Complexity in Environmental Chemistry

Marco Minella
Editor

Complexity in Environmental Chemistry

From Pollutants to Predictive Analytics

Springer

Editor
Marco Minella
Department of Chemistry
University of Turin
Torino, Italy

ISBN 978-3-032-31773-5 ISBN 978-3-032-31774-2 (eBook)
https://doi.org/10.1007/978-3-032-31774-2

This Springer imprint is published by the registered company Springer Nature Switzerland AG
The registered company address is: Gewerbestrasse 11, 6330 Cham, Switzerland

Foreword

We are currently moving away from the climate stability and planetary resilience established in the Holocene and entering a new epoch, the Anthropocene, thus named because of the impact of human activities at global level. Though not recognized by sedimentologists, the Anthropocene is an undeniable cultural concept that acknowledges the power of human actions and their destabilizing effects on the complex, closely linked Earth systems that govern and keep the planet Earth stable: the Climate System, Biosphere Integrity, Biogeochemical Flows, Global Freshwater Resources, Land Use, Ocean pH, and Stratospheric Ozone.

Formal acceptance apart, the Anthropocene is an acknowledgment of how the planet is increasingly governed, and already profoundly transformed, by human actions involving flows of energy, matter, and entropy that have exceeded planetary biophysical boundaries, generating global environmental problems of such complexity that the risk of reaching tipping points is real.

This new epoch began in the 1950s, with the Great Acceleration, a period of strong economic and population growth, an upsurge in industrial activity, and the advent of intensive agriculture (the Green Revolution) based on chemical fertilizers and pesticides that completely changed the relationship between agriculture and the planet. The mid-twentieth century also marked the increased use of fossil fuels, the exploitation of non-renewable resources, the erosion of renewable resources, and ecosystem damage.

The exponential growth of all human activities, prevalently based on non-renewable resources, was supported by progressive technological development, causing global-scale perturbation of many Earth systems in a very brief period of time. The first of these systems affected was the Climate System, which together with Biosphere Integrity, oversees all the others. The release of molecules like carbon dioxide, methane and nitrogen oxides into the atmosphere has upset the Earth's fine energy balance, because atmospheric emissions occurred too fast for their much slower absorption by natural mechanisms. Climate warming is a problem of mismatch between human and natural timescales, between historical time and biological time.

The use of fossil fuels was accompanied by the development of petrochemistry and the synthesis of molecules hitherto unknown to the biosphere, highly stable molecules that persist in the environment, giving rise to the unforeseen phenomena of bioaccumulation and biomagnification in the food web. In "creating" these new molecules, some of which are toxic and immortal (e.g., the so-called forever chemicals), humans distanced themselves from the principles of nature, including circularity and renewability. They did so with little knowledge and great superficiality, under the pressure of markets unconcerned with assessing interactions with ecosystems and the environment, since these aspects are neglected in economic evaluation.

The pollution derived from the environmental release of new substances, whether deliberate, as in the case of agricultural pesticides, "accidental" (e.g., dioxins), or "inevitable," as in the case of various combustion processes (e.g., NOx), is defined very generically and reductively as *chemical pollution*. The definition of chemical pollution offered by Will Steffen and colleagues gives a complete and exhaustive picture, introducing a new, exclusively man-made Earth subsystem with global-scale effects: "Novel Entities: new substances, new forms of existing substances and modified life forms." These include new chemical compounds, new engineered materials (e.g., nanomaterials) and genetically modified organisms hitherto unknown on planet Earth. To these are added elements such as heavy metals and radioactive elements, extracted from deep in the Earth's crust, released in great quantities and mobilized in the biosphere.

In the biophysical context of the planet, the great variety and quantity of Novel Entities released into the environment by humans generates complex, differentiated, chemical, physical, and biological interactions. Research must necessarily be conducted at different scales, without ignoring the whole, i.e., systemically. Since "the whole is greater than the sum of its parts," more research into environmental networks and the development of a multidisciplinary approach are necessary. Monitoring and assessment become essential to measure the mitigation of damage and to demonstrate the threat to planetary resilience posed by these Novel Entities.

Complexity is the key word of this volume, which offers a fundamental epistemological framework, with a wide range of instruments and methods aimed at understanding the mechanisms that pollutants activate in different matrices at different scales. Since a lack of full scientific certainty surrounds the increasing release of Novel Entities, researchers must continuously develop new risk-assessment models to evaluate the often irreversible environmental, social, and health effects. An example is the development of the Chemical Footprint, an indicator of systemic toxicity or a measure of the total input of multiple pollutants that ecosystems can withstand without damage.

However, we must recognize that in the absence of scientific certainty about the short- and long-term environmental effects and dynamics of these increasingly complex Novel Entities, their synthesis and release must be limited to prevent them from entering the matrices of life and spreading through the food web. This can be done by invoking the Precautionary Principle regarding Novel Entities, as highlighted in the report "Planetary Health Check 2025 – A Scientific Assessment of the State of the Planet": "The amount of human-made chemicals and particles as well as human-modified life forms introduced into the environment without adequate testing is beyond safe levels. Following the precautionary principle, no human-made entities should be released without adequate safety testing." Applying this principle means recognizing the biophysical limits of the planet and consequently the limits of human action.

Ecodynamics Group,
Department of Earth,
Environment and Physical Sciences
University of Siena,
Siena, Italy

Nadia Marchettini, nadia.marchettini@unisi.it

Valentina Niccolucci

Preface and Something More

This book offers a comprehensive overview of the topic of complexity in the environmental sciences and, in particular, in environmental chemistry. The topic of complexity is a central topic in the philosophy of science, in epistemology, and, more generally, in modern and contemporary science. Nowadays, the questions posed by complexity science (i.e., the study of phenomena emerging from a collection of numerous interacting objects) in environmental studies can be addressed with new tools, such as those developed for the management of multivariate and big (chemical) data (e.g., new chemometric tools and innovative artificial intelligence algorithms). Furthermore, the better understanding of environmental (chemo) dynamics requires an increasingly broad view of natural phenomena and of the interactions among different environmental compartments.

The first section of this book addresses the topic of complexity in the environment from a general point of view, especially in terms of its definition, the uneasy description of what pollutants are and the complexity of the climate system. Furthermore, analytical tools (e.g., innovative advanced separation techniques) capable of providing information on the complex composition of natural matter are presented, together with an in-depth description of the statistical tools to be used to address complex issues in chemistry (from the basic principles of chemometrics to the application of artificial intelligence algorithms for environmental monitoring).

The second section of this book describes the application of several different approaches to solve complex environmental problems, with the final goal of providing readers with practical suggestions to address environmental issues (e.g., the photochemical transformation of pollutants in the environment, the study of the reactivity of organic compounds in the atmosphere, the speciation of elements in aqueous environment, and the study of long-term trends of water chemistry in complex ecosystems).

The main target audience of this book consists of students of science and scientists (not only chemists) working in the field of environmental chemistry; for them, this book provides not only an overall view of the topic of complexity in their field of research/study but also a toolbox to address complex problems. Furthermore, students (Master's and doctoral students) of environmental sciences (and, in particular, of chemistry) will find here a handy book useful for a better understanding of the complexity of nature and the interactions between humans and the environment. In order not to be in conflict with nature and to achieve truly sustainable

Fig. 1 (**A**) Giotto di Bondone, Last Judgement in the Scrovegni Chapel (1306), Padua (Italy); (**B**) Pablo Picasso, Guernica (1937); (**C**) diagram of the Earth's natural water cycle (Howard Perlman, USGS and John Evans, USGS). In (**A**) and (**B**) the lines and arrows represent the lines of force in the two paintings

development, understanding nature and its complexity is essential. The authors of this book hope that their efforts can be useful for a better understanding not only of the environment (and, in particular, of the environmental chemistry) but also of the role of humanity on our planet.

Before moving on to the body of the book and reading the contents of each chapter, I would like to introduce the main topic in a more naive way, with the sole purpose of giving readers a first idea of what complexity is and why environmental systems are intrinsically complex.

This introduction to environmental complexity, "at first sight," could begin in the Italian city of Padua in the early Fourteenth century, when Giotto di Bondone painted his *Last Judgement* in the Scrovegni Chapel (1306). In this large fresco (Fig. 1A), the amount of information presented is impressive and constitutes a sort of summary of medieval theology. The number of details in the painting is considerable, but there is no real complexity, mainly because there is a clear division of space dedicated to the souls who, based on their behavior in life and on clear rules, are placed in Heaven or Hell. Furthermore, the lines of force in the fresco (see arrows in Fig. 1A) are straight and all in the same direction, without any kind of intersection. This painting can exemplify how the presence of a lot of information does not automatically mean the existence of a complex system, especially when the different parts do not interact.

A scientific equivalent of Giotto's fresco could be the Newton's law of universal gravitation (Isaac Newton, *Philosophiae Naturalis Principia Mathematica*, first published July 5, 1687), which describes the gravitational force (F) that attracts two bodies of masses m_1 and m_2 placed at a distance r (G is the universal gravitational constant).

$$F = G\frac{m_1 \times m_1}{r^2}$$

This law is essential for describing the motion of planets around a star, such as the motion of planets in the Solar systems around the Sun. Here too, there are many elements, but each element interacts simply with the others based on a well-defined law. Each planet interacts with all the other elements in the system, but its behavior can be described as the sum of its individual interactions. There is no complexity in this very complicated system.

At the same time, it is essential not to equate complexity with the concept of uncertainty. For example, consider Heisenberg's uncertainty principle (1927), which states that for an electron around a nucleus, the uncertainty of the particle's position (Δx) and of its momentum (Δp) are greater than or equal to $\hbar/2$ (where $\hbar = h/2\pi$ is the reduced Planck constant).

$$\Delta x \Delta p \geq \frac{\hbar}{2}$$

This principle is the basis of quantum mechanics and the concept of atomic/molecular orbitals. Knowing the electron's exact position does not allow us to know its energy and vice versa. We are far from the idea of complexity: in the microscopic world, we must accept that we cannot know the values of all variables without uncertainty, but this does not mean that the system described is complex.

The same concept is visually clear when considering, for example, the painting technique known as *Pointillism*, in which small dots of (usually) primary colors are applied to form an image of the desired color. Knowing the color of individual dots prevents us from knowing the actual color of the painted image, but at the same time, knowing the color of the image (possible only by observing the painting from a distance) provides no indication of the color of the individual dots. Uncertainty and complexity are not the same concept.

To visually appreciate complexity, I suggest looking at some Twentieth-century works, such as Pablo Picasso's masterpiece *Guernica* (1937), in which the Catalan artist depicts the bombing of Guernica, a town in the Basque Country (northern Spain), by Nazi Germany and Fascist Italy during the Spanish Civil War. The reason for the complexity of this painting lies in the strong interaction of each element, clearly visible in the complex lines of force highlighted on the painting in Fig. 1B. Not only does each element interact with the others, but the painting as a whole is far from being the sum of its parts. The sensation one experiences in front of this painting cannot be explained by observing and describing the individual parts; rather, it is the result of the overall interaction between each element, which, by interacting, generates something greater than their simple sum.

The same occurs in the environment, where each element (e.g., atmosphere, biosphere, hydrosphere, etc.) interacts and influences the behavior of the others in a truly complex way. Examples include the biogeochemical cycles (see the water cycle diagram in Fig. 1C and compare it with *Guernica*), in which elements and compounds continuously circulate between the abiotic environment and biotic (living) organisms. A change in a single compartment, for example, as a result of human activities, can have dramatic consequences on the others, since all the elements interact strongly and in a nonlinear manner. This is a truly complex system.

As the Editor of this book, I wish to express my gratitude to all the authors who contributed to its creation. Without their expertise and professionalism, this book would never have seen the light of day.

The contents of almost all the chapters of this book were presented as lectures during a Doctoral Summer School dedicated to environmental complexity, organized by the Department of Chemistry of the University of Turin (Italy) as part of the project *Chimica 4.0 Molecole e Materiali per la società di domani*, funded by the Ministry of University and Research (MUR) under the "Departments of Excellence 2023-2027" Program (CUP: D13C22003520001). Without this financial support, for which I am sincerely thankful, the Summer School would not have been organized and, consequently, this book would likely not have been written.

Furthermore, I wish to express my gratitude to Springer and, in particular, to Dr. Sofia Costa and Mr. Pradeep Rajasekaran, who provided important guidance in editing and finalizing this book.

Torino, Italy Marco Minella

Contents

Complexity in Environmental Chemistry: From the Definition of the Problems to the Tools to Address Them

Epistemology of (Environmental) Complexity: A Matter of Relations

Elena Ghibaudi

A barometric low hung over the Atlantic. It moved eastward toward a high-pressure area over Russia without as yet showing any inclination to bypass this high in a northerly direction. The isotherms and isotheres were functioning as they should. The air temperature was appropriate relative to the annual mean temperature and to the aperiodic monthly fluctuations of the temperature. The rising and setting of the sun, the moon, the phases of the moon, of Venus, of the rings of Saturn, and many other significant phenomena were all in accordance with the forecasts in the astronomical yearbooks. The water vapor in the air was at its maximal state of tension, while the humidity was minimal. In a word that characterizes the facts fairly accurately, even if it is a bit old-fashioned: ***It was a fine day in August 1913***

Robert MUSIL, The man without qualities, 1913.

Abstract

The present chapter aims at offering an overview of the main issues related with the general discourse on complexity and to comment on how these issues are declined and finds applications in the disciplinary fields of environmental chemistry and environmental science. This topic will be addressed under a philosophical and epistemic perspective.

Keywords

Complexity · Emergence · Relational ontology · Epistemology · Downward causation · Deterministic chaos

E. Ghibaudi (✉)
Department of Chemistry, University of Torino, Torino, Italy
e-mail: elena.ghibaudi@unito.it

M. Minella (ed.), *Complexity in Environmental Chemistry*,
https://doi.org/10.1007/978-3-032-31774-2_1

Chapter Learning Goals

- The fundamental concepts of the discourse on complexity from a philosophical viewpoint.
- The main features of complex systems.
- The methodological implications of the complex perspective.
- Some examples of how the complex perspective applies to environmental chemistry.

This epigraph is the opening paragraph of one of the most famous and important novels of the twentieth century. It offers us a significant example of the difference between the analytical description of a system's state and the explicitation of the meaning of the state itself. The sum of countless pieces of information relating to a specific state risks having limited significance, if they are not framed within a context of meaning: the final sentence, although much less specific than the previous ones, has the advantage of coordinating the preceding information into a meaningful "whole," that is, to bring out knowledge from a complex set of information.

Meteorological knowledge is complex knowledge: this is why this quote, which relies on this kind of description, offers an excellent starting point for introducing some brief considerations on what is meant by *complexity thought*, as well as on the nature of complex systems and their properties, from an epistemological point of view.

This contribution will be organized into four sections: (i) a brief overview of the discourse on complexity from a philosophical viewpoint and its related epistemic and cultural issues; (ii) the description and comment of some properties that are typical of complex systems; (iii) a discussion of the issue of environmental complexity within the general discourse on complexity; (iv) some concluding remarks on the manifold implications of environmental complexity.

1 The Complex Approach to Reality

What is complexity? Several authors have pointed out that any definition of complexity is inevitably complex and, consequently, elusive. However, we can resort to etymology to resolve a possible initial misunderstanding regarding the terms *complex* and *complicated,* that are often confused or considered as synonyms. *Complicated* derives from the Latin *cum plicum*, "with folds": it refers to the image of a sheet of paper folded several times, which can be unfolded, made legible, simplified. A complicated problem may be linearized and reduced to simpler issues. Conversely, *complex* derives from *cum plexum*, "with knots," intertwined, woven together. The consequences of such weave are dramatic: it creates meaning precisely in the way it connects the elements, just as the meaning of a tapestry is not given by a set of threads, but depends on the specific relationships that unite them. If we separate the threads, the system dissolves.

This immediately leads us to wonder what a system is and whether there are simple systems and complex systems or not. Luigi Cerruti [1], following John L. Casti [2], identifies *decomposability* as a discriminating criterion: "Decomposability is a property of simple systems. The components of a simple system interact weakly, so we can disconnect one or more of the components, and the system will behave more or less as it did before. A suitable example is a ship. It can be a highly complicated system, but from the point of view of decomposability, it is not complex; any ship can suffer major damage without dramatically changing her overall seafaring capacity. A complex system with its high degree of interaction is very sensitive to these changes. Disconnecting or damaging any part/component of the system will produce drastic changes in the system's behaviour." This position recalls the distinction, made by von Foerster [3], between *trivial* and *non-trivial machines*: the former are determined synthetically, independent of history, analytically determinable and predictable; whereas, the latter are determined synthetically, dependent on history, analytically indeterminable and unpredictable.

According to other points of view, the very notion of *system* is inherently complex, since a system is structured as a whole, characterized by very precise relationships between its parts, and these relationships are constitutive of the system itself [4, 5]. In fact, there is no commonly accepted definition of what a *complex system* is, and the perspectives and opinions on the subject are manifold (see [6] and ref. therein).

Nevertheless, one aspect that all discussions on complexity have in common is their focus on the relationships that characterise any system. The fundamental assumption underlying this perspective is that "keys to understanding the system are contained in patterns of relationships and interactions among the system's agents" [7], since it is the interdependencies and feedback loops between the elements that create the whole. Consequently, by fragmenting the system into its components, the analytical method destroys precisely what it is trying to understand.

This position has two immediate consequences: a philosophical one and a methodological one.

On the philosophical level, adopting a complex perspective implies adhering to a conception of reality based on relational ontology. This means attributing a foundational character to relationships and interactions with respect to individual entities: "Instead of perceiving objects and concepts as independent units that subsequently interact, relational ontology posits that the very significance of something is constituted by its network of connections and involvements within a broader system" [8]. This marks a significant move beyond substance-based ontologies, insofar as entities are conceived as individuals that are not fundamentally self-subsistent: they are constituted by and exist within networks of relations.

Three implications of relational ontology are the following: (1) the *principle of interdependence:* entities are neither self-sufficient nor autonomous. They are part of a broader relational network; hence, (2) the *context* becomes central to their understanding. The sense of an entity is always shaped by its context because its properties and behaviors are not just intrinsic, but are expressed and understood within specific relational contexts. (3) As relationships are dynamic, ever-evolving processes, *reality is dynamic* and its descriptions must account for this inherent dynamism [8].

If the complex approach to reality stems from the critique of ontological reductionism, the consequences are substantial, as they affect the epistemic approach that has characterized all modern, Newtonian and Cartesian, science [9]. In fact, ontological reductionism translates into epistemic reductionism, i.e., a conception of knowledge according to which knowledge of a system can be traced back purely to knowledge of its constituent parts. On the contrary, by embracing the complex view, the Newtonian ideal (that conceives of the universe as a simple and orderly machine, where each phenomenon is linearly traceable to its cause in time) is abandoned, and the mechanistic ideal of knowledge is criticized: "The model of complexity contrasts the mechanistic idea of nature as a simple, linear machine that can be broken down into elements, with the idea of nature as an organized system whose parts interact dynamically. [...] From the point of view of complexity, knowledge is never knowledge of things, isolated objects, but *always of systems of relations*" [10].[1]

If reality is no longer thought of as a linear system, i.e., a system governed by linear causality, that can simply be broken down into its parts, which can be analyzed in isolation from their context, then the mode of investigation needs to change as well. If reality is thought of as "a network of relationships, in which the parts have meaning in relation to the organization of the whole" [10, 11], then it has to be investigated according to approaches that account for such relationships.

The role and posture of the knowing subject change accordingly, as compared to the Newtonian and Cartesian investigative approach; in fact, the observer is part of the same network he is exploring: "The knowing subject [...] is not a pure, external observer, without place or time, [...] but it is linked by a complex connection to the world it observes" [10]. Silvana Borutti remarks that "from the point of view of complexity, observing and knowing does not mean contemplating an object from an external and privileged point of view" [10]. Thus, the so-called *God's eye view* on nature is rejected, because unrealistic: a scientist is not the recorder of an objective reality from which he is excluded. Knowledge is an active relationship between subject and object. This implies that the knowledge developed by the subject is hermeneutic, interpretative. It is also the expression of the socio-economic-cultural environment, in which the knowing subject is immersed, and by which it is influenced [10, 12].

Back to the initial question (what is complexity?), it becomes clear that it consists of an *epistemological rethinking* [13]. The Newtonian and Cartesian view has accustomed us to look at the universe as a causally determined mechanism, following a linear causality. According to this view, "complication appears as a negative notion: it is linked to our cognitive limitations" [14] and, in principle, it could be overcome by an ideally infinite point of view (the God's eye view). In the universe conceived according to this canon, simple, deterministic systems are the rule, and complex systems (such as living organisms or ecosystems) represent the exception

[1]All quotations from bibliographic sources in Italian or French have been translated by me, unless differently specified.

that escapes complete understanding due to the unattainability of the omniscient point of view. Conversely, according to Isabelle Stengers, the idea of complexity overturns this perspective: in fact, it rejects *a priori* the infinite point of view, because "the complex description transforms the loss of validity of our simplifications into an intrinsic property" [14]. By virtue of this vision, "simple" systems become special cases of the more general case, represented by complex, irreversible systems, etc. The incomplete determinability of a system no longer derives from a theoretical limit of knowledge that could, in principle, be overcome; it rather becomes an intrinsic feature of the system itself. Consequently, the notion of complexity acknowledges implicitly the intrinsic limits of the cognitive act we call science: according to Morin, the complexity thought marks "the overcoming of the limits of universalist abstraction, which eliminated singularity, locality, and temporality" [15].

2 What Is a Complex System?

If defining complexity is challenging, we can try to define the features of a complex system. Edgar Morin was one of the first scholars to point out that a complex system is both *more* and *less* than the sum of its parts. The limitation stems from the very existence of the organization: "Organization imposes constraints that inhibit certain potentials found in the various parts" [15]. On the other hand, the organization brings out functions and properties that do not manifest themselves in the individual components of the system: "the organized whole is more than the sum of its parts, because it brings out qualities that would not exist without such organization" [15]. If we adopt an informational perspective, we could say that the system contains new information that is not found in the individual parts [9].

Two of the most intriguing definitions of complex system are due to the biomathematician Robert Rosen, who interprets complexity from the epistemic viewpoint, i.e., from the perspective of someone who wonder how our knowledge on complex systems is built. According to Rosen "a system is complex when it is not possible to describe it with a unique formal model, and the models in use are not derivable from each other." As an alternative, Rosen states that "a system is complex when its behavior as a whole is not derivable from the properties of its parts" [16].

The first definition "emphasizes the intractability of a complex system from a unique point of view" [17]: it does not deny the usefulness of the reductionist (analytical) method of investigation, nor the effectiveness of this or that formal model. At the same time, it acknowledges the need for plural investigative approaches, that are mutually irreducible, since a single formal model will never be able to provide an exhaustive and complete interpretation of the system's behavior. In this way, Rosen provides an indirect, operational definition of complexity: complexity is seen as a function of the number of ways in which we can interact with the system. This view has been further refined by Don Mickulecky, one of Rosen's pupils, who states that complexity is "the property of a real world system that is manifest in the inability of any one formalism being adequate to capture all its properties. It requires that we find *distinctly different* ways of interacting with systems. Distinctly different in the sense that when we make successful models, the formal systems needed to describe each distinct aspect are NOT derivable from each other" [18]. Mickulecky's clarification is interesting in that it emphasizes that a complex system is, by necessity, an *open system from the epistemic point of view*, i.e., a system that lends itself to a multiplicity of cognitive approaches that cannot be reduced to one another.

This awareness has a straightforward consequence: the cognitive demands of complex systems require openness towards various disciplinary approaches. We can go so far as to say that the complex approach to reality *elevates multidisciplinarity to a method* [12] and promotes a *symphonic conception* of knowledge [19].

The second definition of complex system proposed by Rosen focuses on the system's behavior and refers implicitly to emergent properties, i.e., those peculiar properties of a system (considered as a whole) that originate from the organization of the system itself and are not recognizable at lower levels of organization. In general, any property of a system that cannot be traced back to a description of the system in terms of individual entities is considered as *emergent*. Therefore, the notion of emergence implies the presence of strong interactions between the system's components [12]. Based on this definition, it becomes clear that emergent properties are *markers of ontological complexity*, as they represent an innovation that stems from a different ontological level as compared to the one which the system's components belong to [20]. From the ontological viewpoint, a complex system is *multidimensional*, i.e., it is characterized by multiple levels of organization that are substantially distinct. Hence, any investigation of complex systems "must recognize this multidimensionality" [21]. In other words, *ontological complexity inevitably translates into epistemic complexity*.

Actually, in the field of philosophy, there is extensive debate on the nature of emergent properties, with a distinction being made between strong and weak emergence, depending on whether emergence is attributed ontological or purely epistemological significance [22]. We will not enter into this debate, but will simply remark, along with Walliser [23], that the concept of emergence has strong *heuristic potential* since it allows linking together many methodological problems that are usually treated in isolation. Further, the notion of emergence allows highlighting the outbreak of "collective behaviors or structures that cannot be predicted from the individual elements" [24].

2.1 Properties of Complex Systems

We have stressed that emergent properties originate from the system's organization: the organization of the system is "the configuration of relationships between the parts of the system and between these parts and the whole" [5]. It is a dynamic intertwining of relationships between the system's components, which follows nonlinear logic (feedback and circularity mechanisms), and may even be responsible for self-organization phenomena [5, 9, 25].

The meaning of the organization of a system and its implications are well described by Giovanni Villani [5], who remarks the distinction between "system" and "aggregate" (a bunch of unorganized entities): a molecule is not equal to a bunch of atoms, an organism is not a heap of cells, etc. Now, the functions and reactivity of a molecule can only be explained by considering the relationships between the component atoms as well as the role of the chemical environment; similarly, the properties of an ecosystem are the result of interactions between its multiple and varied components in a specific set of physical conditions. A molecule, an organism, an ecosystem may be thought of as a *functioning whole*, whereas a heap, an aggregate are not a whole, since they lack the dynamic intertwining that characterizes the previously mentioned systems. That's why *emergent properties* are typical of *organized* systems, not of aggregates.

It's worth emphasizing that the system's organization is not a static scheme, given once and for all, and therefore immutable. On the contrary, the organization of a complex system is characterized by intense dynamism which derives from the continuous exchange of stimuli and information that occurs both internally (i.e., between the system's components) and at the boundary between the system and the environment. Within this dynamic description, the system is identified as "a repository of information, which resides at different levels and in different degrees of content" [9]. Therefore, the organization of a complex system is understood in a *processual* sense: it is the result of a dialogical mode that involves a *principle of complementarity* and a *principle of antagonism* [15]. In other words, the system's organization *self-creates* incessantly, according to an autopoietic process [25]. It is the result of an interplay between the parts and the whole that requires continuous adjustments and regulations: just think of the Earth's atmosphere, a thermo-fluid dynamic system in continuous interaction with the oceans, the biosphere, the cryosphere, and the Sun as a source of thermal energy, or think of the metabolism of a living organism, which guarantees the maintenance of the system's identity despite the incessant changes occurring within it.

How can change and permanence be reconciled? How can a system change continuously while retaining its own identity? To answer these questions, Maturana and Varela [25] developed their *theory of autopoiesis* that considers the complexity of systems from the point of view of their dynamic unity: every complex system is characterized by *organizational closure,* i.e., the invariant characteristic of a complex system is its unity. According to Morin [15], a principle of *recursive organization* applies: a complex system is generated by the interrelationships between its components, that have a certain stability or regularity. They produce an organizing totality that retroacts on the components themselves, modifying them. As an example, just consider that society is made up of individuals in relation to each other; these interactions produce an organizing wholeness that retroacts on individuals to co-produce them as human persons [4, 15].

This mention of retroaction leads us to discussing the problem of the type of *causality* that operates within a complex system, namely non-linear systems. Causal relationships within a complex system cannot be traced back to a linear deterministic model, which conceives reality as the result of separate and summable elements, whereby each phenomenon can be traced back to its cause in time, and the entire chain of effects that occurs—starting from precise initial conditions—can be predicted. In other words, complex problems cannot be solved by applying the *principle of superposition*, according to which the general solution can be obtained as the sum of simpler solutions [9]. Causal relationships within a complex system are more complex and exclude the possibility to achieve a firm prediction of the effects that act on the system itself. In particular, complex systems can be interpreted in the light of:

- *Circular explanations*. This originated in cybernetics and describes the feedback of the effect on the cause that provoked it, or rather *the circular transformation of the effect into a further cause* [5]. According to Norbert Wiener [26], the father of cybernetics, the feedback mechanism presupposes the integrated action of the system and the environment: the system acts on the environment, which reacts on the system, creating a circularity that defines limits (negative feedback) and produces a stable, self-regulating interaction. The statement according to which "the parts of the system are meaningful in relation to the organization of the whole" should be read in this perspective. This type of causality generates constraints that are functional to the subsistence of the system's organization.
- *Teleonomic explanations*, according to which phenomena can be explained in relation to the end or function toward which they tend [12]. In discussing this issue, Villani [5] remarks that (i) Finalism should not be confused with the presence of a consciousness capable of establishing the system's goals; (ii) The goal is, in all respects, a *systemic property* that distinguishes unstructured from structured matter. Most important, in the complex perspective, the goal becomes an *inherent property* of the system, that self-regulates itself.

These kinds of explanations may be related to two distinct types of causation that are called:

- *Upward causation:* this expression designates causal processes that lead from lower level entities to higher level ones, so that the lower level can be seen as the cause, and the higher level as the effect. For example, in chemistry, the macroscopic properties of a substance are thought of as the result of its microscopic structure.
- *Downward causation*: this expression indicates causal influence exerted by higher-level organizational structures on lower-level parts of the system. In open, complex systems, downward causation is achieved through feedback mechanisms.

The bidirectional causality typical of complex systems challenges the reductionist view, according to which causation flows solely upward, from microscopic to macroscopic levels. It also suggests that wholes can influence their parts in ways *not fully predictable* from the parts alone. Hence, one consequences of this conception of causality, which is more open than the simplistic idealization of linear determinism, lies in the role assigned to uncertainty and unpredictability in the behavior of systems: in fact, the very concept of complexity encompasses uncertainty [19, 27, 28].

This uncertainty (which should be understood in opposition to the virtually absolute predictability that characterizes linear deterministic causality) has a dual nature: (i) *epistemic*, in that the complex view excludes the possibility of "total and absolute knowledge" for the reasons discussed above [19]; therefore, a completely exhaustive knowledge of a system is inconceivable; (ii) *structural*, i.e., intrinsic to the system itself as a consequence of the system's extreme sensitivity to boundary conditions. The complex system is, in fact, subjected to perturbations that can reorient its evolution in an unpredictable manner starting from given initial conditions. It is the principle of *order from noise* postulated by von Foerster [29] that states that some form of order may arise out of local interactions between parts of a disordered system, since random perturbations let the system explore a variety of states, hence increasing the probability of self-organization [15]. In other words, the principles of deterministic chaos come into play, which seek to understand how the behavior of a system subjected to deterministic laws can be difficult to predict because the system's future states are very sensitive to the system's current state, and how this indeterminacy may entail an adaptive behavior, based on the reciprocal relationships between the system and the environment.

In fact, uncertainty (or, rather, the epistemic openness of complex systems) is a necessary condition for creativity to manifest itself. This issue, developed extensively by Ilya Prigogine in his numerous writings, is directly linked to the dissipative nature of complex systems. Prigogine [28, 30] reinterprets the second law of thermodynamics and the meaning of entropy in a creative key, stating that the notion of entropy should not be associated with the idea of decay. On the contrary, entropy "affirms the existence of a fundamental instability in the dynamic systems that surround us." It is precisely because of this instability—which is expressed at both the macroscopic and microscopic levels—that systems can transition from one structure to another and evolve, also in relation to the surrounding conditions. From this perspective, according to Prigogine, the second principle of thermodynamics states the reality of history and of the time arrow.

This leads us to clarify and re-evaluate the role of probability and uncertainty. In fact, it is precisely the distance from equilibrium that forces us to resort to a description of the dynamic (i.e., unstable) system in statistical terms. The system is not unpredictable *tout court*, but rather *predictable within well-defined limits*. This is not due to our insufficient knowledge of the system, but is a condition that arises from the very nature of the system, i.e., from the presence of interactions between the parts (resonances) that are integral and irreducible parts of the system itself.

The probabilistic description can therefore be read positively, as an expression of the *space of possibilities* of a system, whose limited predictability should not be understood in a negative perspective, as the result of a cognitive impossibility, but rather in a positive way, as a characteristic that is essential for preserving the space of freedom of the world and the evolutionary nature of reality [12]. Mark Taylor [31] remarks that "The unpredictability of a process does not mean that it is directionless. Self-organizing systems can be conceived as teleonomic processes, tending towards increasing levels of complexity."

In Prigogine's words: "The new vision that emerges today is therefore a balanced description of two alienating representations: that of a deterministic world and that of an arbitrary world subject only to chance. Laws do not govern the world, but neither does it obey chance" [27]. In this perspective, the role of chance is also clarified: the system does not evolve randomly, but evolves thanks to chance, or rather to its own freedom. In this way, chance loses the character of pure arbitrariness that we find in Monod [32], and becomes an instrument of the indeterminacy that appears indispensable for ensuring creativity in a world in perpetual flux. It is no coincidence that Prigogine's conclusion is that "The world, in order to be creative, *must be* indeterminate" [27].

3 The Specificity of Environmental Complexity

In the previous sections, we examined the topic of complexity from a philosophical point of view, identifying a series of aspects typical of complex thinking and a complex view of reality, which could be summarized through the following keywords: non-linearity, system's organization and self-organization, emergent properties, non-linear causality, deterministic chaos, dynamism, adaptability, interdisciplinarity.

Faced with this mode of thinking, a scientific investigator may wonder how to translate it into concrete and effective operational modes. However, if we explore the developments that have taken place in many disciplinary fields over the last 50 years, we realize that they have already been strongly influenced by the complexity thought, which has encouraged both the development of new ways of describing the systems under study, and the elaboration of methodological approaches consistent with this new perspective. In other words, the complexity thought is already at work in many disciplinary fields. It finds expression both within each discipline, and in promoting the development of new interdisciplinary fields. In fact, in recent decades we have witnessed the raise of a number of new epistemic approaches, inspired by the discourse on complexity: e.g., systems biology (an interdisciplinary approach that investigates biological systems by focusing on the mutual interactions between biological components like genes, proteins, cells, etc.), systems chemistry (which looks at complex networks of interacting molecules, acting under thermodynamic and/or kinetic control, whose collective behavior generates emergent properties), the physics of chaotic systems, data science, etc. Crossing the boundary between the natural and physical sciences, on one side, and economic, social, and humanistic disciplines, on the other, we remark that complex thinking has inspired new theoretical and investigative approaches even in the fields of economics, social studies, demography, anthropology, linguistics, etc. Despite the variety of disciplines and types of systems investigated, all these fields share a research approach focused on the notions of *dynamic network* and *relational ontology*. The application of complexity thinking has also promoted methodological developments, namely the emergence of investigative approaches or techniques capable of providing information on the relational aspects of the systems under study, or on their emergent properties. This implies, for example, the possibility of monitoring multiple variables in real time, in order to record the structural/organizational changes that may occur in a complex systems (just think of high-resolution microscopy techniques, dynamic imaging of organs and organisms, satellite monitoring of meteorological systems, detection of transient species in chemical systems, etc.), or to follow the behavioral changes brought about by external stimuli or changed boundary conditions; or, again, the development of dynamic models and the simulation of dynamic behaviors, based on a plethora of variables taken into consideration simultaneously. The application of multivariate analysis, the availability of high computing power, and access to AI, the statistical approach to systems as is typical, for example, of chemometrics: all these methods are commonly applied to the study of complex systems, and their development or enhancement is heavily indebted to the complexity thought.

Let's turn to the topic discussed in the present book: environmental chemistry. In what sense this field of scientific inquiry is indebted to the complexity thought? And how is environmental complexity configured?

Environmental chemistry is defined as "a multidisciplinary study of chemical, biological, and integrated processes of the environment, at large. Environmental chemistry draws on a myriad of concepts from chemistry, biology, statistics, geology, and environmental science to assist in defining the sources, reactions, transport, effects, and fates of chemical species in various environmental matrices" [33].

This description immediately highlights two aspects of environmental chemistry that are typical of a complex vision: the multi- and trans-disciplinary approach (a *myriad of concepts* that contaminate and fertilize each other) and the processual and dynamic nature of the object of study ("*reactions, transport, effects, and fates*" are all terms that refer to processes). The same source emphasizes that environmental chemistry is concerned with characterizing the "complex interactions" that occur between the natural environments (air, water, soil, and land, as well as biota) and the man-made environment, according to a perspective that fits into the broader context of environmental science [33]. Such investigation calls into question ecosystems (functional units whose identity is defined by the "dynamic interactions between plants, animals, microorganisms, and abiotic factors (soil, water, nutrients, and light)" [33]). Here we find the vocabulary and concepts encountered in the philosophical introduction of this chapter: the study of environmental chemistry does not aim so much at characterizing entities isolated from their context, but rather at characterizing *complex interactions between different agents*. These interactions are *dynamic* and have a *functional significance*, as they contribute both to the maintenance of those functional units we call ecosystems and to the maintenance of a homeostatic mechanism involving the biosphere and the other components of the geosphere (i.e., the lithosphere, hydrosphere, and atmosphere). In summary, environmental chemistry deals with large systems of interacting chemical species; environmental systems are dynamic in nature, as they vary both spatially and temporally across a wide range of matrices. The underlying idea is clearly non-reductionist: the system is conceived as a network of relationships rather than as a sum of entities; it is precisely these relationships that define the (dynamic) structure of the system and become the object of investigation.

With regard to the *symphonic conception of knowledge*, which is equally typical of the complexity thought, this is evident from the emphasis placed on the fact that environmental chemistry encompasses "targeted areas of study including atmospheric chemistry, biogeochemistry, chemical toxicology, marine chemistry, soil and sediment chemistry, water chemistry, astrochemistry, fire chemistry, chemistry of energy, Earth and geochemistry, and climate change" [33]. All these fields of investigation are characterized by specific cognitive practices, none of which can be considered exhaustive: only their combination may hope to provide an adequate overview of the processes and systems dealt with by environmental chemistry.

It is therefore clear that environmental chemistry embraces complex thinking in all respects. With Cerruti [20, 34], we may conclude that *environmental chemistry is complex from a double viewpoint*: ontological (the environment is understood and conceived as a complex system) and epistemological (the cognitive approaches typical of this field are interdisciplinary).

One of the characteristics of the complex view lies in the fact that, thanks to its inherently interdisciplinary approach, it allows global (planetary) issues to be addressed. It is no coincidence that climate change was mentioned in the previous description of environmental chemistry, as it is perhaps the most important challenge facing humanity today. Other sources highlight how environmental chemistry is considered important at a societal level as it is linked with pollutants and their impact on the environment, reduction of contaminants, and environmental management.

This leads us directly to the topics addressed by the present book, which tackles several issues related with the environmental fate of pollutants (namely, the photochemical transformation of pollutants in the environment, the study of the reactivity of organic compounds in atmosphere, the speciation of elements in aqueous environment, the study of long-term trends of water chemistry in complex ecosystems such as the subalpine lakes, etc.) as well as the methodologies suitable to investigate them (namely, QSAR, chemometrics, speciation models, Artificial Intelligence algorithms, etc.)

It's worth underlining that the very definition of pollutant is a complex issue if examined within the frame of relational ontology. In fact, in the complex perspective, the ontological status of substances (or chemical species) is dynamic and reflects the net of relations in which they are involved. Even though the intrinsic properties of a substance are constant, relations may be acquired and lost, depending on the substance's location. Specific environments may provide the substance with distinct meaning and open operational possibilities as, in distinct regions, the relational properties of the substance change: they may either improve or be depleted, depending on the network of relationships wherein the substance is involved. Hence, the ontological status of substances changes accordingly, and may result improved or impoverished [35]. These considerations are particularly relevant for all those substances (or chemical species) that are generically defined as pollutants, since their ontological status depends strictly on the environmental context in which they are found.

3.1 A Quick Look Back to History

Although the discourse on complexity has emerged as a form of thinking in relatively recent times, the global perspective that characterizes it has much more distant origins.

Historically, the first disciplines to develop this perspective, which then influenced all the various branches of environmental studies, were ecology and geophysiology. It is therefore worth briefly reviewing the contributions made by the influential figures of Vladimir Vernadskij (1863–1945), Alfred Lotka (1880–1949), and James Lovelock (2019–2022).

Vladimir Ivanovich Vernadskij can be considered the founder of the research tradition that focuses on the interaction between the lithosphere, hydrosphere, and biosphere (this latter concept was formulated by H.E. Suess in the nineteenth century and developed by Vernadskij himself in his famous essay *La biosphère*, 1929 [36]). The basis of his studies was the insight that the Earth and the life present on it constitute a single self-regulating system, capable of ensuring conditions suitable for the survival of living organisms, thanks in part to the exchange of energy provided by sunlight [37]. In light of this vision, which we would today describe as complex, Vernadskij denounced the fragmented scientific view of geological phenomena that was typical of his time, and conceived of them as *planetary phenomena*. This gave rise to his conception of the structure of the Earth as a dynamic mechanism, whose parts form a harmonious and indivisible whole, and whose study cannot therefore ignore this overall vision [37]. It then became clear to Vernadskij that the investigation of such a system must transcend the traditional boundaries between individual disciplines. It required a new approach, unencumbered by disciplinary barriers that are "formal, external, superficial fact" [36]: an approach focused on the theoretical problems to be addressed and resolved. It is precisely on the basis of the data at his disposal that Vernadskij came to assert that the existence of a planetary mechanism that includes the biosphere is "in harmony with all empirical data and necessarily follows from their scientific analysis" [36, 37].

The vision developed by Vernadskij led him to conclude that one of the primary objectives that science as a whole must pursue is the study of the *flow of matter and energy* that is established between the environment and living bodies and that incessantly feeds vital processes [37].

It is difficult not to recognize in this operating model one of the cornerstones of what is now called "ecological complexity." The corresponding entry in the *Encyclopedia of Complexity and Systems Science* [38] states: "Ecological systems are large ensembles of macroscopic matter that capture energy, store it, and use it to circulate and rearrange matter within the system. The term 'use energy' aptly describes the material flows within an ecological system because information-based processes determine both the material structure of the system and its function."

The same source recalls that one of the first attempts to develop a comprehensive theoretical structure for ecology was Lotka's concept of "physical biology" [39], which integrates ecological complexity with the broader framework of theoretical physics.

A further example of the path that led to the development of the ideas we now recognize as belonging to complexity thought comes from James Lovelock, a physical chemist who, in 1972, in an article published in the Journal of Atmospheric Environment, wrote the first formulation of what would become the Gaia hypothesis.

The geophysiological model that Verdnaskij had foreshadowed in his writings is explained and confirmed in the studies of Lovelock, who in 1987 published an article in Nature together with other colleagues [40], which included the adjective *geophysiological* in the title of the work. In this paper, the four authors presented the now famous "CLAW hypothesis" (from the initials of their names) regarding a possible feedback mechanism involving dimethylsulfide emitted from the oceans, capable of affecting the Earth's radiative balance and global warming. The CLAW hypothesis therefore supported the ability of marine phytoplankton to influence the global climate. An interesting aspect of this study is its interdisciplinary nature, as evidenced by the different disciplinary backgrounds of its authors: Charlson was a chemist working in the field of atmospheric chemistry; Lovelock was a chemist who became a physical chemist; Andreae had studied earth sciences and was an oceanographer; Warren was a meteorologist, specializing in the study of cloud albedo. Cerruti points out that "an *interaction scheme* emerges from their professional qualifications: atmosphere (Charlson), ocean (Andreae), and clouds (Warren) were connected by the syncretistic way of thinking of a physical chemist" [1]. Only later Lovelock attained a more complete formulation of the Gaia hypothesis [41], which can be summarized as follows: "The whole system of life and its material environment is self-regulating at a state comfortable for the organisms." He commented on this, noting that Gaia is a top-down system's view of the Earth, according to the hard science view of a physical chemist with an interest in control theory [41].

The Gaia hypothesis and the view of the Earth as a geophysiological system are a striking example of the application of complexity thinking in science. In fact, this model has all the features that identify complexity on a theoretical level:

- It appeals to a global systemic dimension.
- It highlights the dynamic nature of the system.
- It uses explanatory models that refer to non-linear causality and downward causation.
- It predicts and verifies mechanisms of self-organization and self-regulation.
- It requires a strongly interdisciplinary approach.

Each one of the three scientists mentioned here can be considered precursor of the school of thought that we now call "environmental complexity": in fact, this field of research cannot help but refer to the systemic models they proposed and developed; even some of the topics covered in this volume are influenced by them, either explicitly or implicitly.

4 Conclusions

In conclusion, the complexity thought has ancient roots and extends its influence across countless disciplines.

It has radically changed both our way of interpreting reality as well as our conception of Nature. Today, reality is no longer conceivable as a "catalogue of objects" determined in themselves and immersed in an environment unrelated to them; on the contrary, reality is thought of as a system within which entities participate in a dynamic set of interactions that contribute to determining them, in the ontological acceptation [5, 10]. Nature is no longer conceived as the mechanism of a clock, revealed and governed by deterministic laws and subject to linear causality, i.e., a fully predictable system. The vision of Nature that science gives us leaves room for probability and recognizes the role of uncertainty, understood as a necessary condition for the expression of that dynamic and creative element that allows us to apply a narrative dimension to Nature itself [28].

As a result, our conception of knowledge is also influenced: "the challenge of complexity makes us renounce forever to the myth of total clarification of the universe" [15], leading us to recognize the intrinsic limits of any cognitive approach. Consequently, it highlights the need to articulate multiple perspectives and integrate multiple views through which to interpret reality. It is exciting to discover what a great scientist like Mendeleev wrote, even though he was immersed in a cultural climate profoundly distant from the one we experience today. Regarding the seek for truth, in his last university lecture (in 1890), he stated: "The truth is not concealed from men, it is amongst us, diffuse and distributed all over the world. One can look for it everywhere: in chemistry, mathematics, physics, history, literature" (cited in [42]). What is striking about Mendeleev's vision is its breadth, which led him to bring together science and humanities, in the belief that science must dialogue with other disciplines in order to allow human knowledge to grow. Most of all, for Mendeleev, science is not "the last word" on things: "Science is certainly and only a new search for truth, but from this it does not follow at all that it claims to reach some form of 'last word'" (cited in [42]). These are statements that we can now inscribe into what has become known as complexity thought.

A further aspect of the complex view of knowledge lies in the fact that complexity thinking highlights the limits of reductionism (which seeks to understand the whole, based solely on the qualities of the parts), but equally rejects the simplistic approach of holism (which ignores the parts in order to understand the whole) [15]. In fact, a further implication of the complexity thought is the renunciation of thinking of reality (and its physical description, i.e., science) as an undivided whole. Complexity implies the assumption of the *constitutive multiplicity of reality*, a point of view that leads to a reevaluation of the qualitative and subjective dimension.

Once again, we would like to quote the surprising statements of Mendeleev, who takes a stance against the reducibility of Nature to unity: "A unit in nature is even unthinkable. [...] in everything, in everything one feels the aggregate, the complexity, the mass of units, the collective" (cited in [43]).

Precisely for this reason, the distinctive features of the complex approach to reality, on a methodological level, are inter- and trans-disciplinary: with this, we mean that disciplines are not limited to working side by side on the same problem, but influence each other by working "inside the same problem," integrating and influencing each other. This is especially evident in the theoretical approach to environmental complexity, which is not limited to taking into account the contributions of different scientific disciplines, but—in order to propose truly effective solutions to the problems it faces—extends its field to more distant disciplines. For example, concerning environmental sustainability, Nuta et al. [44] remark that "Sustainable transformation requires the operational integration of environmental data, public health outcomes, and socioeconomic indicators" because "environmental sustainability depends on understanding the mutual feedback loops among ecological, health, and economic systems." In fact, due to the interconnectedness of natural ecosystems, human health, and socioeconomic development "research should examine system dynamics that connect local exposure, global supply chains, and transboundary flows, providing a fuller picture of how environmental risks propagate" [44]. This also requires new operating methods that integrate seemingly unrelated fields of investigation in order to achieve "integrative, data-informed, and justice-centered environmental governance, grounded in cross-sector collaboration and empowered by digital transformation" [44]. In the conclusion to their recent work, Nuta et al. state that "global environmental health is a *shared responsibility*" [44].

This remark leads us to a last critical reflection regarding the concept of complexity: the complex conception of reality implies an *ethos*.

An ethos, in turns, implies commitments consistent with such ethos. John Cairns [45], an ecotoxicologist, has denounced the *cognitive dissonance* that often manifests itself in very practical ethical choices: "individuals can simultaneously advocate robust economic growth and environmental protection, despite a preponderance of scientific evidence that economic growth is causing massive environmental damage" [45]. Following this line of thought, Cairns reminds us of our responsibilities as human beings, which imply a commitment to ensure that ecosystems' functions and integrity are maintained: "If humankind is dependent upon the biospheric life support system, humankind's primary responsibility should be to nurture it and protect its health and integrity. *Ideally, all humans should participate in this endeavor*" [34].

In front of the great challenges that humanity must face—from climate change to environmental sustainability, from the need to move from an economy of exploitation to an economy of care and sharing of the resources available to all of us—the complexity thought may help promoting a *collective ethos* that we may summarize in the words of two great figures of science and philosophy.

The first one is Heinz Von Foerster, for whom an ethically virtuous behavior can be summed up in the ethical imperative: "*Act always so as to increase the number of choices*" [46], that may be rephrased as: always opt for situations that preserve and expand future possibilities. The second one is Hans Jonas who—in his renown text "The Imperative of Responsibility"—invites us all to act "*in such a way that the consequences of our actions are compatible with the permanence of authentic human life on earth*" [47].

Questions

A. Does the complex perspective rely on deterministic or probabilistic approach in the description of natural systems?
B. How does the complexity thought affect the notion of causality?
C. What does "emergence" mean in the context of complexity?
D. According to the complexity thought, what is the role of relations within a complex system?
E. What is the role of interdisciplinarity in the complex perspective?

Glossary

- Ontology: It is the philosophical study of being; it aims at explaining the nature of reality.
- Epistemology: It is the study of the nature and grounds of knowledge; it aims at explaining the nature of our experience of reality.
- Causality: it is the theoretical investigation of the relationship between something that happens and the reason for it happening.
- Teleonomy: the condition of having a fundamental (intrinsic) purpose.

Further Readings

Scientific readings*: a few scientific reference texts/websites on the subject of complexity*

Kauffman, S. A. *At Home in the Universe: The Search for the Laws of Self-Organization and Complexity*, 1995, ISBN 0195095995

Lovelock, J. Gaia: A New Look at Life on Earth, 2000, ISBN 978-0-19-286218-1.

Prigogine, I., Stengers, I. *Order Out of Chaos: Man's New Dialogue with Nature*, 1984, ISBN 10: 0553343637

Capra, F., Luisi, P.L. *The Systems View of Life,* 2014, ISBN 978-1-107-01136-6

Damasio, A. *The Strange Order of Things: Life, Feeling, and the Making of Cultures,* 2018, ISBN 9780307908759

(continued)

Ceruti, M. *La fine dell'onniscienza,* Edizioni Studium, Roma, 2014 (available only in Italian)

WEB site of the Santa Fe institute for studies on complexity: https://www.santafe.edu/

***Literature readings**: a few literary masterpieces that exemplify the complex view*

Borges, J. L. Luis. *Ficciones*, Editorial Sur, Buenos Aires, 1944 (english version: *Fictions,* ISBN 10: 0802130305)

Calvino, I. *Le città invisibili,* Einaudi, Torino, 1972 (english version: *Invisible cities,* ISBN 10: 0099429837)

Cortazar, J., *Rayuela,* 1963 (english version: *Hopscotch,* ISBN 9780394752846)

Wallace, D.F. *Infinite Jest*, 1996, ISBN-10 0316920045

Bibliography

1. Cerruti L (2005) The inherent complexity of chemistry. In: Proceedings of the International meeting 2004: Complexity in the living: a problem-oriented approach. Consiglio Nazionale delle Ricerche, Roma
2. Casti JL (1994) Simple and complex models in science, Written version of a keynote lecture presented at the joint meeting of the German and Austrian societies for systems research. Cottbus, Germany
3. Von Foerster H (2007) Cibernetica ed epistemologia: nuove prospettive. In: Bocchi G, Ceruti M (eds) La sfida della complessità. Bruno Mondadori, Milano, pp 88–116
4. Morin E (1992) Method: towards a study of humankind, Volume 1, The Nature of Nature. Peter Lang, New York
5. Villani G (2008) Complesso e organizzato. Franco Angeli, Milano
6. Gentili PG (2021) Why is complexity science valuable for reaching the goals of the UN2030 agenda? Rendiconti Lincei Scienze Fisiche e Naturali 32:117–134. https://doi.org/10.1007/s12210-020-00972-0
7. Anderson RA, Crabtree BF, Steele DJ, McDaniel RR (2005) Case study research: the view from complexity science. Qual Health Res 15(5):669–685. https://doi.org/10.1177/1049732305275208
8. Sustainability Directory. https://climate.sustainability-directory.com/term/relational-ontology/. Accessed 2 Apr 2026; The original text has been adapted, according to the Creative Common license CC BY 4.0. (Free for Reuse with Attribution)
9. Ambrogi P, Floriano MA, Ghibaudi E (2009) Sistemi complessi e didattica delle scienze: l'esperienza di SPAIS2008. La Chimica nella Scuola 2:140–146. https://www.societachimica.it/sites/default/files/chimind/pdf/2025-06/cns_2009-2.pdf
10. Borutti S (2023) Complessità e critica dell'epistemologia. InCircolo 15:12–32. https://www.incircolorivistafilosofica.it/miscellanea/
11. Villani G (2009) Complessità sistemica. Un concetto generale di notevole importanza chimica. La Chimica nella Scuola 2:109–114. https://www.societachimica.it/sites/default/files/chimind/pdf/2025-06/cns_2009-2.pdf
12. Popolla F, Ghibaudi E (2010) Il recupero dei concetti di teleonomia e complessità in ambito scientifico nella prospettiva del dialogo tra scienza e teologia. Archivio Teologico Torinese 16(2):333–351

13. Morin E (2005) Restricted complexity, General complexity. In: Proceedings of the Colloquium "Intelligence de la complexité: Épistémologie et pragmatique". Cerisy-La-Salle, France., English translation by Carlos Gershenson
14. Stengers I (2007) Perché non può esserci un paradigma della complessità. In: Bocchi G, Ceruti M (eds) La sfida della complessità. Bruno Mondadori, Milano, pp 37–59
15. Morin E (2007) Le vie della complessità. In: Bocchi G, Ceruti M (eds) La sfida della complessità. Bruno Mondadori, Milano, pp 25–36
16. Rosen R (1985) Anticipatory systems. Pergamon Press, Oxford
17. Cerruti L (2003) Il riduzionismo. Prospettiva storica e problemi attuali. In: Atti del XII Congresso della Divisione di Didattica Chimica. CNR, Pisa, pp 5–30
18. Mikulecky DC (2001) The emergence of complexity: science coming of age or science growing old. Comput Chem 25:341–348
19. Morin E (2002) Le complexus, ce qui est tissé ensemble. In: Benkirane R (ed) La complexité, vertiges et promesses. Le Pommier, Paris
20. Cerruti L (2004) La chimica e le ricerche sulla complessità. Sistemi complessi e unità autopoietiche. La Chimica nella Scuola 2:37–44. https://www.societachimica.it/sites/default/files/chimind/pdf/2025-06/cns_2004-2.pdf
21. Morin E (1999) Seven complex lessons in education for the future. UNESCO
22. O'Connor T (2021) Emergent properties. In: Zalta EN (ed) The Stanford encyclopedia of philosophy (Winter 2021 Edition). https://plato.stanford.edu/archives/winconnor2021/entries/properties-emergent/. Accessed 2 Apr 2026
23. Walliser B (2009) Emergent phenomena. In: Gérin M, Maurel MC (eds) Origins of life: self-organization and/or biological evolution? EDP Sciences, France, pp 95–104
24. Complexity Institute. https://www.complexityinstitute.it/bussola-della-complessita/. Accessed 2 Apr 2026
25. Maturana HR, Varela FJ (1987) The tree of knowledge: the biological roots of human understanding. New Science Library, Shambhala, Boston
26. Wiener N (1968) Cybernetics, or control and communication in the animal and the machine. The MIT Press, Boston
27. Prigogine I (1997) The end of certainty: time, chaos, and the new laws of nature. Free Press, New York
28. Prigogine I (2002) La fin des certitudes. In: Benkirane R (ed) La complexité, vertiges et promesses. Le Pommier, Paris
29. Von Foerster H (1960) On self-organizing systems and their environments. In: Yovitz MC, Cameron S (eds) Self-organizing systems. Pergamon Press, pp 31–50
30. Prigogine I (2007) L'esplorazione della complessità. In: Bocchi G, Ceruti M (eds) La sfida della complessità. Bruno Mondadori, Milano, pp 155–169
31. Taylor M (2001) The moment of complexity: emerging network culture. The University of Chicago Press, Chicago
32. Monod J (1971) Chance and necessity: an essay on the natural philosophy of modern biology. Knopf, New York
33. Hanrahan G (2012) Introduction to environmental chemistry. In: Key concepts in environmental chemistry. Elsevier, Amsterdam, pp 3–30
34. Cerruti L (2009) Breve storia dell'ecotossicologia. Osservazioni epistemologiche e riflessioni etiche. In: Atti del XIII Convegno nazionale di Storia e Fondamenti della Chimica. Accademia dei XL, Roma
35. Ghibaudi E, Cerruti L (2017) Chemical substance, material, product, goods, waste: a changing ontology. Found Chem 19:97–123. https://doi.org/10.1007/s10698-017-9281-8
36. Vernadskij VI (1929) La biosphère. Librairie Felix Alcan, Parigi
37. Tagliagambe S (1994) Uno scienziato ai confini tra più saperi. In: Vernadskij V, Pensieri filosofici di un naturalista, Teknos, Roma
38. Maurer BA (2009) Ecological complexity. In: Meyers RA (ed) Encyclopedia of complexity and systems science. Springer Science + Business Media, New York
39. Lotka AJ (1925) Elements of physical biology. Williams and Wilkins, Baltimore

40. Charlson RJ, Lovelock JE, Andreae MO, Warren SG (1987) Oceanic phytoplankton, atmospheric sulphur, cloud albedo and climate: a geophysiological feedback. Nature 326:655–661
41. Lovelock J (1992) The evolving Gaia theory. Paper presented at the United Nations University, Tokyo
42. Tagliagambe S (1992) Introduzione. In: Mendeleev D, Sullo spiritismo, Bollati Boringhieri, Torino
43. Gordin M (2018) The chemist as philosopher. In: Scerri E, Restrepo G (eds) Mendeleev to Oganesson. Oxford University Press, New York
44. Nuta F, Mustata Wilson G, Boateng GO, Nuta AC (2025) Environmental degradation, health, and socioeconomic impacts. Front Environ Sci 13:1732268. https://doi.org/10.3389/fenvs.2025.1732268
45. Cairns Jr J (2008) Putting the eco in ecotoxicology. SETAC annual meeting, 2008, Tampa, FL. http://www.johncairns.net/Commentaries/SETACcomm.pdf. Accessed 2 Apr 2026
46. Von Foerster H (1984) Observing Systems. Intersystems Publications, Seaside, CA
47. Jonas H (1985) The imperative of responsibility: In Search of an ethics for the technological age. The University of Chicago Press, Chicago

Environmental Complexity Makes Intriguing the Definition of Pollutants

Claudio Minero

Abstract

A web search for what a pollutant is provides numerous definitions and concepts. All these definitions are reasonable, even if they only answer part of the question and require a general framework to understand their meaning. These different definitions arise from the fact that the environment is a complex system and that its relationships with pollutants are highly nonlinear, with loop-back interactions. The system's components do not respond instantaneously to each other's changes, but only after a certain time lag and often on multiple internal time scales. In a very concise way, pollution is the disturbance in the chemical or biological state of environment due to emissions that produce adverse effects. An appropriate definition of the environment is provided, based on the risk assessment on defined receptors and the analysis of the environment as a complex chemical reactor. This includes the dispersion and specific interaction/transformation of chemical species in environmental compartments, each of which is characterized by distinct chemical and physical processes. Seven criteria for defining a pollutant are discussed.

Keywords

Pollution · Complex system · Criteria for definition

C. Minero (✉)
Department of Chemistry, University of Turin, Torino, Italy
e-mail: claudio.minero@unito.it

M. Minella (ed.), *Complexity in Environmental Chemistry*,
https://doi.org/10.1007/978-3-032-31774-2_2

Chapter Learning Goals

- Various web definitions of pollutants
- The environment as a complex system
- Types of emissions
- Effects of pollution at different time and size scales
- The risk and its relationship with different receptors
- Environmental compartments and their exchanges
- Specific interactions of chemical species within a compartment
- Seven criteria for defining a pollutant

1 Introduction

A web search on what pollution is and how a pollutant (the agent that creates pollution) can be defined yields many definitions, each defining only a few aspects. Pollution is defined as "the presence in or **introduction into the environment** of a substance which has harmful or poisonous effects." [1] In reference to air pollution, the WHO defines pollution as "contamination of the indoor or outdoor environment by any chemical, physical, or biological agent that modifies the **natural** characteristics of the atmosphere." [2] Along the same lines is the EPA definition [3], also cited by Wikipedia under "Pollution," which states that pollutants are "Any substances in water, soil, or air that degrade the **natural** quality of the environment, offend the senses of sight, taste, or smell, or cause a health hazard. The **usefulness** of the natural resource is usually impaired by the presence of pollutants and contaminants."

The now obsolete IUPAC definition of an air pollutant [4] is " A substance, gaseous material, or aerosol that has been introduced into the air (either by human activity or by natural processes) in sufficient concentration to produce a measurable effect on humans, animals, vegetation, or materials (monuments, etc.): SO_2, NO_2, H_2S, CO, hydrocarbons, etc." This definition introduces the concept of sufficient concentration, or the need to define a **threshold concentration** to classify a pollutant. The threshold concentration does not have a univocal meaning because it depends on the receptor, be it humans, animals, vegetation, or materials. It is nevertheless important when it comes to legally establishing concentration limits that must not be exceeded to protect the health of citizens. Following this reasoning, it can be assumed that a pollutant can be **defined by law**. Different legal systems respond differently to social and political needs. Legislative implementation can sometimes compensate for the lack of "scientific evidence" in the definition of a pollutant, as in the case of human toxicity. It should be noted that even **undefined mixtures of chemical compounds** can be legally defined as pollutants, and therefore a pollutant is not a single chemical substance. For example, the State of California, as well as other national and supranational regulations (such as the EU), legally define PM_{10} as a pollutant above a certain concentration, to avoid endless legal disputes over how and why PM_{10} should be considered a pollutant. [5]

To complicate the definition, IUPAC introduces the definition of air **contaminant** (in atmospheric chemistry) [6] as "A substance, gaseous material, or aerosol, which is present in an air mass at levels greater than in clean air. An air contaminant has been commonly added by anthropogenic activity." Here, the definition of contaminant is related to "clean air" as reference, a concept like that of "natural quality" reported above. The difference between a contaminant and a pollutant would therefore lie, at least in some interpretations, in the fact that a contaminant becomes a pollutant when its concentration exceeds a threshold above which adverse effects are observed.

The Merriam-Webster Dictionary defines pollution as "the action of making an environment **unsuitable or unsafe for use** by introducing chemical or manufacturing waste" and the pollutant as "something that pollutes, especially, something (such as chemical or manufacturing waste) that makes an environment unsuitable or unsafe for use."[7] In this regard, often new substances, new forms of existing substances, and modified life forms, including chemicals and other new types of engineered materials or organisms not previously known (**new entities**) to the Earth system, are included.[8] In the framework of the planetary boundary (PB) approach, that aims to define a safe operating space for human societies to develop and thrive, based on our evolving understanding of the functioning and resilience of the Earth system, Steffen et al. [9] renamed the "chemical pollution" boundary to "novel entities" (NE), as well as naturally occurring elements mobilized by anthropogenic activities. They argued that the anthropogenic introduction of novel entities into the environment constitutes a matter of global concern when such entities display **persistence**, **mobility** leading to widespread distribution, **accumulation** in organisms and environmental compartments, and the potential to adversely affect critical Earth system processes or subsystems. The definition given by Britannica is "the addition of any substance (solid, liquid, or gas) or any form of energy (such as heat, sound, or radioactivity) to the environment **at a rate** faster than it can be dispersed, diluted, decomposed, recycled, or stored in some harmless form." [10] The cited page is highly informative both on the type of pollutants and on the type of pollution of the various environmental compartments (air, water, soil) and on the historical evolution of the concept of pollution. On water pollution it adds "the release of substances ... to the point that the substances interfere with beneficial use of the water or with the natural functioning of ecosystems."

From this brief overview, numerous definitions and concepts emerge that are reasonable, reflect only part of the problem, and require a general framework to understand what their meaning is. These different definitions arise from the fact that the environment is a complex system, and that its relationships with pollutants are highly nonlinear.

A complex system is composed of multiple components or subsystems that can interact with each other (through feedback) in a nonlinear manner. Well-known examples include the climate system, social and economic systems, and even simple ecosystems. In general, a problem is linear if it can be broken down into a set of independent subproblems. A system is linear if it responds proportionally to the inputs. The principle of superposition of effects is then said to apply to that system.

Natural systems and their interactions are essentially nonlinear. However, to simplify initial investigations or for practical purposes, linearity assumption is often used first.

A nonlinear system is more complex the more parameters are required to describe it. Therefore, the complexity of a system is not an intrinsic property, but always refers to its description, and therefore depends both on the model used in the description and the variables considered. The greater the quantity and variety of relationships between the elements of a system, the greater its complexity, provided that the relationships between the elements are nonlinear. The environment is characterized by a notable variety of elements that interact with each other in a nonlinear fashion, and therefore even their interactions with a single constituent, considered as pollutant according to the definitions described above, are a complex problem.

In complex systems, time plays a crucial and often counterintuitive role. A characteristic feature is the presence of temporal lags or delays between a cause and its effect within the system. Unlike linear systems where the response is direct (immediate cause-effect), in complex systems many dynamics are subject to significant delays. For example, an environmental ecosystem can react slowly to changes: the release of CO_2 today could produce marked climate effects only after decades due to accumulation in the atmosphere and oceans (involving sea level rise or ocean acidification). The system's components do not respond instantaneously to each other's changes, but only after a certain **temporal lag**. This means that the system can accumulate latent tensions or imbalances that only become visible later, making it more difficult to connect causes and effects. A complex system often has multiple internal time scales. Different processes occur on different clocks: some components react in milliseconds, others in months or years. This temporal heterogeneity can create inconsistencies because assessments based on rapid signals may not account for slow but cumulative effects. This delay between cause and effect makes it difficult to perceive the urgency of the pollution problem (and not only), but also to define what a pollutant is.

Rockström and colleagues reported that "Although Earth's complex systems sometimes respond smoothly to changing pressures, it seems that this will prove to be the exception rather than the rule. Many subsystems of Earth react in a nonlinear, often abrupt, way, and are particularly sensitive around threshold levels of certain key variables. If these thresholds are crossed, then important subsystems, such as a monsoon system, could shift into a new state, often with deleterious or potentially even disastrous consequences for humans."[11] In this paper the planetary boundaries due to the chemical pollution, like the amount emitted to, or concentration of persistent organic pollutants, plastics, endocrine disrupters, heavy metals and nuclear waste in the global environment, or the effects on ecosystem and functioning of Earth system are reported as still to be determined.

In complex systems, sudden transition phenomena are extremely frequent. These phenomena, called "**critical transitions**," are a distinctive aspect of complex systems [12] and are also known as "bifurcations," or points where a small variation in an internal or external parameter can lead the system from a stable state to a drastically different one. In the presence of such bifurcations, the system no longer reacts linearly, but rather through sudden and often irreversible changes. For example, a gradual increase in pollution (the persistence of certain compounds in the environment) can exceed a critical threshold (tipping points) that rapidly transforms an ecosystem from healthy to degraded, making it difficult or impossible to return to its original state. Critical thresholds remain invisible until they are crossed and are partly predictable by the slowing of dynamic resilience (critical slowing down), which signals that the system is losing its capacity to absorb perturbations and is approaching a structural threshold, beyond which regime change is inevitable [13]. Therefore, any change that, whether within a limited or broader ecosystem, could lead to a possible critical transition, must be considered pollution.

In summary, the main concepts that emerge from both the various definitions of pollution and the behavior of the complex systems described above, and whose meaning/relationships must be examined to properly address what pollution is, are the following: #1 Natural environment / environmental compartments or media; #2 Introduction / presence of substances/energy in the environment; #3 Adverse change / undesirable environmental effects / degradation of natural quality; #4 Novel entities; #5 Rate of release, increasing rate of production, rate of transformation or persistence or mobility among environmental compartments; #6 Risk potential; #7 Usefulness of the natural resource. The analysis that follows attempts to refine these concepts.

2 Essentials for Pollution Definition

2.1 Natural Environment

If according to the EPA definition [3] that a pollutant is any substance in water, soil, or air that degrades the natural quality of the environment, the natural environment needs a description, but if we look at the natural environment, we can see mountains, rivers and waterways, urban settlements, cultivated fields, clouds and rain, air, wind, and roads. What is the natural environment? It does NOT exist unless we mean the wild environment, but where is this still found? Man has always interacted strongly with the environment, because it is a resource from which to obtain food, wood, minerals, and water. Various human societies, after the development of agriculture, have modified the soil, moved waterways, flooded fields (see rice fields), built villages and then cities, dug quarries for stone materials, cut down forests for the need to supply wood to create cultivable fields, for construction, for industry, for heating and furniture. Here, it is interesting to note that the ancient Romans conducted extensive deforestation in the Italian Apennines to build their naval fleets. In the First Punic War—264-241 BC—the Romans built more than 1110 quinquereme ships, for which wood and pitch was obtained from the conifers of the Sila Mountains. [14] The environment therefore has no precise definition, and a vague term cannot be used for referring to it.

The definition of environment is "surrounding." It is what surrounds a thing. "It is everything that isn't me," to quote a definition attributed to A. Einstein. It is where we live.

The complex system "environment," can be divided into the following: • The physical environment, which includes physical or non-living elements such as soil, water, and air. It also includes climatic factors such as solar radiation, rainwater, precipitation, humidity, pressure, and winds. • The biotic environment that includes plants, trees, animals, mammals, underwater life, humans, and microorganisms such as bacteria and fungi. • The anthropic environment that includes cities, agricultural and mining sites, indoor environments (in homes), and outdoor environments (urban air, etc.).

These subsystems are further classically divided into **environmental compartments**, such as the atmosphere, biosphere, soil, surface water (lakes and rivers), groundwater, boundary zones such as wetlands and estuaries, seas, and oceans. Each of these subsystems has its own physical, chemical, and biological processes, which can be identified as specific. Environmental Chemistry discipline develops along these lines. However, understanding the dynamics of environmental compartments requires knowledge of their interactions, that is, their related processes of exchange of matter and energy. For example, living things are highly dependent on one another; the biotic environment strongly modifies the physical environment; physical and chemical factors strongly influence both the physical and biotic environments. Therefore, environmental studies include expertise in physics, geophysics, geography, pedology, hydrology, chemistry, biology, history, economics, and more. What a pollutant is derives from the interaction it has with these compartments, and how it is exchanged among these environmental compartments.

Finally, as already highlighted above, the environment is an economic resource for human civilizations, providing water, food, renewable and non-renewable resources, minerals, energy, and recreational use. Since the Earth is a closed system, with necessarily limited material resources (minerals, elements, energy, living and fossil organic matter), it is not an infinite economic resource. This observation gives rise to the issue of environmental sustainability and the circular economy, which, although very important, are outside the scope of this chapter, which focuses on the definition of pollutants. However, some aspects of environmental sustainability are relevant to pollution, and vice versa.

By reducing the system to environmental compartments, it is tempting to reduce the definition of pollutants to entities that modify an **ecosystem** (not the natural environment, which has no definition), that is, a subset of environmental compartments limited to a particular environmental configuration. The concept of ecosystem has long been known in ecology. It consists of three main actors: 1) producers, who photosynthesize with sunlight and thus produce organic material using the pool of inorganic materials; 2) a first trophic level of consumers that utilize photosynthesized biota (herbivores); a second trophic level of consumers that feed on herbivores; 3) decomposers (bacteria and fungi) that consume both producers and consumers and release inorganic and organic matter in the pool usable by producers. The ecosystem model implies a closed system of chemical and biological cycles in a kinetic steady state because it is open to the input of energy from sunlight. It is in fact a system with a closed structure composed of a chain of spontaneous processes that, once they have occurred, maintain the system in a steady state. An ecosystem (or ecological system) is formed by the interaction of the local environment and biological organisms. The biotic and abiotic components are linked through nutrient cycles and energy flows. By definition an ecosystem does not generate waste. Ecosystems are dynamic entities: they are subject to periodic external perturbations and are always recovering from past perturbations. The ability of a system to absorb perturbations and reorganize during change, while essentially maintaining the same function, structure, identity, and feedback loops, is defined as ecological resilience. Consequently, the kinetic steady state is not constant indefinitely, as an ecosystem is controlled by external and internal factors. External factors such as climate are important. It follows that an ecosystem is a snapshot of the interactions of its components over a narrow time frame. Therefore, identifying a pollutant through its impact on an ecosystem—that is, through the variations induced in the steady state of its chemical and biological populations—is valid only within the ecosystem's time frame of stability.

The interaction of various ecosystems with human activity weakens the utility of the ecosystem concept even further. The industrialized production system has an open structure because it uses copious amounts of energy, exceeding that recovered and reintroduced into the system. It redistributes inorganic matter (for example, primary production is moved from cultivated fields to areas of use, such as urban settlements, and the original ecosystem is depleted of nutrients). It also experiences significant upstream exploitation through the indiscriminate use of natural resources for production purposes. It creates artificial materials and transfers individual components across space and time. Furthermore, downstream exploitation occurs due to the exceeding of natural absorption capacity, resulting in **waste production**. Consequently, the temptation to reduce the definition of a pollutant to something that modifies an ecosystem proves highly unproductive because there is essentially no reference state. This definition must be reformulated by returning to the specific environmental compartments of each environment, noting the variations that have occurred over relatively short periods of time or over historical series, due to the introduction into them of substances or energy that cause disturbances in the chemical or biological state of the environmental compartments.

The environment is not just a backdrop for economic activity, but an integral part of it, providing essential resources and services. These include the use of mineral resources, including rare metals of current technological importance, fossil fuels, renewable fuels, food, and freshwater. These resources are finite, and their use for economic production and consumption invariably generates environmental impacts. Conversely, changing the usability of environmental assets has direct economic value, not only for their original "aesthetic" aspect (usability), but also for their sustainability. This reasoning is in agreement with what is quoted in the EPA definition [3], which states that pollutants are "Any substances in water, soil, or air that ... offend the senses of sight, taste, or smell, or cause a health hazard. The **usefulness** of the natural resource is usually impaired by the presence of pollutants and contaminants." This sentence is in agreement also with what was previously quoted by the Merriam-Webster Dictionary that defines pollution as "the action of making an environment **unsuitable or unsafe for use** by introducing chemical or manufacturing waste." [7]

The most obvious example is water, both surface and groundwater. Various sources of aquifer deterioration exist due to agriculture, with the use of pesticides, fertilizers, and livestock farming with the discharge of manure and livestock waste. The latter induce direct and indirect environmental changes through interaction with the soil, such as the induced reduction of Mn(IV) and Fe(III) from soils, increasing the concentration of Mn(II) and Fe(II) in groundwater. Water contamination can be accidental, from wells, surfaces, or landfills, or intentional (injection of radioactive materials into wells). Classic examples include NAPLs (non-aqueous phase liquids, which are less dense than water and therefore found above the water table, for which remediation with wells is still possible) or DNAPLs, which are denser than water and can accumulate in aquifer fractures, leading to solubilization for decades and are not amenable to removal with wells.

Water has a **direct economic value** because it is essential for agriculture and for human consumption of drinking water. For this reason, the unpleasant taste, mainly due to the dominant cations (Fe(III) > 50 ppb, Cu^{2+} > 2.5 ppm, Zn^{2+} > 5 ppm), and the unpleasant, although non-toxic, odor caused by additional contaminants must be considered critical for water use. Some odors are produced by blue-green algae, and actinomycetes, such as earthy/musty odors (due to geosmin - OTC 4 ng/L, where OTC is the odor Threshold Concentration), 2-methylisoborneol -OTC 9 ng/L, and septic/swampy odors (dimethyltrisulfide -OTC 10 ng/L) present in the water distribution system.

Air also has direct environmental economic value, not only due to odor but also due to visibility, which influences human perception of the environment and quality of life (fog, clear skies, safety of car and air traffic). Soil has clear environmental economic value as it is used for agriculture. Soil productivity can be affected by acid rain and climate change (temperature, rainfall amount).

Pollution can also alter *the appearance of the landscape*. This is not a secondary aspect since the environment is where we live and where recreational activities and tourism develop for the sake of nature. Deterioration of the environment, even due to impacts that are not particularly harmful to health, but which limit its usability (for example macro plastics and odors), constitutes damage that must be considered in the definition of pollutant.

2.2 Emissions

Disturbances in the chemical or biological state of environmental compartments, which we can currently consider pollution, are caused by both natural and anthropogenic emissions. Natural events include volcanic emissions and wildfires, which are a major source of global air pollution, with significant impacts on air quality and climate. [15]. Physically, materials mobilized during and after wildfires often contain a mixture of ash, dust, and other particles that can alter the surfaces and soils on which they are deposited. Increased vertical and horizontal particle fluxes due to wind (i.e., aeolian transport) can occur for months after the fire. Both black carbon and PAHs are deposited. Ozone, sulfur dioxide, VOCs, and PM downwind of fires can have negative effects on plants, including reduced stomatal conductance, oxidative stress, and reduced CO_2 assimilation [16].

A significant source of emissions of various substances into the environment is due to human activities.

The most palpable pollution is air pollution, which is not recent. Pliny the Elder (61 AD) reported that "as soon as I left the heavy air of Rome and the row of smoking chimneys emitting all sorts of vapors and soot into the air below, I immediately felt a benefit." Since the beginning of engineering development (Italy-Germany 1500) and the invention of the steam engine (Watt 1769), there has been a sharp increase in energy consumption, as well as smoke and ash emissions from boilers, furnaces, and later steam engines (e.g., locomotives).

The release of constituents into the air that are not present there, or present at much lower concentrations, is primarily due to emissions from combustion processes, traffic and transportation, domestic heating, industrial boilers, and waste incineration. Emissions also occur in the form of vapors, such as unburned fuels, solvents, and household emissions. Many of these emissions are unavoidable, but they can be reduced with measures related to environmental sustainability and related remedies, or the circular economy. A well-known example is the shift in energy sources (from oil and coal to solar photovoltaic or wind power) to reduce CO_2 emissions. The same applies to water contamination (treatment of civil and industrial waste) and soil (waste storage).

Source of emissions of various substances into the environment due to human activities include mineral extraction, manufacturing, industrial processes, as well as transportation and storage accidents. Sometimes, dispersion into the environment is intentional, as is the case of herbicides, pesticides, and fertilizers. Dispersal into the environment is inevitable, even for consumer goods that become waste at the end of their useful life. Among these are novel entities (NEs), that is, new substances, new forms of existing substances, and modified life forms, including "chemicals and other new types of materials or engineered organisms not previously known to the Earth system,"[8] as well as natural elements (e.g., heavy metals) mobilized by anthropogenic activities in new ways, new forms, or at substantially faster rates by anthropogenic activities than natural ones, xenobiotic organic chemicals, persistent organic pollutants (POPs), plastics, etc. For these NEs, there is no "natural variability" against which a control variable can track change on human timescales, nor is there a biophysical precedent for identifying environmental thresholds. For these entities, the "before" reference state is "absence," so the effects of NEs can be assessed or estimated: from measurements in the appropriate environmental compartments of their chemical concentration or presence; from direct biological effects, such as the decline of top predators (e.g., following the widespread use of DDT), and the resulting ecological cascades; by the induced physical effects, such as the reduction of albedo with effects on the climate (e.g., for black carbon particles); by the chemical reactions they can induce in the environmental system after release, which are often unknown a priori (e.g., the reduction of stratospheric ozone by CFCs in polar stratospheric clouds). As already cited [9] the introduction of novel entities to the environment is of concern at the global level when these entities exhibit persistence, mobility, with consequent widespread distribution and accumulation in organisms and the environment.

Some chemicals are produced for specific uses, while others are unwanted byproducts, including waste or combustion products such as toxic gases and particles from industrial emissions and fuel combustion. More than 160 million chemicals are known, of which approximately 40,000–60,000 are commercially available. As of December 2020, the EU's REACH (Registration, Evaluation, Authorization and Restriction of Chemicals) regulation registered approximately 23,000 chemical substances. Of these, over 12,000 were registered as non-intermediates (intermediates have reduced safety data requirements) with annual production exceeding 1 t, and approximately 2400 substances were or are undergoing evaluation and analysis, for which environmental impact estimates are not yet available. 6000 of these substances account for over 99% of the total volume of chemicals traded globally. In 2017, the chemical industry was the world's second-largest manufacturing industry. The trend is growing, with chemical sales expected to nearly double by 2030.[8] **Production volume** is directly linked to the release of some chemicals, both immediate, such as pesticides that are intentionally applied and chemicals that end up in the waste stream (e.g., personal care products), and time-lag, where the final release is more difficult to quantify (e.g., CFCs used as blowing agents in polystyrene insulation are released during building renovations, dyes and other plastic additives are released as plastic ages). Chemical production may be responsible for the unavoidable amount and rate of global release, as production is fed into highly complex and difficult-to-quantify global supply chains.

It follows that large-scale production, persistence, mobility (and thus accumulation in organisms and the environment, and biomagnification in the food chain) must be considered elements of the definition of a pollutant, in accordance with the definition already reported in [9, 10].

All these emissions are primary emissions, meaning they originate from a source. However, transformation products of primary emissions may be present in the environment, which in turn cause disturbances in the chemical or biological state of environmental compartments (see below).

The definition of a pollutant must therefore consider the emission, whether unavoidable, accidental, malicious, or intentional, as an important factor. The abatement of emissions and their inventory are important criteria for pollution control.

2.3 Adverse Effects

A chemical species becomes a hazard, and therefore can be considered a pollutant, when it poses a risk to all entities existing in the environment, which is everything that forms the environment and interacts with it. The concept of risk due to pollution has expanded to include not only ecological impacts but also economic indicators, underscoring the multidisciplinary nature of current environmental challenges. The technical details of risk analysis are reported in the specialized literature [17]. In short, risk is the probability of suffering harm from a hazard, and risk assessment is a procedure for its reduction, elimination, and rational management. Web searches, including those using AI, provide many interesting concepts not listed here. This procedure can be informal and qualitative (subjective, personal life), or rigorous and quantitative (required for environmental legislation). The steps for risk assessment are (1) hazard identification, which for environmental risks is often not easy (extrapolation from bioassays and animal toxicity, epidemiological studies, etc.); (2) dose assessment. This requires defining the microenvironments (e.g., indoor, groundwater, and, for biota, assessing bioavailability); (3) assessing the probability P(i) of an undesirable impact as a function of the dose released. This requires defining the receptors in the microenvironments, i.e., organisms, organs within an organism, population subgroups (children, the elderly), etc.; and (4) the methodology for statistical extrapolation (e.g., the log probit model, the multistage model, with the aim of establishing an acceptable level of risk for the entire population, e.g., maximum acceptable concentration limits to avoid health risks, etc.). Finally, the risk impact is assessed as N(t) x G(t) x P(i,t), where N is the exposed target population and G is the extension over generations. Here, it's clear that discovering the relationships between dose and response of the various compartments is difficult due to the complexity of the environment. For example, for global warming, it's difficult to assess the dose (heat released/induced by numerous processes) and its relationship with specific compartments and the sum of their effects, which is nonlinear.

Risk is assessed at the receptor and therefore depends on where the pollutant is detected or measured. This concept is very important, since between the point of emission and the risk assessment point lies the environment, with all the interactions, transformations, and dispersions caused by the environment (see below). At the risk assessment point, analytical instruments may be located, which, like other receptors, respond only to the *residual concentration*, that is, the concentration of chemical species that have survived the environmental transformations following the point of emission. This objective limitation of measurements and the effect on generic receptors can be reduced with the appropriate choice of *monitoring strategies*, which consider the problem as a whole and define both the assessment/sampling sites and the duration and timing of observations/measurements (as well as the choice of appropriate analytical techniques and their calibration).

This analysis clarifies that concentration is not the primary parameter for defining a pollutant, as for the IUPAC definition [4], although it plays an important role in human and ecosystem toxicity. The opposite is true. Risk analysis derives **threshold concentrations** that are adopted by regulatory bodies to protect receptors. The various receptors to which a pollutant can cause harm (subject to risk assessment) are (1) humans, with regard to acute/chronic toxicity; (2) specific environmental systems, such as vegetation, agriculture, forestry, surface or groundwater, urban air quality, and so on, with the potential for bioaccumulation and alteration of biodiversity; and (3) materials and cultural heritage. These include the potential degradation of metals (a key role in technological applications), fabrics, rubber, and stone (marble, etc.).

From these receptors, further criteria for defining a pollutant arise: (1) biological effects at low concentrations because they can endanger human health, other living systems, or ecological systems; (2) effects on materials and goods produced or usable in human society, such as stone building materials or materials with historical/archaeological/cultural value, and technological materials.

Point emissions create an immediate disturbance in the vicinity of the source on chemical and biological populations. A classic educational example is shown in Fig. 1, where A and B show the variation in chemical populations (concentration) and C and D show the variation in biological populations. [18] If one were to consider the river as an ecosystem, even one located near the emission point, one would also have to consider external disturbances, such as seasonal variations in temperature and flow due to rainfall or groundwater withdrawals.[19] This is a good example of an adverse effect, as the title of this section suggests, or of a degradation of "natural quality" because a direct comparison can be made between "before" and "after" in terms of local loss and reduction in biodiversity.

Local-scale emissions cause effects that can be identified through measurement, particularly for chemical species. Emissions into the aqueous environment occur through an array of pathways, such as surface runoff from urban or agricultural regions, direct discharge of raw or treated wastewater from municipal, hospital, and industrial wastewater treatment plants, and landfill leachate. Their effects are observable through alterations in the chemical or biological state of environmental compartments. These effects include, but are not limited to:

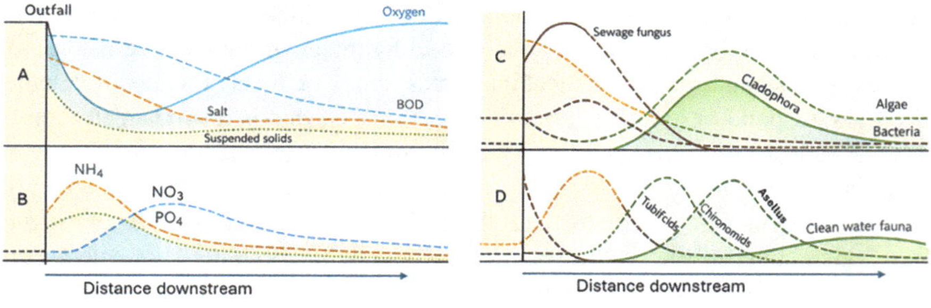

Fig. 1 Schematic representation of the changes in water quality and the population of organisms in a river below a discharge of an organic effluent. A) physical changes; B) chemical changes; C) change in microorganisms; D) change in macroinvertebrates. Near the discharge, there is a significant consumption of oxygen due to the oxidation of DOC, which releases NH_4 (then oxidized to nitrate) and phosphates. Salts are diluted, and suspended solids settle. Sensitive species such as fish and invertebrates that require clean, cold, oxygenated water disappear, while tolerant species such as algae, bacteria, fungi, and invertebrates thrive. When oxygen levels slowly recover, a mix of tolerant and moderately sensitive species reappears, but biodiversity remains lower than upstream. Downstream, the river may return to its original state, but some impacts, such as sediment alteration, may persist. Overall biodiversity and fish populations may decline, and migration routes (such as those of salmon) may be blocked or impeded. Adapted from [18]

- Changes in surface aquatic chemistry. Examples include contamination of water from industrial and domestic wastewater by metals, nutrients (nitrogenous and phosphate-containing compounds that enhance biota development), and acid rain, resulting in sulfate and nitrate deposition in surface waters (rivers and lakes) and a decrease in pH. This in turn increases soil weathering reactions, releasing calcium and magnesium ions, altering the alkalinity and acid/base chemistry of water and soil. [20]
- Contamination of freshwater ecosystems (surface and groundwater) by so-called emerging contaminants (ECs). ECs include a broad group of chemicals and their derivatives, such as pesticides, fertilizers, pharmaceutically active compounds, personal care products, artificial sweeteners, endocrine disruptors (EDCs), and synthetic and natural hormones. [21] ECs can accumulate in food webs, threatening the integrity of entire ecosystems.
- Soil degradation. Contamination by heavy metals, pesticides, industrial waste, and plastics reduces soil fertility and structure, poisons crops, contaminates water, damages ecosystems by killing beneficial microbes, and poses a serious threat to food security and human health throughout the food chain. Effects include nutrient and organic matter depletion, soil erosion, acidification, and desertification, with changes in soil pH. Not all these effects are immediately measurable, and non-local emissions may also be the cause.
- Accumulation in lake and river sediments, with possible mobilization during flood events, with a related impact on water quality.
- Decreased visibility, which is the clarity or transparency of the atmosphere and the associated ability to see distant objects. The terms haze and smog describe the effects of air pollution on visibility. Visibility affects human perception of the environment, but also air flight and road safety. [22]

However, there are cases where a differential comparison of the state of the environment is not straightforward and can only be conducted through time series analysis, since the environment, as a complex system, responds with time lags (see Sect. 1), which can vary depending on the variables. On a global scale (an environment characterized by considerable complexity), some examples include changes in average temperatures, [23] the greenhouse effect and resulting climate change, and the ozone hole [24].

The emission of greenhouse gases (GHGs) such as CO_2 causes significant global warming: the Earth's average temperature has increased by approximately 1.1–1.2 °C since the pre-industrial era (1850–1900), and the rate of warming is accelerating, especially in recent decades, leading to more extreme heat, heat waves, and widespread climate impacts. The decade from 2015 to 2024 was the warmest on record. [23] This also leads to warming seas, melting mountain and Arctic glaciers, rising sea levels, an increased risk of forest fires, and ecosystem disruption. Chlorofluorocarbons (CFCs) and halons, released from products such as refrigerants and aerosols, reach the stratosphere from the troposphere, where they destroy stratospheric ozone. The ozone layer participates in regulating global climate and temperatures and, by absorbing UV light, protects human health, reducing the risk of skin cancer and eye damage in humans, and helps preserve biodiversity, plant growth, food chains, and biochemical cycles.

In conclusion, the adverse effects, for which a pollutant can be defined as such and caused by emissions, are multiple and require assessment on both a local and global scale, over very different time scales, and across different types of environmental compartments. They must also be defined in relation to the object being injured, namely, people, environmental compartments and their subsets (ecological systems), the inorganic physical environment (soil, water), and the anthropogenic environment (civil and industrial materials, cultural heritage).

2.4 The Environment as a Complex Chemical Reactor

The environment is classically divided into the following environmental compartments (atmosphere, water (fresh surface water and seawater and oceans), sediments, soil, and biomass), and the exchange between these compartments is described in various environmental chemistry textbooks (see, for example [25]). The flows between the various compartments and the typical reaction classes for each are shown in Fig. 2.

According to the definition of Pollution as "the addition of any substance (solid, liquid, or gaseous) or any form of energy (such as heat, sound, or radioactivity) to the environment at a rate greater than that at which it can be dispersed, diluted, decomposed, recycled, or stored in a harmless form'" [10], it is necessary to analyze the location of the release, the properties of the environmental compartments, and their interactions with each other. Given an emission factor (which is a rate) as a representative value that relates the quantity of a pollutant emitted into the environmental compartment to an activity associated with the release of that pollutant, the

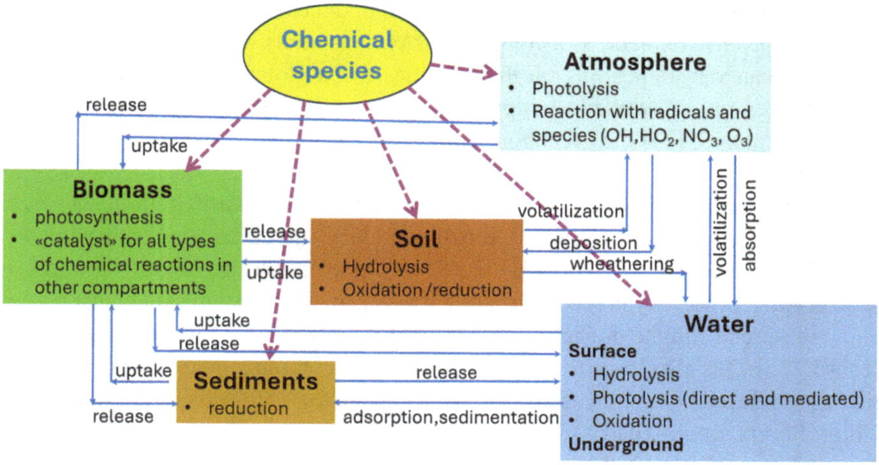

Fig. 2 Emission and main exchange mechanisms and transformation processes of a chemical substance in different environmental compartments

environmental compartment receives a given concentration for a given time (dose) for each of the emission sources. This is the *capacity factor*, or the quantity of pollutants accepted by the environmental compartment. In the absence of environmental reactivity or dispersion/exchange, the environmental concentration of these pollutants will increase steadily and continuously over time.

2.4.1 Dispersion

The release of a species into an environmental compartment is subject to possible dilution in the system, whether aquatic or atmospheric. Atmospheric dispersion is of particular concern because the atmosphere is the most dynamic compartment and can affect even locations very distant from the point of release. The release of a non-volatile and insoluble substance onto the soil causes only local pollution.

The thermal circulation of the atmosphere always follows the pattern shown in Fig. 3 (left). Warm, less dense air tends to rise, and low pressure at the surface induces a horizontal current. These movements are the basis of general circulation. Such movements also occur at the regional or local level and determine the weather. Local winds are sea-land and valley-mountain breezes (see Fig. 3, right). In the case of valley-mountain winds, circulation depends greatly on which side of the valley is warmed by the sun, that is, on the orientation of the valley (E-W, N-S, and intermediate situations). Regional winds can be superimposed on these local circulations, confining the currents in the valley or dispersing them along the valley depending on their direction. This example outlines that local concentrations measured at a given site (at valley or on the N or S slopes) give only partial information on the pollutant impact.

A volume of air released to the ground, if transported upward, expands and cools. Assuming the movement is adiabatic (i.e., without heat exchange, since air is a poor thermal conductor), there is a linear decrease in the temperature T of the air mass with height z. This is called an adiabatic profile (adiabatic lapse rate $-\Delta T/\Delta z = \gamma_d$ of approximately 0.98 °C for every 100 m of height for a dry atmosphere). The surrounding atmosphere, depending on the environmental conditions, can have any profile (from superadiabatic with a minor negative slope, to subadiabatic with a greater negative slope, to isothermal with no variation of T with height z, or with

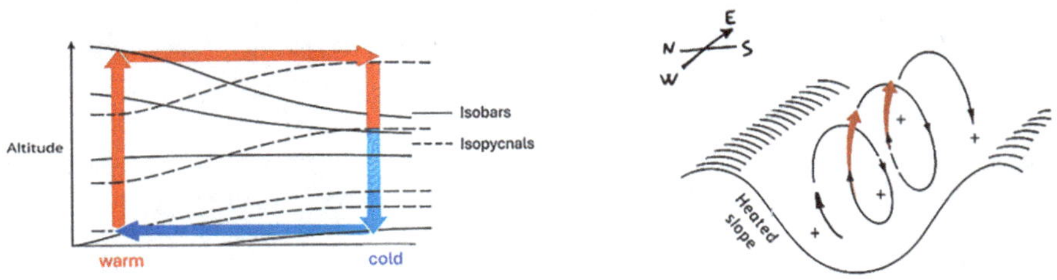

Fig. 3 (Left) Thermal circulation in the atmosphere. In red, warmer air parcels move vertically, while cooler air (in blue) moves horizontally over the ground (wind); (right) a valley ridge wind with the left wall of the valley illuminated by the sun and a regional wind from the W-E

inversion, in which T increases with z and positive slope). These profiles can also be composite, for example with a superadiabatic profile at ground level and an inversion at altitude and can vary during the day depending on conditions of sunlight or fog. The air mass rises until its temperature (adiabatic profile) is higher than the outside temperature (any profile).

An image often reported in textbooks is shown in Fig. 4, which shows the possible atmospheric temperature profile schematically in red dashed lines.

Depending on the atmospheric temperature profile, the air mass can rise indefinitely and be dispersed or rise (if warmer than the surrounding atmosphere) to the inversion height of the atmospheric gradient, where it is confined. The height of the inversion zone defines the volume of atmosphere in which the emitted species can be diluted, and therefore its concentration. Under inversion conditions, the concentration at ground level can increase with increasing emissions, up to critical points of impact on receptors.

It follows that the definition of a pollutant cannot ignore the evaluation of environmental conditions, including potential dispersion, which do not depend on the chemical species itself and its intrinsic properties (see below), but solely on the properties of the environment. Since dispersion changes its concentration locally, a

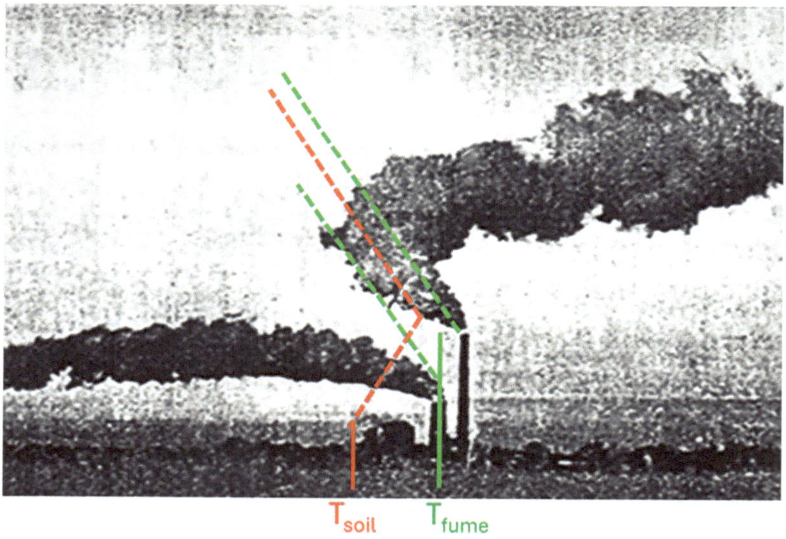

Fig. 4 Smoke plumes from a 150-m-high stack and another 75-m-high stack in Salem, Massachusetts. The photo, taken on a cold February morning, shows the dispersion of fumes at two different heights. The original photo was taken by Ralph Turcotte, a staff photographer for the Beverly (MA) Times. Adapted from [26]. Superimposed on the photograph are a possible thermal profile of the atmosphere in red, and the adiabatic profiles of the smoke in green. There is an inversion at low altitude and an adiabatic above. From the lowest chimney, the fumes, warmer than the air at ground level, rise, cooling until they reach a height with a corresponding atmospheric temperature, where they stop and are confined. In the highest stack, they are always at a higher temperature, so they rise without stopping and disperse

species must also be defined as a pollutant based on the effects it produces as a function of its concentration.

2.4.2 Environmental Dynamics

The fate of various chemical species, including pollutants, depends on the specific properties of each chemical species, both for the exchanges between compartments and the chemical reactions that occur within them. This is the *intensity factor*, which depends on the chemical properties of the compartments combined with the specific chemical properties of the pollutant. *The environment is reactive.* Consider for example the evolution of photochemical smog, which shows how emissions and atmospheric reactivity due to the presence of sunlight modify the concentration of chemical species in the air throughout the day (Fig. 5).

Transformation factors in the environment can be abiotic, i.e., photochemical, hydrolysis, and oxidation/reduction reactions, or biotic, including photosynthesis, hydrolysis, and redox reactions, catalyzed by bacteria, fungi, and generally aquatic and terrestrial flora. Redox processes are determined by the compartment's redox potential (oxidative in contact with the atmosphere, reductive in anoxic environments). Photolysis processes can be either direct (through absorption of light by the species, typically in the UV region, but also in the Visible region for iron complexes) or indirect, i.e., mediated by products formed by photolysis of other species.

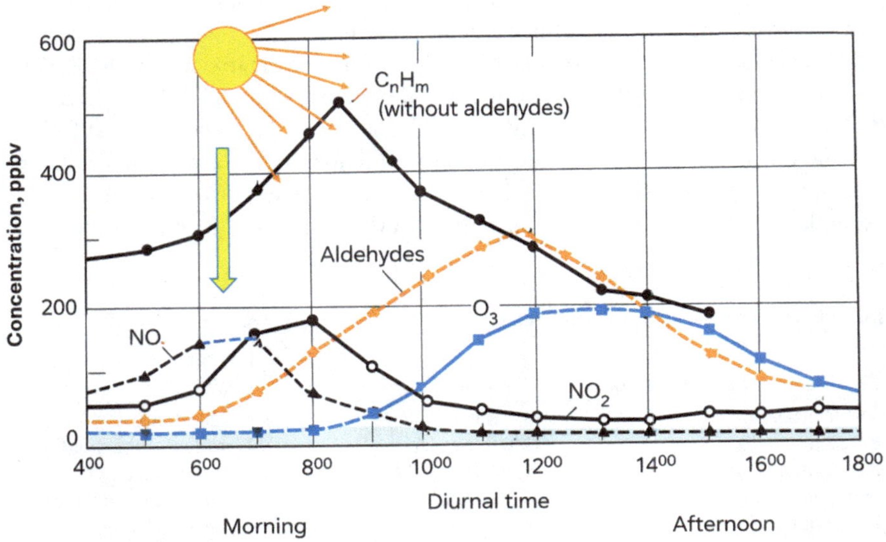

Fig. 5 Diurnal cycle of photochemical smog observed in Los Angeles 1952–1957. During peak traffic hours in the morning, vehicles release significant amounts of nitrogen oxides (NO_x), primarily nitric oxide (NO), and unburned hydrocarbons (C_nH_m, HCs). These emissions are not solely products of cars. HCs undergo oxidation processes facilitated by hydroxyl radicals in the atmosphere, forming peroxy radicals (and aldehydes), which then convert nitric oxide into nitrogen dioxide. This is a key player in forming ground-level ozone through photochemical dissociation, which in turn reacts with residual NO and VOCs, comprising aldehydes. Adapted from [27]

These photolytically produced species are $^1O_2$, from the sensitization of humic acids and natural dyes; superoxide HO_2, from the photolysis of Fe(III) complexes and secondary reactions in the atmosphere; H_2O_2, $^1O_2$, and the OH radical, in water from the photolysis of Fe(III) complexes, nitrate and nitrite anions, from the decomposition of O_3, and photolysis or Fenton reaction of H_2O_2.

For a given chemical species, these reactions lead to its degradation at a given rate, which is translated either as a residence time or as a half-life. The two concepts have different origins: the first from engineering concepts (such as the volume/flow ratio), and the second from chemical kinetics, assuming first-order kinetics in which the concentration of reactive species in the environment is constant. Environmental half-lives can vary greatly, from 2nd to decades, depending on the *persistence* of the species (if not at all reactive to the processes mentioned above).

2.4.3 Species-Specific Properties

Both the reactivity and the distribution of a chemical species in the various environmental compartments depend, in addition to the properties of the compartments (see below), on various specific properties of the species. Table 1 reports how the specific properties of a compound (the measurable chemical-physical parameters) influence its distribution in the various environmental compartments and its transformations, and what are the concepts used in environmental sciences. Each compound has different chemical-physical characteristics in relation to the air, water, biota, soil, and sediment compartments, and therefore is distributed and reacts in the various environmental compartments in different ways.

Some cases are here reported as examples. Consider the toxicity (LD_{50}-Rat) for three tetrachlorodibenzo[1,4]dioxins, (TCDD): the isomers (2,3,7,8), (1,3,7,9) and (6,7,8,9) have LD_{50}s of 4×10^{-5} g/kg, >0.1 g/kg and > 1 g/kg respectively. It follows that the positional isomer is important, and consequently the analytical evaluation must not quantify the total TCDD, but the individual species. Consider as a second example the chemical reactivity of butyl chlorides in water. It is known that the

Table 1 Properties specific to the pollutant

Physical/chemical parameters	Distribution	Environmental significance
Vapor pressure	Air-water	Evaporation
Solubility	Surface-water	Transport at the interfaces
Adsorption isotherm	Sediment-water	Accumulation
Henry constant	Biota-water	Sedimentation and burial
Lipophilicity /distribution coefficient		Bioaccumulation
Transport coefficients	**Molecular transformations**	
Speciation, acid/base equilibria	Hydrolysis	Residence time
Redox potential	Photolysis	Half live
Kinetic constants	Redox	
In water/air		
With major environmental species		
Structure-reactivity correlations		

hydrolysis rate (expressed as $t_{1/2}$ at pH 7) increases in the series ter- (30 s) > sec- (40 days) > normal (1 year). It follows that tert-butyl chloride will never be detectable in the environment, even if emitted (while the corresponding alcohols will instead be detected as reaction products). Finally, consider as a third example the distribution between biota and water of three isomers of tetrachloro-phenols. We obtain that the water/hydrophobic surface partition coefficient (2.9% of organic carbon at pH 6.8) is 10, 20, and 180 respectively for the isomers (2,3,5,6), (2,4,5,6) and (3,4,5,6). This implies that the latter is present in the hydrophobic phase, while the former may still be partially present in water. The environmental fate (in which compartment they are found and how they can be moved) of the different structural isomers is therefore different. For the isomers (2,3,5,6), (2,4,5,6) and (3,4,5,6), the pK_a values are 5.04, 5.40, and 6.35, respectively, which justify their distribution according to their fractions of associated species at a given pH,.

It follows that any chemical species, be it a new entity, will behave in the environment depending on its chemical-physical characteristics. The definition of a pollutant cannot therefore ignore the evaluation of *its distribution properties* in the various environmental compartments, which are characterized by different mobility (air > water > > soil and sediments) and reactivity.

2.4.4 Properties of Environmental Compartments

Each compartment, considered without perturbation of xenobiotic species, has its own chemistry, due to its constitution and the possible exchange with other compartments. Depending on the conditions, this determines that it has a pH and pε (a variable directly proportional to the redox potential E) due to the presence/absence of oxygen and other redox species (Fe(III), Mn(IV), nitrates, sulphates, organic matter), as well as different minerals and inorganic anions and cations. The pH and pε determine the speciation/transformation in environmental compartment of the majority and minority species, including xenobiotic/anthropogenic ones. The majority species are produced by photosynthesis (living organic matter, debris) and in addition to sunlight directly used in photochemical reactions, they are the driving force of all environmental transformations. The analysis for each environmental sector is outside the scope of this text and can be found in excellent textbooks [25].

Here it is important to note that for the soil, water, and sediment compartments, redox reactions have a fundamental role. In these compartments there are various environmental species (O_2, nitrates, Mn(IV), Fe(III), sulphates), with their redox reduction potential (respectively to H_2O, N_2, Mn(II), Fe(II), sulfides), which can oxidize organic matter to CO_2 if available. The scheme in Fig. 6 shows the standard redox potential at pH 7 for the various species mentioned.

Different environmental compartments behave differently due to the different availability of these oxidizing species. Any type of organic matter, natural resources, organic waste, and other pollutants, such as new entities, are eventually oxidized/reduced depending on the compartment's mean redox potential. Potential pollutants can be oxidized or reduced, *forming secondary species*, which in turn are subject to further potential transformations.

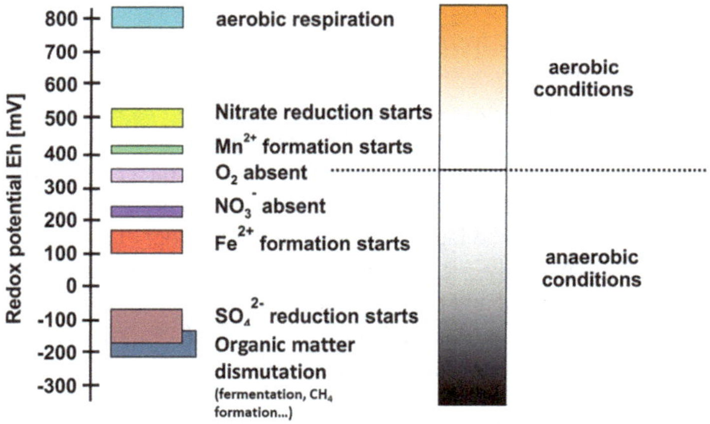

Fig. 6 Redox potentials for environmental oxidants

Take soil, for example. It is in contact with the atmosphere and therefore has a high redox potential. Oxidation reactions will be favored. If the soil is flooded, the present organic matter consumes dissolved O_2 (low solubility and poor diffusivity from air), and the redox potential drops. Oxidizing the organic matter requires a reduction of nitrate (denitrification), if present, or a reduction of MnO_2 (if present), or a reduction of Fe(III), if present in the soil as a constituent in clays, with the release of Fe(II) into the water. The nitrogen cycle in soil (the aforementioned denitrification reaction, or the nitrification reaction, i.e., the oxidation of ammonium released by decomposing biota or introduced into the soil as fertilizer) is not environmentally "neutral," because the two reactions produce byproducts, N_2O and NO, respectively, which are released into the atmosphere. In this sense, even a fertilizer, which can directly be a primary source of water contamination, due to the specific reactivity of one environmental compartment, can induce the production of secondary pollutants in another environmental compartment.

In sediments, conditions are anaerobic, and the oxidation of organic matter (including pollutants) is driven by sulfates (present in seawater, less so in freshwater), which are reduced to sulfides. In the absence of environmental oxidants, the same organic matter can decompose, forming fermentation products (and then CO_2, methane, hydrocarbons, and elemental carbon over increasingly longer periods). In this transformation, the organic matter of sediments and soils becomes increasingly lipophilic, increasing the possibility of partitioning organic molecules, including organic pollutants, which are retained there and made less available to water and biota. In an aquatic system (see Fig. 7 for a stratified lake), these processes are distinguishable as a function of depth and lead to an internal recirculation of elements in their various oxidation states, changing **speciation** and then their **bioavailability** and possible **biotoxicity**.

Each type of environmental compartment behaves like a complex chemical reactor, transforming chemical species and possibly releasing them, also transformed (as environmental transformation products) into adjacent environmental compartments.

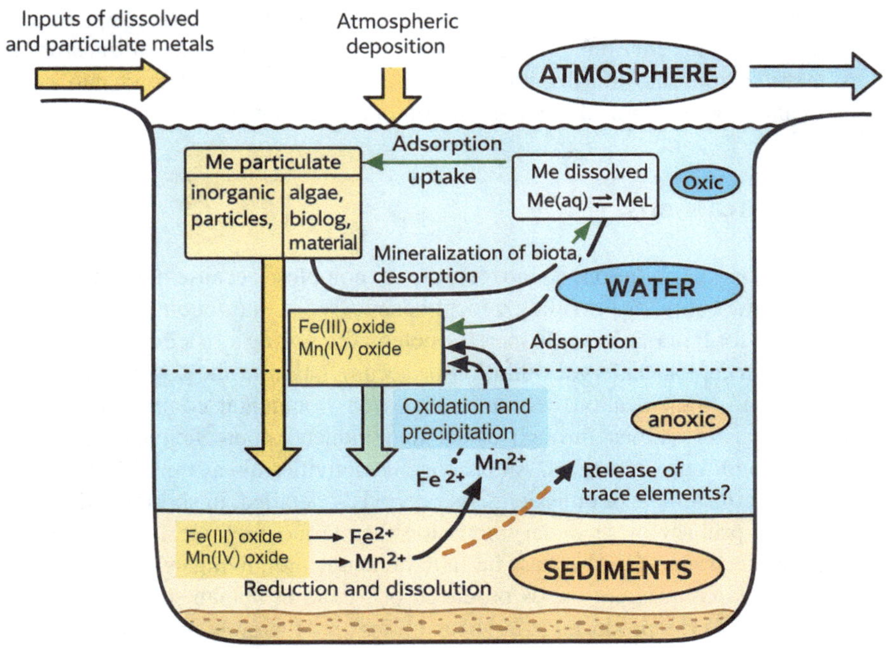

Fig. 7 Cycling of trace elements in a stratified lake. Depending on the redox potential of the environment, the metal is present in its high/low oxidation state, and thus it is subject to different solubility equilibria. The oxic condition is due to the presence of dissolved oxygen, which disappears in the bottom due to its respiration (consumption) by organic matter. The lake can exchange various chemical species with sediments, the atmosphere, and surface water. Adapted from [28]

They are characterized by their own internal dynamics, specific types of chemical reactions, and interrelationships with other compartments. In the atmosphere, consideration must be given to its vertical structure and interaction with the spectrum of sunlight as a function of altitude, the reactions occurring in the stratosphere and troposphere, the reactive species present there, the dry and wet deposition of substances on the soil and in surface waters, and emissions from the soil (outgassing, VOCs) and water (organometallic compounds, dimethyl sulfide, H_2S, sea spray). The biosphere is regulated by the processes of photosynthesis, respiration, and assimilation, which, depending on climatic conditions and water availability, lead to a different net primary production, which in turn evolves into soil organic matter, humic and fulvic acids. The latter have significant effects on the chemical weathering of soil, leading to the transformation of the minerals that constitute them and the dissolved salt load of surface waters. Plants also function as water pumps into the atmosphere, influencing the local water cycle. The lithosphere exchanges elements with the biosphere (nutrients and trace elements) and, in addition to chemical weathering, interacts with surface waters as an ion exchanger. These, in turn, are subject to hydraulic phenomena and are the agents of mechanical soil weathering (in addition to winds). Marine waters, likes winds (horizontal movements of the atmosphere), are subject to currents which are major heat transporters from the tropics to the poles. Marine waters have their own nutrient and sulfur cycles and redox profiles.

In conclusion, the environment, through the diverse and specific reactions that occur in the various environmental compartments, is an active agent of transformation, mobilization, and translocation of the chemical species released into it. Its effects cannot be overlooked in the definition of a pollutant.

3 Conclusions

The definition of pollution is intriguing and complex because any substance, whether a new entity, or providing a local benefit (see N and P compounds, which are nutrients for biota), or locally inert, or potentially posing a risk due to its impact on the final receptor, undergoes dilution or accumulation in the complex environmental system, translocation between various environmental compartments, and chemical transformations through direct and induced photochemical reactions, hydrolysis, and redox reactions. Anthropogenic activities always modify the environment in terms of use or emissions. As previously detailed, in short a pollutant is defined as a primary or secondary emission that 1) has biological effects at low concentrations because it can cause harm to human health, living systems, ecological systems, materials, and goods produced or usable in human society, reducing their usability; 2) It is produced in large quantities (and therefore inevitably released into the environment); 3) It is easily transported and dispersed in the environment; 4) It is persistent; 5) It can lead to substances that induce the transformation of other species (reactive species) or induce physical changes (e.g., aerosols, albedo changes); 6) It reduces the potential for using the environment, both for direct economic value and for recreational use; 7) It can lead to the formation of metabolites, which can themselves be pollutants (secondary pollutants, classified as such on points 1), 3–6) above.

The multiple reactions and spatial-temporal distribution depend both on the specific properties of the compound and on the properties of the environment. Therefore, risk assessment due to the release of substances or energy into the environment, in addition to a precise monitoring strategy, must identify and define receptors, in the broadest sense of the term, also considering the economic value of the environment and its sustainability. The latter are not always easily quantifiable, such as quality of life.

Questions

- Is the natural environment the reference for defining the level of pollution?
- Why does persistence make a chemical species a potential pollutant?
- Is it important to evaluate the distribution of chemical species across environmental compartments, and why?
- How do the specific chemical and physical properties of a chemical species help define a pollutant?
- This statement is true/false: a species that is not persistent and not toxic is not a pollutant. There are at least four reasons to support the right answer.

Glossary

- NAPL: Non Aqueous Phase Liquid
- DNAPL: Dense Non Aqueous Phase Liquid
- VOC: Volatile Organic Carbon
- PAH: Polycyclic Aromatic Hydrocarbon
- LD_{50}: Lethal Dose, 50% is the amount of a substance—administered all at once—required to kill half of a test population, usually rats or mice
- PM_{10}: Particulate matter with diameter 10 μm or smaller

Bibliography

1. 3.b Pollution, Oxford's English dictionary - Oxford Languages - Oxford University Press. https://www.oed.com/dictionary/pollution_n?tab=meaning_and_use#29518781
2. World Health Organization. https://www.who.int/health-topics/air-pollution#tab=tab_1
3. US EPA. https://web.archive.org/web/20220816015617/https://sor.epa.gov/sor_internet/registry/termreg/searchandretrieve/glossariesandkeywordlists/search.do;jsessionid=uJCB5Gu4w83V2Rvwe9ICdwSYPN92FIBDbphBhn8YAqz9GhWQ2_kk!918770883?details=&vocabName=Environmental%20Issues%20Glossary&filterTerm=pollution&checkedAcronym=false&checkedTerm=false&hasDefinitions=false&filterTerm=pollution&filterMatchCriteria=Contains
4. PAC, 1990, 62, 2167 (Glossary of atmospheric chemistry terms (Recommendations 1990)) on page 2172. https://doi.org/10.1351/goldbook.A00194
5. https://ww2.arb.ca.gov/resources/california-ambient-air-quality-standards
6. PAC, 1990, 62, 2167 (Glossary of atmospheric chemistry terms (Recommendations 1990)) on page 2182. https://doi.org/10.1351/goldbook.A00190
7. Pollution, Merriam-Webster.com Dictionary, Merriam-Webster. https://www.merriam-webster.com/dictionary/pollution
8. Persson L et al (2022) Outside the safe operating space of the planetary boundary for novel entities. Environ Sci Technol 56(3):1510–1521. https://doi.org/10.1021/acs.est.1c04158
9. Steffen W et al (2015) Planetary boundaries: guiding human development on a changing planet. Science 347:6223. https://doi.org/10.1126/science.1259855
10. https://www.britannica.com/science/water-pollution
11. Rockström J, Steffen W, Noone K et al (2009) A safe operating space for humanity. Nature 461:472–475. https://doi.org/10.1038/461472a
12. Scheffer M, Bascompte J, Brock W et al (2009) Early-warning signals for critical transitions. Nature 461:53–59. https://doi.org/10.1038/nature08227
13. Ma Z, Zeng C, Zhang YC et al (2025) Predicting critical transitions with machine learning trained on surrogates of historical data. Commun Phys 8:258. https://doi.org/10.1038/s42005-025-02172-4
14. (in Italian) Vittoria Calvani e Andrea Giardina, in Le vie della storia, Bari, Editori Laterza, 1979, 1986, vol. 2, pp. 52–58, cited in https://www.ariannaeditrice.it/articolo.php?id_articolo=18501
15. Holder AL, Sullivan AP (2024) Emissions, chemistry, and the environmental impacts of wildland fire. Environ Sci Technol Lett 11(9):903–907. https://doi.org/10.1021/acs.estlett.4c00612
16. Paul MJ, LeDuc SD, Boaggio K, Herrick JD, Kaylor SD, Lassiter MG, Nolte CG, Rice RB (2023) Effects of air pollutants from wildfires on downwind ecosystems: observations, knowledge gaps, and questions for assessing risk. Environ Sci Technol 57(40):14787–14796. https://doi.org/10.1021/acs.est.2c09061

17. Anenberg SC, Belova A, Brandt J, Fann N, Greco S, Guttikunda S, Heroux M-E, Hurley F, Krzyzanowski M, Medina S, Miller B, Pandey K, Roos J, Van Dingenen R (2016) Survey of ambient air pollution health risk assessment tools. Risk Analy 36:1718–1736. https://doi.org/10.1111/risa.12540
18. Hynes HBN (1960) The biology of polluted waters. Liverpool University Press, Liverpool. Chap 9, page 94
19. River flow and abstraction document, downloadable from https://wildfish.org/news/document-library/
20. Skjelkvåle BL, Stoddard JL, Jeffries DS, Tørseth K, Høgåsen T, Bowman J, Mannio J, Monteith DT, Mosello R, Rogora M, Rzychon D, Vesely J, Wieting J, Wilander A, Worsztynowicz A (2005) Regional scale evidence for improvements in surface water chemistry 1990-2001. Environ Pollut 137:165–176
21. Ruhí A, Acuña V, Barceló D, Huerta B, Mor J-R, Rodríguez-Mozaz S, Sabater S (2016) Bioaccumulation and trophic magnification of pharmaceuticals and endocrine disruptors in a Mediterranean river food web. Sci Total Environ 540:250–259. https://doi.org/10.1016/j.scitotenv.2015.06.009
22. Hyslop NP (2009) Impaired visibility: the air pollution people see. Atmospheric Environ 43(1):182–195. https://doi.org/10.1016/j.atmosenv.2008.09.067
23. NOAA, Global Climate Report Annual 2024. https://www.ncei.noaa.gov/access/monitoring/monthly-report/global/202413
24. Latva-Pukkila LN, Kyrölä E (2014) Analysing time-varying trends in stratospheric ozone time series using the state space approach. Atmos Chem Phys 14:9707–9725. https://doi.org/10.5194/acp-14-9707-2014
25. Schlesinger WH, Bernhardt ES (2013) Biogeochemistry (Third Edition) - An Analysis of Global Change. Elsevier. ISBN: 978-0-12-385874-0
26. Tan F, Tam V, Middleton DG (2022) A review of application of integral atmospheric jet dispersion model to flammable hazards. Fabic Newslett. https://doi.org/10.20944/preprints202103.0756.v1
27. Dameris M (2014) Human Impact. In: Fabian P, Dameris M (eds) Ozone in the Atmosphere. Springer, Berlin, Heidelberg. https://doi.org/10.1007/978-3-642-54099-8_5. chp 5, page 103 fig 5.13; original figure in Leighton P.A., Photochemistry of air pollution, 1961, Academic Press, NY and London, figure 59 at page 273
28. Sigg L, Stumm W (2016) Aquatische Chemie. VDF Hochschulverlag AG an der ETH Zürich, Zollikon. ISBN: 978-3728137678

The Climate System as a Complex Dynamical System: Drivers, Variability, and Climate Change

Elisa Palazzi

Abstract

The climate system is a highly complex and dynamical system resulting from interactions among the atmosphere, hydrosphere, cryosphere, lithosphere, biosphere and, more recently, the anthroposphere. These components interact with each other by continuously exchanging energy, momentum, and matter, giving rise to non-linear processes and feedback mechanisms that control climate variability across a wide range of spatial and temporal scales.

This chapter introduces the fundamental concepts needed to understand how the climate system works and how it is changing, subject to natural and anthropogenic climate forcings and to internal variability. A clear distinction is first made between climate and weather, and between long-term statistics in defining climate change and internal variability in shaping year-to-year and decadal fluctuations. Observational evidence of global warming is then presented, highlighting not only the increase in global mean surface temperature, but also concurrent changes in the cryosphere, sea level, biosphere, and extreme events.

The physical basis of climate change is discussed through the global energy balance of the Earth and the greenhouse effect. Attention is given to the anthropogenic amplification of the natural greenhouse effect, which has led to a persistent radiative imbalance and a measurable increase in global temperature since the pre-industrial period. The chapter further examines the main natural and anthropogenic climate forcings, including solar variability, orbital changes, volcanic eruptions, greenhouse gas emissions, aerosol particles, and land-use changes, and introduces the concept of radiative forcing as a useful framework to compare the relative contribution of the different forcing factors on the energy balance.

E. Palazzi (✉)
Department of Physics, University of Turin, Torino, Italy
e-mail: elisa.palazzi@unito.it

M. Minella (ed.), *Complexity in Environmental Chemistry*,
https://doi.org/10.1007/978-3-032-31774-2_3

Finally, the chapter presents climate models as essential tools for understanding the climate system working and for projecting future climate evolution. The hierarchical structure of climate models, the use of model ensembles, and the main sources of model uncertainty are discussed. Together, these elements provide a coherent framework for interpreting past, current, and future climate variability and change.

Keywords

Climate system · Climate change · Radiative forcing · Greenhouse gases · Climate variability · Feedbacks · Climate models · Model uncertainties

Chapter Learning Goals

After reading this chapter, you will be able to:

- Describe the climate system as a complex system composed of multiple interacting components.
- Distinguish between weather and climate and between climate variability and climate change.
- Explain the physical basis of the Earth's energy balance and the greenhouse effect.
- Identify the main natural and anthropogenic climate forcings and their role in driving past and current climate change.
- Understand the concept of radiative forcing and its use in comparing different drivers of climate change.
- Explain the role of feedback mechanisms in amplifying or dampening climate responses to a given forcing.
- Recognize the purpose, structure, strengths and limitations of climate models, including the main sources of uncertainty in climate projections.

1 The Climate System and Its Complexity

The climate system qualifies as a complex system, characterized by multiple and mutual interactions among its components [1]. They include the atmosphere, the hydrosphere, the cryosphere, the lithosphere, the biosphere, and, increasingly, the anthroposphere. Together, they form an interconnected system in which energy, momentum, and matter are continuously exchanged across a wide range of spatial and temporal scales. The atmosphere is the thin gaseous envelope surrounding the Earth and plays a central role in regulating the planet's energy balance through radiative, dynamical, and thermodynamical processes. The hydrosphere comprises all liquid water bodies, including oceans, lakes, rivers, and groundwater, and acts as a major reservoir of heat and carbon. The cryosphere, which includes glaciers, ice sheets, sea ice, seasonal snow cover, and permafrost, strongly influences climate through its high albedo (reflectivity) and its interaction with both the atmosphere and the ocean. The lithosphere and land surface provide the physical boundary conditions for atmospheric and hydrological processes, while the biosphere encompasses ecosystems and living organisms that actively interact with the climate system through biogeochemical cycles. Finally, the anthroposphere represents human activities, which have become a dominant driver of climate change since the industrial era. Indeed, it is from the Industrial Revolution onward that, also in response to population growth, human modifications of the environment have produced tangible, and in some cases irreversible, effects on ecosystems, air and water quality, human health, biodiversity, nutrient cycles, as well as climate. Some authors [2] have also proposed the term "Paleoanthropocene" to refer to the period between the earliest, barely detectable anthropogenic environmental changes already present during the Holocene and the Industrial Revolution, but the effect on the climate is not comparable to the post-industrial one.

The interactions among the climate system components give rise to non-linear processes and to feedback mechanisms, whereby a perturbation in one component can propagate through the system and its effect can be either amplified (in positive feedbacks) or damped (in negative feedbacks) by subsequent responses. As a result of these circular rather than linear cause-effect chains, the overall behaviour of the climate system cannot be fully understood or predicted by considering its individual components separately.

In climate science, and in particular in climate models, the inherent complexity of the system is often addressed through a reductionistic approach, in which individual components and their processes are represented within separate computer models (or modules), such as atmospheric, oceanic, cryospheric, or land-surface models. These model components are then coupled together to form an integrated representation of the climate system. While such an approach necessarily involves simplifications, it provides a practical framework for studying the functioning of the climate system and for investigating its response to natural and anthropogenic perturbations—in the past, present, and future conditions [3].

1.1 Weather, Climate, Climate Variability, and Climate Change

A correct understanding of climate variability and change requires a preliminary distinction between weather and climate, two concepts that are often confused in everyday language but refer to different, though related, aspects of the atmosphere and of the climate system.

Weather describes the instantaneous (or averaged/cumulated over very short periods) state of the atmosphere at a given place. It is characterized by atmospheric variables such as air temperature, liquid and solid precipitation, atmospheric pressure, humidity, and wind speed, and it can change rapidly over hours or days. Because of the chaotic nature of the atmosphere, weather forecast is possible only over short time horizons (1 week-10 days, depending on the background atmospheric conditions and the spatial resolution of the forecast). In contrast, climate refers to the statistics of several variables describing the atmosphere, ocean, land surface, and cryosphere, i.e., the climate system components, over long time periods. According to the World Meteorological Organization (WMO), climate is typically defined using statistics calculated over periods of 30 years or more [4]. Climate is a statistical description that includes mean values, variability, and the frequency and intensity of extreme events. As in the popular adage sometimes attributed to the American writer and humourist Mark Twain, "weather is what you get, climate is what you expect."

Related to the distinction between weather and climate is the concept of climate variability and climate change. Even in the presence of a long-term warming trend, the climate system exhibits natural fluctuations on interannual to decadal time scales. These variations arise from internal processes within the coupled atmosphere–ocean–cryosphere system, such as the internal mode of variability referred to as ENSO (El Niño Southern Oscillation). As a result, individual years or even decades may be cooler or warmer than the preceding ones, despite an underlying long-term increase in global temperature. Understanding the coexistence of long-term climate change and short-term climate variability is essential for correctly interpreting observations and for avoiding misconceptions about the nature of global warming.

2 Global Warming and Observed Climate Change

Multiple observational datasets provide clear evidence that global warming is unequivocal. Instrumental records show a persistent increase in the global mean surface temperature since the late nineteenth century, with the most recent decades being the warmest in the observational record. This long-term warming trend emerges clearly when short-term internal variability is averaged out. An example is provided in Fig. 1, where global surface air temperature anomalies with respect to the 1850-1900 pre-industrial level are shown as annual averages since 1967 (left) and as 5-year averages since 1850 (right), based on several global temperature datasets. The analysis performed by the Copernicus Climate Change Service (C3S) at the beginning of 2025 confirmed that 2024 was the first calendar year with the global average temperature exceeding 1.5°C above pre-industrial levels (left panel of Fig. 1), making it the warmest year on record, surpassing 2023. The year 2024 saw record-high greenhouse gas concentrations, record ocean temperatures, and extreme weather, including floods, heatwaves, and wildfires. While climate scientists monitor multi-year averages to assess long-term trends, individual years are sometimes quite exceptional, and 2024 marked a crucial point in global warming.

Global warming is not limited to an increase in surface air temperature but involves widespread changes throughout the climate system. The amounts of snow and ice have declined in most regions of the world, with glaciers retreating, Arctic sea ice extent decreasing, and permafrost thawing. At the same time, global mean sea level has risen, driven by both the thermal expansion of heated seawater and the melting of continental ice. Extreme high sea-level events have become more frequent, increasing the exposure of coastal regions to flooding. Changes in temperature extremes provide further evidence of a warming climate. Cold extremes have become less frequent and less intense, while warm extremes and heatwaves have increased in

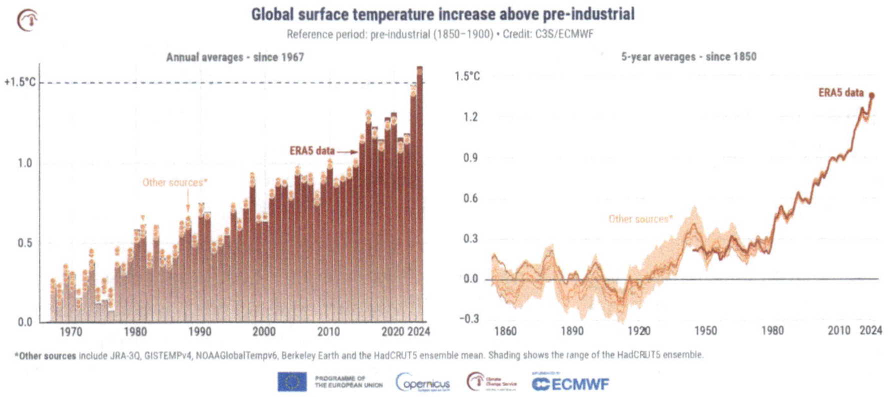

Fig. 1 Global surface air temperature increases above the 1850–1900 pre-industrial reference period, based on several global temperature datasets shown as annual averages since 1967 (left) and as 5-year averages since 1850 (right). Credit: C3S / ECMWF (from: https://climate.copernicus.eu/)

frequency and intensity. In many regions, the number and the intensity of heavy precipitation events have also increased, reflecting changes in the hydrological cycle. As the atmosphere warms, its capacity to hold water vapor increases (Clausius-Clapeyron law), leading to an intensification of the hydrological cycle.

These physical changes have profound impacts on natural and human systems. Ecosystems are increasingly stressed by changing temperature and precipitation patterns, shifts in seasonality, and the increased occurrence of extreme events. Observed impacts include changes in species distribution, phenology, and ecosystem functioning and services, as well as biodiversity loss. The relationship between climate change and biodiversity is complex, and most biodiversity losses are probably associated with habitat fragmentation and loss, land-use change, and introduction of alien invasive species. Also climate change matters, especially in hot-spots like mountains, but could be just an exacerbating factor rather than a main cause. Human systems are also affected, through impacts on water resources, agriculture, health, infrastructure, and economic activities.

Understanding the climate change drivers requires an examination of the basic physical mechanisms in the climate system, starting from the Earth's energy balance and the processes that regulate the exchange of radiation between the Earth surface, atmosphere and space, as discussed in the following section.

3 Earth's Energy Balance and the Greenhouse Effect

The climate system is fundamentally driven by the balance between the incoming energy received from the Sun and the outgoing energy emitted back to space by the Earth. Any persistent imbalance between incoming and outgoing energy results in a change in the temperature of the system, and thus in its climate.

As shown in Fig. 2, the Earth receives energy from the Sun in the form of shortwave solar radiation (on average, 340 W/m^2). A fraction of this incoming radiation is reflected back to space by clouds, atmospheric particles, and the Earth's surface (particularly snow and ice). This fraction is quantified by the planetary albedo and is about the 30% of the incident radiation (about 100 W/m^2). The remaining part of the solar radiation, 70%, is absorbed by the atmosphere, oceans, and land surface, warming the planet. To maintain energy balance, the Earth must emit an equivalent amount of energy back to space in the form of longwave (infrared) radiation, leaving the atmosphere at its top (Top of Atmosphere - ToA). Greenhouse gases trap the infrared radiation emitted by the surface in the atmosphere and warm the lower atmosphere and the Earth's surface. Taking these processes into account, the net radiative balance of the atmosphere would be negative, whilst that of the surface positive: to close the budget, the surface and the atmosphere exchange latent (85 W/m^2) and sensible (20 W/m^2) heat.

In a more simplified representation, the Earth can be treated as a blackbody emitting infrared radiation according to the Stefan–Boltzmann law, a physical law stating that the total energy radiated per unit area by a blackbody is proportional to the fourth power of its absolute temperature. If the Earth had no atmosphere (and thus no greenhouse effect), this balance would lead to a global mean surface temperature of −18 °C. However, the observed global mean surface temperature is about +15 °C, indicating the presence of an additional warming mechanism, represented by the greenhouse effect. Certain atmospheric gases, known as greenhouse gases, are transparent to incoming solar radiation but absorb and re-emit infrared radiation emitted by the Earth's surface. The most important greenhouse gases include water vapor (H_2O), carbon dioxide (CO_2), methane (CH_4), nitrous oxide (N_2O), ozone (O_3), and several industrial gases such as chlorofluorocarbons (CFCs) and sulphur hexafluoride (SF_6). By absorbing outgoing infrared radiation and re-emitting part of it back toward the surface, these gases warm the lower atmosphere and the Earth's surface and cool the stratosphere. The greenhouse effect is a natural and fundamental process that makes the Earth habitable. However, changes in atmospheric composition can modify the global energy balance. Increases in greenhouse gas concentrations enhance the greenhouse effect, reducing the amount of infrared radiation escaping to space and leading to a net gain of energy within the climate system. Human activities, particularly the burning of fossil fuels and changes in land use including deforestation, have significantly increased the atmospheric concentrations of several greenhouse gases from the beginning of the industrial period. This has resulted in an amplified greenhouse effect and a persistent positive energy imbalance. Current estimates of Earth's energy imbalance are indeed around 0.6 W/m^2 (see green arrow in Fig. 2) according to recent measurements from satellites and ocean heat content studies. This means Earth is absorbing more energy than it radiates back to space, which is the main driver of ongoing global warming. The amount 0.6 W/m^2 might seem small, but when multiplied by the entire surface area of Earth, it's equivalent to trillions of watts of extra energy being trapped every second. The excess energy since the 1970s has mainly gone into warming the ocean (91%), followed by the warming of land (5%), the melting of ice sheets and glaciers (3%) and the warming of the atmosphere (1%), as shown in Fig. 3.

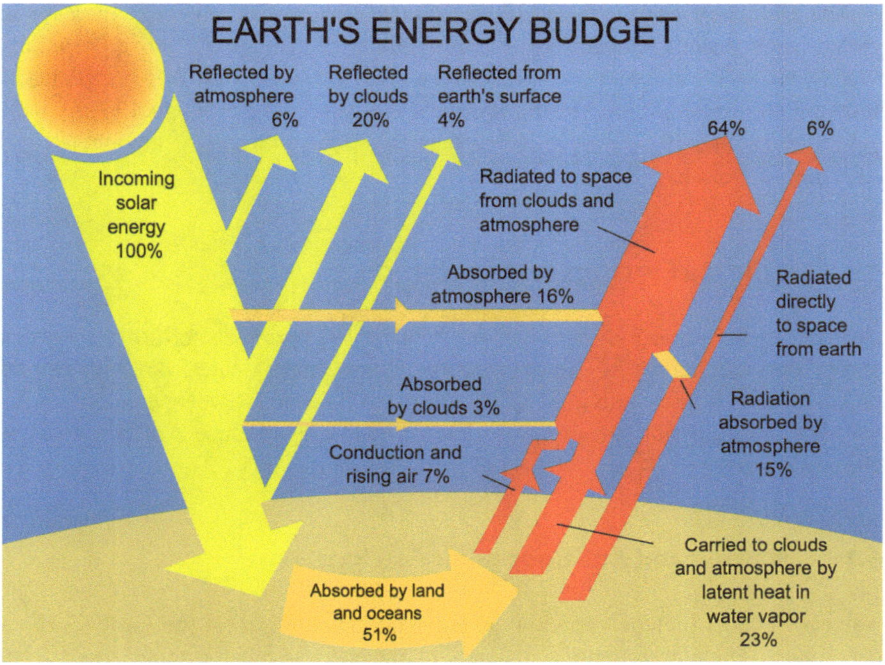

Fig. 2 Schematic diagram of the global mean energy balance of the Earth. Units Wm^{-2}. From [5]

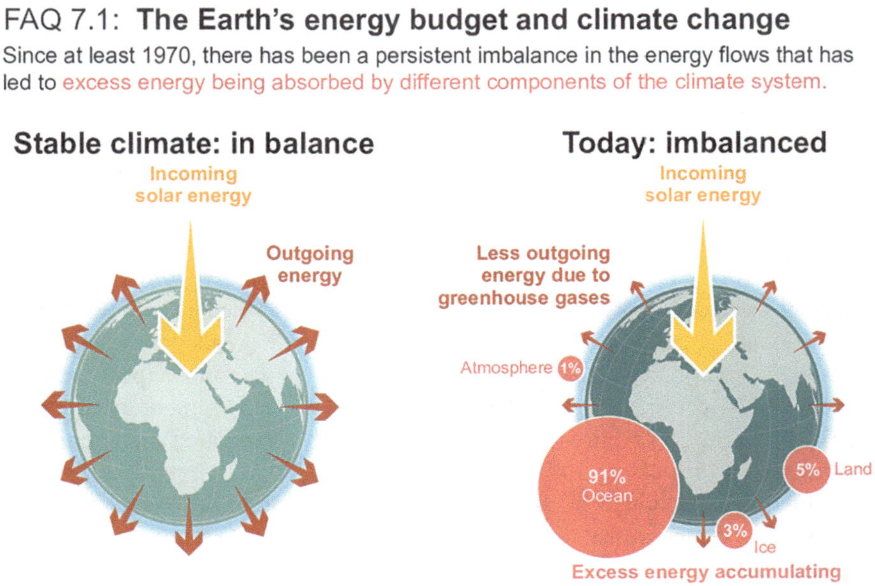

Fig. 3 FAQ 7.1 Fig. 1 in IPCC, 2021: Chapter 7. In: Climate Change 2021: The Physical Science Basis. Contribution of Working Group I to the Sixth Assessment Report of the Intergovernmental Panel on Climate Change [6]

The additional energy stored in the climate system has led to an increase in global mean surface temperature of approximately 1.3 to 1.4 °C above pre-industrial levels when averaged over recent five-year periods, with different datasets showing slight variations, nearing the Paris Agreement's 1.5 °C goal [7] as individual years (like 2024) exceed it. Further warming is expected if greenhouse gas emissions continue.

4 Climate Forcings and Feedbacks

Changes in the climate system can be driven by factors that perturb the Earth's energy balance, known as climate forcings. They can originate from both natural processes and human activities. By altering the balance between incoming solar radiation and outgoing infrared radiation, these forcings induce a response of the climate system.

4.1 Natural and Anthropogenic Forcings

Natural forcings include variations in solar output, changes in the Earth's orbital parameters, volcanic eruptions, impacts of meteorites. Solar forcing arises from changes in the Sun's energy output, which occur over multiple time scales. The most regular pattern is the approximately the 11-year solar cycle (known as Schwabe cycle) associated with variations in sunspot activity. Longer-term modulations, such as the century-scale Gleissberg cycle, also affect solar irradiance. However, assessments by the IPCC indicate that changes in solar brightness since the pre-industrial period have contributed only a very small fraction of the observed global warming [8]. Orbital forcing results from predictable variations in the Earth's orbit around the Sun, known as Milankovitch cycles. Changes in eccentricity, axial tilt and precession modify the geographical and seasonal distribution of solar radiation over time scales of tens to hundreds of thousands of years. These processes have presumably played a fundamental role in driving glacial–interglacial cycles in the past but cannot explain the rapid warming observed over the last century. Volcanic eruptions represent another important natural forcing. Explosive volcanic eruptions inject large amounts of gases and particles into the stratosphere. While volcanic ash rapidly settles out of the atmosphere, sulphur-containing gases form sulphate aerosols that efficiently scatter incoming solar radiation. This leads to a temporary cooling of the Earth's surface, typically lasting a few years, as observed after major eruptions such as Mount Pinatubo in 1991.

Anthropogenic climate forcings are primarily associated with human activities. The most important are emissions of greenhouse gases, including carbon dioxide, methane, and nitrous oxide, which amplify the natural greenhouse effect. Additional anthropogenic forcings arise from emissions of aerosols and aerosol precursors, as well as from changes in land use and land cover, including deforestation. Aerosols can exert both cooling and warming effects depending on their composition, size, and interaction with radiation and clouds, while land-use changes modify surface albedo, evapotranspiration, and carbon fluxes.

4.2 Radiative Forcing and Climate Feedback

To compare the magnitude of different climate forcings and assess their effects on the climate system, the concept of radiative forcing is used. Radiative forcing is commonly defined as the change in the net radiative flux at the ToA or at the tropopause resulting from a given perturbation, after allowing for stratospheric adjustment. A positive radiative forcing corresponds to an increase in the energy retained by the Earth system and leads to warming, whereas a negative radiative forcing results in cooling (more energy leaves the systems compared to the incoming energy).

In response to a radiative forcing, the climate system adjusts through changes in temperature, circulation, and other variables. These adjustments modify the radiative fluxes and give rise to climate feedbacks. A feedback occurs when the outcome of a process influences its own cause, either amplifying or reducing the initial perturbation. Positive feedbacks amplify the initial change, while negative feedbacks tend to stabilize the system. A well-known example is the ice–albedo feedback, a positive one. As temperatures rise, snow and ice cover decrease, reducing the surface albedo and allowing more solar radiation to be absorbed. This additional absorption leads to further warming and further ice loss, reinforcing the initial temperature increase. Besides the ice-albedo feedback, other positive feedbacks include the water vapour feedback (warming increases atmospheric water vapour and water vapour is a greenhouse gas, which amplifies warming); the permafrost carbon feedback (thawing permafrost releases CO_2 and CH_4, these greenhouse gases further enhance warming); methane hydrate (clathrate) feedback (warming oceans may destabilize methane hydrates in sediments, and the released methane further increases greenhouse forcing); ocean stratification feedback (warmer surface waters reduce vertical mixing, less heat and carbon are transported to deep oceans, increasing surface warming). Feedback involving clouds can be both positive and negative: some cloud changes (e.g., fewer low clouds, more high thin clouds) can increase warming by trapping more heat; an increase in low, thick clouds raises Earth's albedo, more incoming solar radiation is reflected back to space which produces a cooling effect.

Feedback mechanisms play a crucial role in determining the overall sensitivity of the climate system to external forcing. In response to a given radiative forcing ΔQ, the variables characterizing the state of the climate system will change, modifying the radiative fluxes (ΔR). The radiative forcing ΔQ causes other elements of the climate system to respond, either dampening or amplifying the initial radiative-induced warming. This response is quantified by the radiative feedback parameter (λ). ΔR, ΔQ and λ are related to each other through the following equation:

$$\Delta R = \Delta Q + \lambda \Delta T \tag{1}$$

This is closely linked to the concept of "equilibrium climate sensitivity," which measures how much the Earth's global average surface temperature will eventually increase when the applied forcing ΔQ is the instantaneous doubling of atmospheric carbon dioxide (CO_2), once the climate system has fully adjusted to the change (ΔR = 0). In this case,

$$\Delta T = -\frac{\Delta Q}{\lambda}. \tag{2}$$

Major climate assessments [8] estimate the equilibrium climate sensitivity to be likely between about 2 °C and 4.5 °C, with a best estimate around 3 °C, and very likely between about 2.0 °C to 5.0 °C, though uncertainty remains due largely to cloud feedbacks.

5 Climate Models and Climate Projections

Understanding the behaviour of the climate system and its response to external perturbations relies heavily on the use of climate models. Climate models are mathematical representations of the climate system based on physical, chemical, and biological principles. They describe the evolution of key variables defining all climate system components through sets of dynamical and thermodynamical equations that express conservation of mass, momentum, and energy, and several empirical formulas and parameterizations. Because the equations are too complex to be solved analytically, they are solved numerically, resulting in simulations that are discrete in space and time. In climate models, resolution refers mainly to how finely the Earth is divided in space (horizontal and vertical grids) and time. It varies by model type and purpose.

Climate models differ in their level of complexity and in the number of processes and components they represent, forming a model hierarchy [3]. Simple models, such as energy balance models or radiative-convective models, provide insight into fundamental mechanisms governing the climate system. More complex models, like the global climate or earth system models, explicitly simulate atmospheric and oceanic circulation, land-surface processes, cryosphere dynamics, and biogeochemical cycles. Climate models are used to study the response of the climate system to perturbations, such as in climate sensitivity experiments, and to produce climate projections. Projections are simulations of future climate conditions driven by prescribed trajectories of greenhouse gas emissions and land-use changes, known as emission scenarios. The latest emissions scenarios used in climate research are the Shared Socioeconomic Pathways (SSPs) adopted in IPCC AR6 [9]. They combine socioeconomic storylines with radiative forcing levels by 2100. In addition to future projections, climate models are also employed in hindcast and attribution experiments, in which past climate changes are simulated to assess the relative contributions of natural and anthropogenic forcing to observed trends.

Climate projections are not based on a single model simulation, but on ensembles of simulations, which allow different sources of uncertainty to be explored

[10]. One source of uncertainty arises from internal climate variability, which is generated spontaneously by the climate system even in the absence of any change in radiative forcing. This uncertainty can be explored making use of multi-member ensembles. Multi-member simulations use different versions of the same climate model, i.e., run from a variety of different starting conditions. A second source is model uncertainty, which includes structural uncertainty, related to different ways of representing physical processes, and parametric uncertainty, associated with the choice of parameter values for unresolved sub-grid processes or empirical formulations, for example in the description of ecosystem processes. Model uncertainties can be separated into three different factors: (i) limited theoretical understanding of a given process, such as how aerosols affect cloud formation; (ii) indeterminacy/inadequacy in model parameters; (iii) model inadequacy owing to the fact that models are missing or are approximating certain climate processes. This uncertainty is quantified through multi-model ensemble simulations, like the commonly used Climate Model Intercomparison Project (CMIP) experiments. A third source is scenario uncertainty which derives from the uncertainty of future demographic changes, economic development, land-use, and technological changes. These changes, along with the environmental changes in the future climate itself, will determine future anthropogenic emissions of greenhouse gases and aerosol particles. The relative importance of the scenario uncertainty grows over time within a climate simulation.

The relative importance of these sources of uncertainty varies with the variable of interest, the space and time scales involved in the climate prediction, and the lead-time of the projection. For time horizons of a decade or two, the dominant sources of uncertainty on regional scales are model uncertainty and internal variability, while for time horizon of many decades or longer, the dominant sources of uncertainty at regional or larger spatial scales are model uncertainty and scenario uncertainty [10]. By considering ensembles that sample these different uncertainties, climate projections provide a range of plausible future climate outcomes rather than a single prediction and associates to the model mean a model uncertainty, as a measure of the inter-model spread. Recognizing and communicating climate uncertainty is therefore not a weakness of climate science, but a fundamental aspect of a rigorous and transparent scientific assessment.

6 Conclusions

This chapter has presented an overview of the climate system, emphasizing its nature as a complex system. The climate system complexity is central to understanding both natural climate variability and long-term climate change. Observational evidence unequivocally shows that the Earth is warming, and that this warming is accompanied by widespread changes in all the climate system components: the cryosphere, sea level, extreme events, and ecosystems. These changes cannot be interpreted as isolated phenomena but as interconnected responses of the climate system to a persistent perturbation of its energy balance, driven by changes in atmospheric composition caused by human activities (Fig. 4).

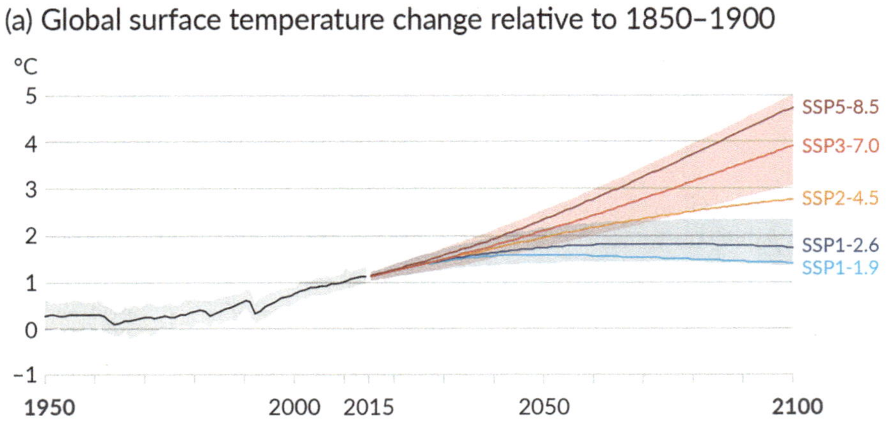

Fig. 4 Global surface temperature changes in °C relative to 1850–1900. These changes were obtained by combining Coupled Model Intercomparison Project Phase 6 (CMIP6) model simulations with observational constraints based on past simulated warming, as well as an updated assessment of equilibrium climate sensitivity (see Box SPM.1). The projections for each of the five scenarios are shown in color. Shades represent uncertainty ranges—more detail is provided for each panel below. The black curves represent the historical simulation. Taken from Panel a of Figure SPM.8 in [11]

The modification of the Earth's global energy balance arises from an amplified greenhouse effect. While natural forcings such as solar variability, orbital changes, and volcanic eruptions have shaped climate in the past, the dominant driver of the observed warming since the industrial era is anthropogenic forcing, primarily due to greenhouse gas emissions, aerosols, and land-use changes. The concept of radiative forcing provides a unifying framework to compare the relative influence of climate change drivers, while climate feedbacks play a crucial role in determining the magnitude of the climate response. Positive feedbacks amplify the effect of climate forcings, negative feedbacks decrease it.

Climate models represent indispensable tools for understanding the climate system and for exploring its response to past and future perturbations. Through models of different kind, scientists investigate climate mechanisms, attribute observed changes, and project future climate behaviour under different emission scenarios. Importantly, the presence of uncertainty—arising from internal variability, model formulation, and future socio-economic pathways—is an inherent feature of climate projections and reflects the complexity of the system and the state of our knowledge, rather than a lack of robustness.

In conclusion, a comprehensive understanding of climate change requires viewing the climate system as an interconnected whole, grounded in physical principles and informed by observations and models. Such an understanding is essential for supporting decision-making, adaptation and mitigation strategies.

Questions

- What makes the climate system a complex system?
- Why is climate defined as a statistical description?
- What is radiative forcing and why is it useful?
- Why can global warming coexist with year-to-year cooling?
- What are the main sources of uncertainty in climate projections?

Glossary

- **Climate system**: The coupled system formed by the atmosphere, hydrosphere, cryosphere, lithosphere, biosphere, and anthroposphere (i.e., human activities), and by the interactions among them.
- **Complex system**: A system composed of many interacting elements whose collective behaviour cannot be understood solely by analysing the individual components (the whole is more than the sum of its parts), often characterized by, among other characteristics, non-linearity, feedbacks and some degree of unpredictability.
- **Weather**: The instantaneous or short-term state of the atmosphere at a given location, described by variables such as temperature, precipitation, pressure, humidity, and wind.
- **Climate**: The statistical description of the climate system, including mean values, variability, and extremes of relevant variables pertaining to all climate system components, typically averaged over periods of at least 30 years, as recommended by the World Meteorological Organisation.
- **Climate forcing:** An external factor, either natural or anthropogenic, that alters the Earth's energy balance by changing the amount of incoming solar radiation or outgoing infrared radiation, thereby driving a response of the climate system.
- **Climate change**: A long-term change in the statistical properties of the climate system, resulting from natural factors, anthropogenic forcing, or both.
- **Global warming**: The long-term increase in the global mean surface temperature of the Earth, primarily driven by anthropogenic greenhouse gas emissions since the industrial era.
- **Greenhouse effect**: The process by which certain atmospheric gases, the greenhouse gases (GHGs), absorb and re-emit infrared radiation emitted by the Earth's surface, warming the lower atmosphere.
- **Greenhouse gases (GHGs)**: Atmospheric gases, such as CO_2, CH_4, N_2O, and H_2O, capable of absorbing infrared radiation due to their molecular structure.

(continued)

- **Radiative forcing**: A perturbation of the Earth's energy balance caused by external factors, expressed as a change in net radiative flux at the Top of the Atmosphere (ToA) or at the Tropopause; positive forcing leads to warming, negative forcing leads to cooling.
- **Climate feedback**: A process by which a change in the climate system triggers secondary effects that either amplify (positive feedback) or dampen (negative feedback) the initial change.
- **Internal climate variability**: Climate variability generated internally by the coupled atmosphere–ocean–cryosphere–land system, even in the absence of external forcing. It manifests ad fluctuations occurring on interannual to decadal time scales.
- **Equilibrium Climate Sensitivity -** The long-term increase in global average temperature expected to occur after the effects of a doubled CO_2 concentration have had time to reach a steady state.
- **Climate model**: A numerical representation of the climate system based on physical, chemical, and biological principles, used to simulate past, present, and future climate conditions.
- **Climate projection**: A simulation of future climate conditions based on assumptions about future external forcing, particularly greenhouse gas emission scenarios.

Bibliography

1. IPCC, 2013: Climate change 2013: the physical science basis. Contribution of working group I to the fifth assessment report of the intergovernmental panel on climate change [stocker, T.F., D. Qin, G.-K. Plattner, M. Tignor, S.K. Allen, J. Boschung, A. Nauels, Y. Xia, V. Bex and P.M. Midgley (eds.)]. Cambridge university press, Cambridge, United Kingdom and New York, NY, USA, 1535 pp.
2. Foley SF, Gronenborn D, Andreae MO, Kadereit JW, Esper J, Scholz D, Pöschl U, Jacob DE, Schöne BR, Schreg R, Vött A, Jordan D, Lelieveld J, Weller CG, Alt KW, Gaudzinski-Windheuser S, Bruhn K-C, Tost H, Sirocko F, Crutzen PJ (2013) The Palaeoanthropocene—the beginnings of anthropogenic environmental change. Anthropocene 3:83–88
3. McGuffie K, Henderson-Sellers A (2014) The climate modelling primer.4th edn. Wiley Blackwell
4. WMO (2025) Climatological standard normals (CLINO) for the period 1991–2020; WMO-No. 847. https://library.wmo.int/idurl/4/53252
5. https://commons.wikimedia.org/wiki/File%3AEarth_energy_budget.svg?utm_source=chatgpt.com#filehistory
6. Forster P, Storelvmo T, Armour K, Collins W, Dufresne J-L, Frame D, Lunt DJ, Mauritsen T, Palmer MD, Watanabe M, Wild M, Zhang H (2021) The Earth's energy budget, climate feedbacks, and climate sensitivity. In: Masson-Delmotte V, Zhai P, Pirani A, Connors SL, Péan C, Berger S, Caud N, Chen Y, Goldfarb L, Gomis MI, Huang M, Leitzell K, Lonnoy E, Matthews JBR, Maycock TK, Waterfield T, Yelekçi O, Yu R, Zhou B (eds) Climate change 2021: the physical science basis. Contribution of working group I to the sixth assessment report of the intergovernmental panel on climate change. Cambridge University Press, Cambridge, UK/New York, pp 923–1054. https://doi.org/10.1017/9781009157896.009

7. United Nations (2015) Paris agreement. UN Treaty Series 110:288–319. https://legal.un.org/avl/pdf/ha/pa/pa_e.pdf
8. Masson-Delmotte V, Zhai P, Pirani A, Connors SL, Péan C, Berger S, Caud N, Chen Y, Goldfarb L, Gomis MI, Huang M, Leitzell K, Lonnoy E, Matthews JBR, Maycock TK, Waterfield T, Yelekçi O, Yu R, Zhou B, IPCC (eds) (2021) Climate change 2021: the physical science basis. Contribution of working group I to the sixth assessment report of the intergovernmental panel on climate change. Cambridge University Press, Cambridge, UK/New York., 2391 pp. https://doi.org/10.1017/9781009157896
9. Riahi K et al (2017) The shared socioeconomic pathways and their energy, land use, and greenhouse gas emissions implications: an overview. Glob Environ Chang 42:153–168. https://doi.org/10.1016/j.gloenvcha.2016.05.009
10. Hawkins E, Sutton R (2009) The potential to narrow uncertainty in regional climate predictions. Bull Am Meteorol Soc 90(8):1095–1107
11. IPCC (2021) Summary for policymakers. In: Masson-Delmotte V, Zhai P, Pirani A, Connors SL, Péan C, Berger S, Caud N, Chen Y, Goldfarb L, Gomis MI, Huang M, Leitzell K, Lonnoy E, Matthews JBR, Maycock TK, Waterfield T, Yelekçi O, Yu R, Zhou B (eds) Climate change 2021: the physical science basis. Contribution of working group I to the sixth assessment report of the intergovernmental panel on climate change. Cambridge University Press, Cambridge, UK/New York, pp 3–32. https://doi.org/10.1017/9781009157896.001

Suspect and Non-target Screening in Environmental Matrices: Strategies for Data Acquisition and Processing

Chiara Cavaliere and Anna Aquilone

Abstract

The continuous introduction of new chemicals into the environment has increased the need for monitoring strategies capable of detecting both known and previously unknown compounds. While concentrations of legacy contaminants, such as persistent organic pollutants, have progressively been reduced through regulatory actions, contaminants of emerging concern (CECs), including pharmaceuticals, personal care products, pesticides, industrial chemicals, and their transformation products (TPs), are now ubiquitously detected in all environmental matrices. Unlike legacy pollutants, many CECs and TPs are more polar, water-soluble, and present at trace concentrations, posing significant analytical challenges. Reliable detection further requires careful optimization of sample preparation, which must maximize the number of extracted analytes while minimizing matrix effects that could compromise sensitivity and quantification.

Mass spectrometry (MS) remains central to environmental analysis due to its sensitivity, selectivity, and structural elucidation capability. In particular, liquid chromatography (LC)-high-resolution MS (HRMS) enables the analysis of polar, thermolabile, and high-molecular-weight compounds, and it is a fundamental tool for both suspect screening (SS) and non-target screening (NTS) approaches. SS focuses on "known unknowns," using precompiled suspect lists to guide data acquisition and interpretation, while NTS addresses "unknown unknowns," aiming for a comprehensive characterization of the chemical composition of samples without prior assumptions.

Advanced HRMS acquisition modes, namely data-dependent and data-independent, generate comprehensive MS/MS datasets, which are subsequently

C. Cavaliere (✉) · A. Aquilone
Department of Chemistry, Sapienza University of Rome, Rome, Italy
e-mail: chiara.cavaliere@uniroma1.it

M. Minella (ed.), *Complexity in Environmental Chemistry*,
https://doi.org/10.1007/978-3-032-31774-2_4

processed through dedicated workflows for feature extraction, retention time alignment, noise reduction, and assignment of putative annotations. Databases, spectral libraries, and in silico prediction tools, including machine learning models for retention time and fragmentation prediction, support compound annotation and prioritization. The distinction between annotation and identification can be explained through the identification confidence levels proposed by Schymanski et al., which provide a tool to clearly communicate the reliability of detected compounds.

Complementary analytical dimensions, such as ion mobility, further enhance compound separation and identification. These integrated approaches allow for robust detection, prioritization, and retrospective analysis of CECs and TPs, supporting more comprehensive environmental monitoring and a deeper understanding of chemical exposure. Initiatives are being made to standardize analytical protocols and reporting practices for emerging chemicals in environmental samples, facilitating consistency and reliability across NTS studies.

Keywords

Contaminants of emerging concern · High-resolution mass spectrometry · Mass spectral libraries · Non-target screening · Suspect screening · Transformation products

Chapter Learning Goals

"What You Will Learn in This Chapter"

- Understand the challenges of monitoring emerging contaminants and their transformation products in the environment.
- Distinguish between target, suspect, and non-target screening approaches and their analytical implications.
- Recognize the role of liquid chromatography-high-resolution mass spectrometry in environmental analysis.
- Comprehend the principles, strengths, and limitations of data-dependent (DDA) and data-independent (DIA) acquisition modes in non-target screening.
- Understand the contribution of databases, spectral libraries, in silico prediction, and machine learning to compound annotation and prioritization.
- Distinguish the concepts of feature, annotation, and identification, with relative confidence levels.

1 Introduction

To date, although regulatory actions at national and international levels are successfully reducing the concentrations of many legacy contaminants, the number of newly introduced compounds continues to grow [1, 2]. More than 350,000 chemicals and substances are currently in use worldwide, and over 204 million are registered in major chemical inventories [2]. This very large number of compounds to which humans and ecosystems are potentially exposed raises increasing concerns and poses significant challenges for environmental monitoring.

Legacy contaminants, such as persistent organic pollutants (POPs) listed in the Stockholm Convention (https://chm.pops.int) and polycyclic aromatic hydrocarbons (PAHs), are characterized by strong hydrophobicity (i.e., high $\log K_{ow}$ values), which confers them a high environmental persistence, and the tendency to bioaccumulate in living organisms and biomagnify along the food chain. In contrast, the defining feature of more recently identified contaminants is not necessarily their persistence, but rather their ubiquitous presence across all environmental compartments [1]. These contaminants of emerging concern (CECs) include pharmaceuticals and personal care products (PPCPs), pesticides, industrial chemicals, plasticizers, and other widely used substances. Furthermore, many CECs, such as pharmaceuticals [3], can undergo transformation reactions, driven by both biotic (such as living organism metabolisms) and abiotic factors (e.g., oxidation, hydrolysis, photodegradation, etc.), leading to the formation of transformation products (TPs) that are generally more polar and water-soluble than their parent compounds [4]. Although most CECs, and particularly their TPs, are commonly present at low concentration levels in the environment, they can still negatively impact ecosystems and human health due to their widespread occurrence, and potential biological activity. For these reasons, TPs should be included in the monitoring programs.

Mass spectrometry (MS) has long played a central role in environmental analysis, due to its versatility, sensitivity, and selectivity, together with its ability to be coupled with separation techniques such as gas chromatography (GC) and, later, also high-performance liquid chromatography (LC). These characteristics have made MS particularly suitable for the quantification of known contaminants and for the structural elucidation of unknown, or rather, previously unexpected compounds, even in complex environmental matrices, such as air, particulate matter, water, soil, sediments, and biota samples.

2 Advantages of Liquid Chromatography-Mass Spectrometry in Environmental Analysis

For the analysis of POPs and other persistent hydrophobic pollutants, GC-MS has long been the technique of choice. However, the growing introduction of new classes of compounds into the market, including PPCPs and many pesticides (which are often more polar than POPs to overcome persistence and bioaccumulation issues), has led to LC-MS progressively replacing GC-MS in many environmental applications. This shift has occurred primarily because LC enables the analysis of polar and thermolabile compounds that are not amenable to GC without chemical derivatization, as well as high-molecular-weight compounds, including biomolecules.

Generally, GC exhibits a higher chromatographic resolution than LC, even if the gap has been narrowed by the advances in ultrahigh-performance LC (UHPLC), using stationary phases with sub-2 μm particle size. However, LC offers a much broader range of stationary-phase chemistries than GC, allowing enhanced selectivity and versatility for separating the structurally diverse compounds commonly encountered in environmental matrices. In reversed-phase (RP) LC, a wide variety of stationary phases are available, and the most used in this field include conventional alkyl-bonded phases such as C18, and C8, as well polar-embedded or polar end-capped C18 phases, specifically designed to improve the retention and peak shape of polar analytes. Also, aromatic stationary phases, such as biphenyl, phenyl-hexyl, and pentafluorophenyl (PFP), provide different selectivity and enhanced retention for more polar compounds [2]. For highly polar compounds that are weakly retained under RPLC conditions, hydrophilic interaction liquid chromatography (HILIC) offers retention mechanisms orthogonal to those of RPLC. In HILIC, retention, selectivity and peak shape are strongly influenced by mobile phase composition and by the chemistry of the stationary phase, which may include silica, amide, amino, diol, or zwitterionic functionalities. Finally, for the analysis of strongly acidic or basic analytes, ion exchange or various mixed-mode stationary phases can be used.

Another important advantage of LC-MS is the possibility to simultaneously analyze a wide range of molecules with different physicochemical properties (e.g., volatility, polarity, and molecular mass) within a single run, thereby significantly improving the number of compounds that can be detected. These features make LC-MS particularly well suited for suspect and non-target (or untargeted) screening in environmental analysis.

Figure 1 illustrates the applicability of electron ionization (EI) and the main atmospheric pressure ionization (API) techniques as a function of compound polarity and molecular mass. In GC-MS, the ability of EI to generate highly reproducible fragmentation spectra enables compound identification by searching against well-established spectral libraries, such as NIST database (https://chemdata.nist.gov/), which contains data on more than 347,100 compounds. In contrast, LC-MS, using soft ionization techniques, lacks such standardized fragmentation, and requires MS/MS experiments to obtain diagnostic ions.

Among API techniques used in LC-MS, atmospheric pressure photoionization (APPI) is generally limited to the analysis of less polar compounds (e.g., PAHs [5] and brominated flame retardants), and its applicability partially overlaps with that of EI, while atmospheric pressure chemical ionization (APCI) is effective for the ionization of moderate polar compounds (e.g., steroids and lipids) [6]. Nonetheless, electrospray (ESI) is by far the most widely used technique in environmental LC-MS applications, owing to its broad applicability to polar contaminants and their TPs. Despite these advantages, ESI is particularly susceptible to matrix effects, which can lead to ion signal suppression or enhancement and adversely affect quantitative accuracy, especially when analyzing complex environmental matrices. In case of severe signal suppression, the less-abundant compounds could not even be

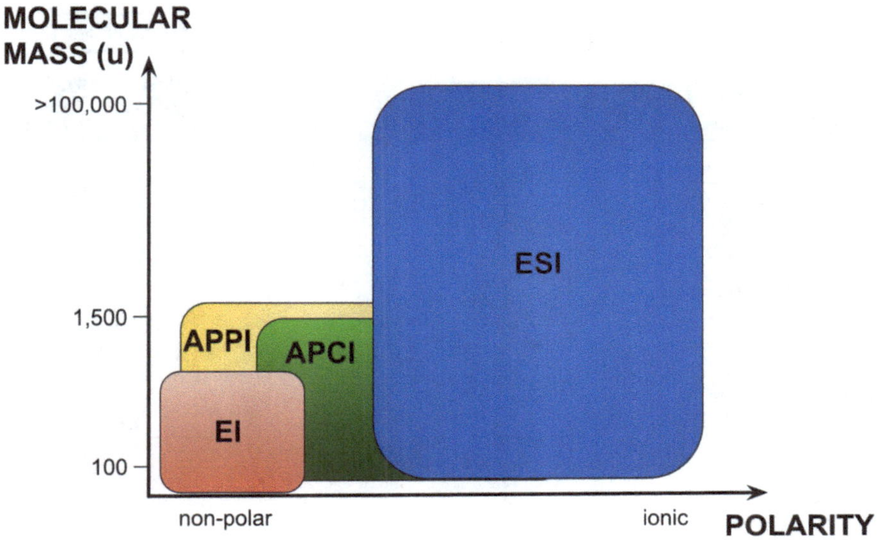

Fig. 1 Main ionization techniques employed in environmental analysis, according to compound polarity and molecular mass

detected at all. Of course, employing different API techniques can provide complementary information on the compounds present in complex mixtures [6, 7]. However, in environmental monitoring, applying such an approach to a single sample would be impractical, particularly given that dozens of samples typically need to be analyzed.

Worth mentioning is the possibility of using EI as an interface for LC-MS systems. This approach offers the advantage of obtaining highly reproducible mass spectra, like classical EI, which are easily interpretable for the identification of unknown compounds. This interface, formerly known as Direct EI (DEI) [8] and now evolved into Liquid EI (LEI) [9], operates in the gas phase under high vacuum and high temperature conditions, using 70 eV electrons. However, since the sample is initially in the liquid phase, LEI enables the ionization of low-to-medium molecular weight compounds that are otherwise unsuitable for GC analysis. Furthermore, compared to ESI, LEI does not suffer from matrix effects [10]. This advantage has led to its successful application in determining both polar and non-polar environmental contaminants in water samples [11, 12].

3 Mass Spectrometry Analytical Approaches

In MS, three analytical strategies are commonly employed, as illustrated in Fig. 2 [13]: target screening, suspect screening (SS), and non-target screening (NTS), the last being also referred to as non-target analysis (NTA) or untargeted analysis.

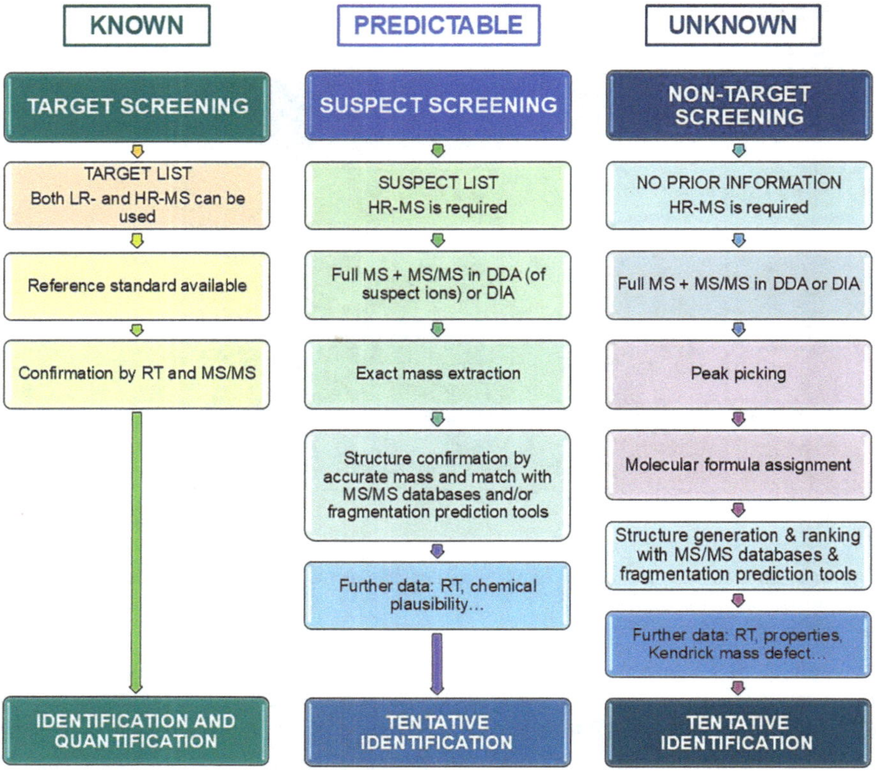

Fig. 2 Mass spectrometry analytical approaches: target, suspect, and non-target screening. Target screening applies to compounds with available reference standards and can be performed using both LR-MS and HR-MS. Suspect and non-target screening are used when reference standards are unavailable or compounds are unknown, relying primarily on HR-MS; tentative identifications can subsequently be confirmed with reference standards

> Unambiguous MS-based compound identification is possible only when a reference standard is available; otherwise, the identification can be considered tentative, unless additional techniques (e.g., NMR) are employed.

3.1 Target Screening

Target analysis requires the availability of an authentic reference standard, which allows to identify and quantify a compound in a sample through comparison of retention time and mass spectral data. This approach commonly relies on low-resolution (LR) mass analyzers, such as triple quadrupoles (QqQ) or linear ion traps. Among these, the QqQ mass spectrometer operated in multiple reaction monitoring (MRM) mode is considered the gold standard, as it provides exceptional sensitivity and selectivity, together with a robust quantitation over a wide dynamic range. Nonetheless, the recent literature indicates that the new generation HRMS mass spectrometers are suitable for quantitative applications, in some cases demonstrating performance that is comparable or even superior to that of LRMS [14, 15].

On the other hand, the major limitation of target methods is that they are inherently restricted to compounds for which standards are commercially available. This constraint is poorly suited to the analysis of complex environmental matrices, in which contaminants from different classes typically occur together at widely varying concentration levels. A notable example is represented by TPs, which are often unknown and, even when known, their authentic standards are not always available [16].

3.2 Suspect and Non-target Screening

Unlike target analysis, SS and NTS approaches enable the expansion of knowledge on the occurrence of CECs in the environment and the associated human exposure without the need for prior analyte selection. These strategies allow the inclusion of compounds for which analytical standards are unavailable or too expensive. When necessary, the purchase or synthesis of reference standards for full confirmation can be performed after the analysis, based on the relevance of tentative identifications, such as detection frequency, signal intensity, statistical significance, and potential ecological and toxicological effects [2].

Both SS and NTS approaches in LC-MS analysis rely on HR mass analyzers. Although the boundary between SS and NTS is not always clearly defined, the two strategies differ in the level of prior knowledge assumed. SS implies at least partial a priori information about compounds that may be present in the sample ("known unknowns"). Generally, a suspect list is compiled and used either during MS data acquisition or data processing, often restricting the search to compounds for which tandem mass spectra are available in spectral libraries.

In contrast, NTS aims at the comprehensive characterization of the chemical composition of a sample without any a priori knowledge of specific analytes [17] ("unknown unknowns"). This approach enables the discovery of completely new or unexpected molecules, such as components of natural organic matter (NOM), novel TPs, or newly emerging drugs of abuse. However, NTS requires broad and unbiased acquisition strategies as well as extensive and complex data processing workflows.

The NORMAN network [2] and the Benchmarking and Publications for Non-Targeted Analysis Working Group [18] (BP4NTA, established in 2018) have focused on harmonizing analytical methods and protocols for the detection of emerging chemicals in the environment, and promoting standardized reporting in SS/NTS studies. In addition, the NORMAN network provides a series of databases and suspect lists for SS and prioritization, whereas BP4NTA provides publicly accessible resources, including lists of analytical and computational tools.

4 Sample Preparation

The main environmental matrices include air, particulate matter, freshwater and marine surface waters, drinking water, influents and effluents of wastewater treatment plant (WWTP), groundwater, soils, sediments, and biota. These matrices are highly complex, as they contain a broad range of chemically diverse natural and anthropogenic substances, with natural components typically occurring at higher concentrations. Therefore, effective sample preparation and, where necessary, clean-up procedures are essential to address such complexity and enable the detection of the analytes of interest at trace levels.

Because of these low concentrations, particular attention must be taken during sample collection that should aim to maximize representativeness of environmental chemical patterns while accounting for spatial and temporal variability and minimize both contamination (particularly critical for ubiquitous CECs, e.g., perfluoroalkyl substances (PFAS)) and compound loss [2].

In target screening, selective sample preparation protocols are employed to assure high sensitivity and accurate quantitation. On the contrary, in SS and NTS, protocols are designed to extract as broad a range of compounds as possible [19] while minimizing matrix interferences, especially when ESI source is employed.

Sample preparation of solid samples involves a solvent extraction step, which is commonly performed using Soxhlet extraction or Quick, Easy, Cheap, Effective, Rugged, and Safe (QuEChERS)-based approaches, often followed by a clean-up step based on solid-phase extraction (SPE), also in the dispersive mode. The most used sorbents are C18, graphitized carbon black (GCB), and primary secondary amine (PSA).

To limit as much as possible sample manipulation and avoid laboratory contaminations, direct injection techniques can be employed. Nonetheless, SPE still remains indispensable in complex samples to improve sensitivity and reduce matrix effects [20]. Indeed, for liquid samples, SPE is the technique of choice, especially for large water volumes. The most used sorbents are hydrophilic-lipophilic balance (HLB) copolymers, weak anion exchangers (WAX), strong ionic exchangers, hydroxylated polystyrene-divinylbenzene copolymers, GCB, and mixed-mode materials [17, 21]. To date, HLB are the most used materials for environmental NTS applications, because of their ability to retain compounds in a wide range of polarity [20, 22]. However, they fail, for example, in the retention of new generation and short-chain PFAS, which are better retained by WAX materials.

Nonetheless, also in environmental analysis, current trends are increasingly oriented toward more sustainable sample preparation strategies, particularly focusing on method miniaturization and the reduction or replacement of toxic organic solvents, in alignment with the principles of green analytical chemistry [19, 23].

Further details on environmental sample preparation can be found in the exhaustive review by Hajeb et al. [22]

5 High-Resolution Mass Spectrometry

As already stated in 3.2 subparagraph, SS and NTS require HRMS instrumentation. First, it is important to introduce the term "resolution" that is one of the most controversial within MS community, as it is sometimes also referred to as resolving power. For the recommended MS terminology, it is advisable to refer to https://www.msterms.org/, containing links to numerous definitions from IUPAC, and other sources (more than 2200 entries in January 2026). According to these guidelines, for an isolated peak corresponding to a singly charged ion of mass m, the resolution (based on the peak-width definition) is expressed as $R = m/\Delta m$, where Δm represents the peak width measured at a specified height. Most commonly, Δm is determined as the full width at half maximum (FWHM), namely the peak width measured at 50% of its maximum height. It is also important to report the m/z value at which the resolution was determined, since in most mass analyzers, resolution varies with m/z (Table 1)

Resolution should not be confused with mass accuracy, which depends on mass analyzer resolution but also on calibration stability. Indeed, mass accuracy (from multiple replicate measurements) or mass error (for a single measurement) is the difference between the experimental m/z (accurate mass) and the theoretical m/z (exact mass), and it can be expressed as absolute mass difference (in mmu) or relative mass difference (in ppm):

$$mass\ accuracy(\mathrm{ppm}) = \frac{m_{\exp} - m_{exact}}{m_{exact}} \times 10^6$$

Table 1 Main characteristics of common mass analyzers. CID: collision-induced dissociation; HCD: higher-energy collisional dissociation; *B:* magnetic field; *k:* field curvature constant; SS: suspect screening; NTS: non-target screening; DIA: data-independent analysis; DDA: data-dependent analysis

Mass analyzer	MS/MS capability	Mass resolution (FWHM)	Resolution proportional to	Mass accuracy (ppm)	Typical application
Quadrupole (Q)	QqQ: In space (CID cell)	~10^3	m/z	50–100	Target analysis
Linear ion trap (LIT)	In time	10^3 - 10^4	m/z	10–100	Target analysis; SS
Time-of-flight (TOF)	QqTOF: In space (CID cell)	~10^4	Constant	<2	SS; NTS (DIA)
Orbitrap (FT-Orbitrap)	Q-Orbitrap: In space (HCD cell); LIT-Orbitrap: In space (HCD cell) or in time	~10^5	$\sqrt{\frac{ze}{m}k}$	<1	SS; NTS (DDA)
FT ion cyclotron resonance (FTICR)	In time	10^5 - 10^7	$\frac{ze}{m}B$	<0.2	NTS

Table 1 shows the main properties of common mass analyzers, listed following increasing resolution. An arbitrary division between high and low resolution is set at 10,000. Orbitrap and TOF instruments are the most commonly used in environmental SS and NTS analyses, whereas the ultra-HR offered by FT-ICR is rarely used in this field, except for NOM studies, due to its high cost, greater operative complexity, and the drawback that chromatographic separation often has to be sacrificed to maximize resolution performance (obtained by longer acquisition times). For a detailed comparison of the physical and analytical features of three high-resolution analyzers, the reader is referred to the paper by Zubarev and Makarov [24].

5.1 Tandem HRMS Acquisition: Data-Dependent and Data-Independent Modes

For MS/MS experiments, TOF and Orbitrap mass analyzers must be preceded by another mass analyzer able to select the precursor ions for fragmentation. Typically, this secondary mass analyzer is a Q, as in QqTOF and Q-Orbitrap instruments, or a LIT, as in LIT-Orbitrap instruments. In practice, after a HR full scan survey acquisition, the precursor ions are filtered at LR by Q (or LIT), fragmented in the collision cell (quadrupole or multipole collision cell for TOF, or higher-energy collision cell for Orbitrap), and the resulting product ions are then analyzed by a HR full scan acquisition.

There are two main distinct strategies for acquiring HRMS/MS data, namely data-dependent acquisition (DDA) and data-independent acquisition (DIA).

In DDA, precursor ions are selected for fragmentation based on their intensity in the survey scan, i.e., for each scan cycle, the top N most abundant ions are selected (selective, intensity-driven MS/MS). This approach produces high-quality MS/MS spectra, but low-abundance ions are missed. Furthermore, the maximum number of ions (top N) that can be selected for MS/MS experiments strictly depends on the mass analyzer and specific model characteristics. Indeed, model scan rate (that can vary considerably, for example, among different Orbitrap instruments), resolution and consequently cycle time setting, ion injection time, and chromatographic peak width are all parameters to consider.

In contrast, DIA fragments all ions regardless of their abundance, either within pre-defined wide m/z windows, as in *sequential window acquisition of all theoretical fragment-ion spectra* (SWATH) mode, or across the entire m/z range without isolation windows, as in *all-ion fragmentation* (AIF) mode, generating comprehensive MS/MS data for both high- and low-intensity species (comprehensive, unbiased MS/MS). The most common DIA approach, especially with TOF instruments, consists in the simultaneous acquisition of mass spectra at high and low collision energies, allowing to alternately acquire spectra of all detectable precursors and their products. Although DIA provides broader coverage and better reproducibility, the resulting spectra are more complex and require advanced computational tools for deconvolution; otherwise the relationships between precursor and fragment ions could remain ambiguous.

Table 2 Key points of data-dependent and data-independent acquisitions

	DDA	DIA (SWATH)	DIA (AIF)
Precursor ion selection criteria	Abundance	*m/z* window	None
Q1 mode	Narrow pass (0.7–4 *m/z*)	Medium pass (10–25 *m/z*)	Wide pass (e.g. up to 1000 *m/z*)
MS/MS spectra quality	High	Deconvolution needed	Deconvolution needed
MS/MS spectra reproducibility	Low	High	High

The possibility of acquiring tandem mass spectra in both DDA and DIA modes would provide the maximum amount of information and should be considered, but in most cases, it is in contrast with analysis time and availability [2].

Table 2 briefly summarizes the key features of the two acquisition approaches, namely DDA and DIA.

It is important to note that, in DDA, the precursor ion isolation is carried out in narrow pass Q1 mode, i.e., a nominal mass range. Therefore, coeluted isobaric ions will also be isolated, leading to mixed fragmentation spectra [2].

Differently from target LRMS-based methods, NTS approach, in both DDA and DIA modes, offers the possibility of a retrospective analysis of compounds not initially considered.

6 Ion Mobility Spectrometry

Although MS can provide extensive structural information on compounds, often it fails in differentiating isomeric species or providing detailed structural conformation. In LC-HRMS, compounds are separated according to their chromatographic retention time, accurate *m/z* value, and peak intensity. Ion mobility spectrometry (IMS) provides a fourth separation dimension by discriminating ions according to their drift time (in milliseconds) required to travel the IMS tube or cell under an electric field in the presence of a drift gas [25, 26]. The drift time is a function of the ion collision cross section (CCS), which depends on its size, shape, and charge [2], as well as on the nature of the drift gas (e.g., air, nitrogen, helium). Integrating CCS values as an additional orthogonal parameter significantly increases confidence in compound structural annotation. This is particularly advantageous for the analysis of contaminants in complex environmental matrices, where co-eluting isobaric interferences and structural isomers are frequently encountered [25].

Although experimental CCS databases, such as METLIN-CCS [27], are currently limited to a few thousands of compounds (but the number is rapidly expanding) [28–30], CCS values can also be computationally predicted [25], supporting SS and NTS workflows [31, 32]. Indeed, the availability of CCS values (which are reproducible even using different IMS devices) [25, 28] supports compound confirmation, reduces false positives, and enhances the reliability of retrospective analysis in environmental monitoring.

To date, various IMS technologies and devices are available on the market [25, 26]. In particular, IMS is a well-established technology coupled with QTOF mass analyzers. Furthermore, in combination with DIA, IMS can resolve highly complex spectra [2], and reduce background noise [25]. In environmental analysis, this allows increased sensitivity and selectivity for less abundant contaminants present at trace-levels.

Recently, ion mobility systems for Orbitrap analyzers have also been proposed; however, their use is still rather limited to the structural characterization of biomolecules [33, 34].

7 Data Processing and Compound Annotation and Identification

The LC-HRMS acquisitions generate thousands of tandem mass spectra per sample, making manual data interpretation unfeasible without the support of dedicated software and computational tools. This is especially challenging in monitoring programs involving a large number of samples to be processed simultaneously.

It is estimated that less than 20% of metabolites are identified in the NTS, while ~80% of signals remain uninterpreted in the recorded mass spectrum. This limitation is mainly due to the incomplete coverage of existing mass spectral libraries. For instance, as of January 2026, mzCloud (https://www.mzcloud.org/), which is one of the largest HR mass spectral libraries, contains more than 16,5 million spectra from 32,330 compounds, whereas chemical databases such as PubChem (https://pubchem.ncbi.nlm.nih.gov/) and ChemSpider (https://www.chemspider.com/) list more than 120 million unique compounds each.

The mismatch between the number of available reference spectra and the chemical space potentially present in environmental sample limits confident compound annotation in NTS workflows.

7.1 Data Processing Workflows

Many data processing workflows are implemented in vendor software, such as Compound Discoverer® from Thermo Fisher Scientific, MetaboScape® from Bruker, MassHunter® from Agilent, and UNIFI® from Waters. Nonetheless, various open access software tools are also available, e.g. MS–DIAL, enviMass, MZmine3, and XCMS [2].

For certain software, a previous raw data format conversion is needed. Then, all samples from the same batch (technical replicates, experimental replicates, different samples), as well as procedural blanks, and quality control (QC) and quality assurance (QA) samples, must be submitted together (classifying them properly for relative quantitation and statistical analysis) and subject to the same processing workflow.

The three main sources of the spectrum complexity are (1) noise signals, which requires blank subtraction; (2) signals of isotopic distribution, which requires deisotoping; and (3) presence of multiply adducts (e.g., $[M + H]^+$, $[M + Na]^+$, $[M + K]^+$, $[M + NH_4]^+$, $[M + ACN + H]^+$, $[M + MeOH+H]^+$, $[M-H_2O + H]^+$), which requires signal grouping. Other steps could be spectral deconvolution, and gap filling.

An initial mass list of the *m/z* values found in each MS scan across the LC run exceeding a user-defined threshold (noise level) is generated. Chromatogram is built by extracting the ion current for each detected mass, and the retention times are aligned. It is fundamental to reduce the spectrum complexity without altering the information useful to identify the ions in the spectrum.

The main goal is to turn the highly complex LC-MS raw data into a list of features. A *feature* refers to a distinct signal detected in the mass spectrum that is associated with a specific compound or ion in the sample, and it includes *m/z*, retention time, and intensity. A feature is not necessarily a single compound or ion, but it could represent an isotopic variant, an adduct, or a fragment ion (https://nontargetedanalysis.org/reference-content/methods/data-processing-and-analysis/).

In NTS, a very large number of features can be obtained, and their full elucidation could be unfeasible. However, prioritization strategies (e.g., specific fragments in MS/MS spectra, or specific Kendrick mass defects [35]) can be applied to filter out the non-relevant signals for the objectives of the study [2].

7.2 Suspect Lists

To date, several public and high-quality suspect lists are available [2]. They can be found on NORMAN website (NORMAN Suspect List Exchange, https://www.norman-network.com/nds/SLE/), and the CompTox Chemicals Dashboard (https://comptox.epa.gov/dashboard/chemical_lists). Other suspect lists can be built from REACH registry, and PubChem.

Suspect lists can be also used for prioritization in NTS.

7.3 Statistical and Chemometric Analyses

Processing software can perform statistical and chemometric analyses for supporting data interpretation and finding potential relationships between samples and or feature/compounds. Indeed, after feature extraction and compound annotation or identification, dataset complexity requires multivariate approaches to extract useful information, and find trends, relationships, and discriminatory variable across groups. Unsupervised methods (e.g., principal component analysis and hierarchical clustering) can provide an initial data overview, facilitating the visualization of compositional differences in chemical profiles across sample sets, and reducing dimension prior to more targeted modeling. Supervised approaches (e.g., partial least squares discriminant analysis) can then be applied to identify which feature or annotated compound is responsible for differences between classes, serving as potential marker compounds or discriminant feature [36]. Therefore, statistical analysis can point out correlations, trends, and potential causal relationships that are not evident from compound lists alone. Importantly, chemometric methods support prioritization of detected features for further identification and validation, particularly when thousands of signals are present and only a subset can be fully characterized [37].

7.4 Databases and Tandem Mass Spectral Libraries

A compound database is a structured collection of chemical information that can be used for compound annotation, whereas a spectral library is a repository of mass spectra and fragmentation mass spectra for direct spectral matching to support compound identification (https://nontargetedanalysis.org/reference-content/methods/data-processing-and-analysis/).

The main spectral libraries are mzCloud, (https://www.mzcloud.org/), MassBank (https://massbank.eu/MassBank/), and MassBank of North America (MoNA, https://mona.fiehnlab.ucdavis.edu/). Automated spectral searches are possible only with libraries that are integrated into the software, such as mzCloud in Compound Discoverer, while other require manual downloading or separate processing steps, as they are not directly accessible for automated searching.

A typical processing workflow of Compound Discoverer, including mzCloud spectral library and mass list searches, is shown in Fig. 3.

There is a fundamental difference between compound annotation and identification. A compound can be putatively annotated based upon physicochemical properties and/or spectral similarity with public/commercial spectral libraries, whereas compound identification necessarily requires the match of retention time and tandem mass spectrum with a reference standard.

For a feature, molecular formula is assigned based on accurate mass (*m/z*) measurement, isotopic pattern, observed adducts, and product ion spectrum. Compound annotation is also achieved by database, spectral library, or structural similarity searching. In silico fragmentation can be also used for product ion spectrum comparison.

7.5 Prediction Tools and Machine Learning

The SS and NTS strategies increasingly rely on computational approaches to improve compound annotation. For example, to encompass the limited number of product ion mass spectra reposited in the libraries, in silico predictive MS fragmentation has been proposed, such as MetFrag (https://msbi.ipb-halle.de/MetFrag/) [38].

To refine candidate selection, TP suspect list can be generated using prediction tools such as EnviPath [39] that is commonly employed for microbial transformation predictions, and BioTransformer [40] that can be employed for both microbial and human transformations [2]. Other tools are available to calculate mass of potential TPs.

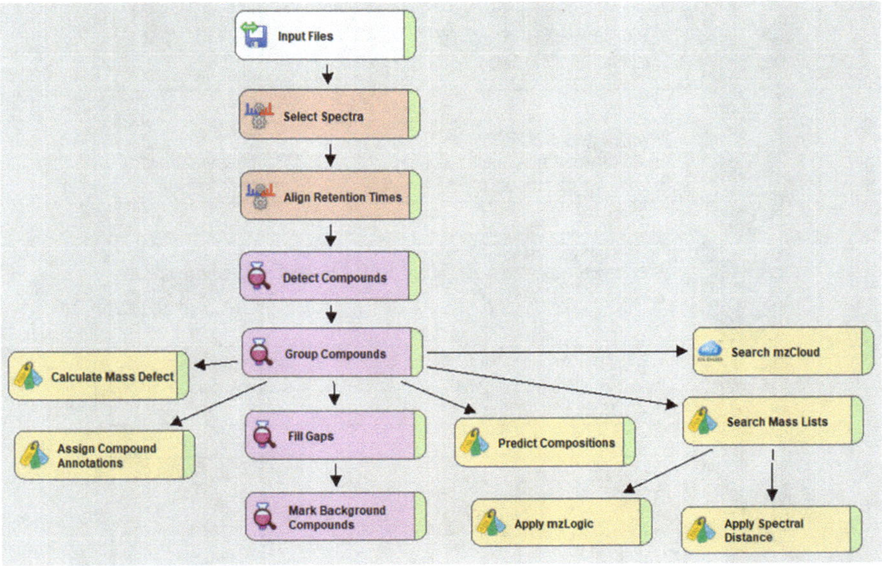

Fig. 3 Typical Compound Discoverer processing workflow with the main nodes

Retention time (RT) predictions based on molecular structures, commonly implemented through machine learning-based quantitative structure-retention relationships (QSRR), provide additional evidence by linking structural features to chromatographic behavior. Unlike GC, LC RTs are influenced by column type, mobile phase, flow rate, gradient, and temperature. Nonetheless, RT predictions by QSRR models can be used to discard candidates [41].

More broadly, machine learning is playing an increasingly important role in MS-based methods. For further discussion, the reader is referred to the review of Beck et al. [42].

7.6 Identification Confidence

To harmonize the communication of the reliability of compound identifications obtained through SS/NTS strategies, Schymanski et al. [43] proposed a five-level confidence scheme, which has since been widely adopted by the scientific community and recommended by the NORMAN network. This system provides a structured framework to report the degree of confidence (increasing from level 5 to level 1) associated with detected compounds, as illustrated in Fig. 4.

Level 1 represents the highest confidence, where the structure is confirmed using a reference standard, including matching retention time and MS/MS spectra. If reference standard is not available, level 2, corresponding to probable structure, can be obtained only if supported either by match with a spectral library (2a) or diagnostic fragmentation patterns (2b).

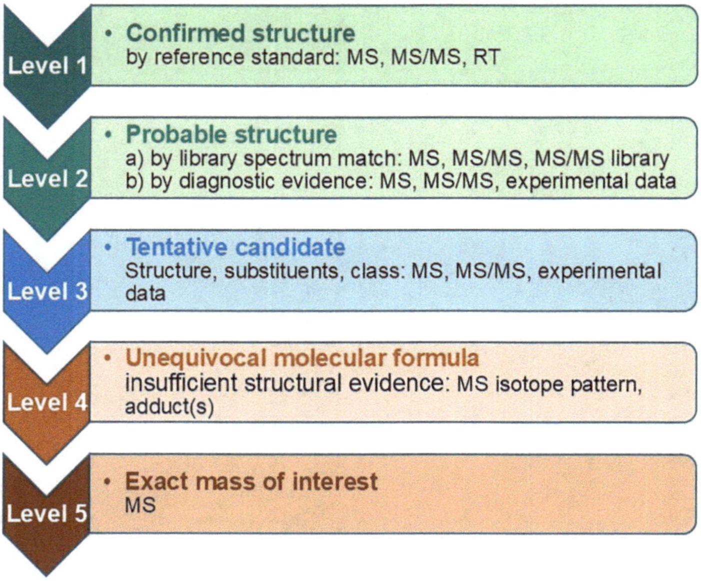

Fig. 4 Identification confidence levels in high resolution mass spectrometric analysis, as proposed by Schymanski et al. [43]

The intermediate level 3 corresponds to tentative candidates, where multiple possible structures fit the observed data, such as accurate mass and some fragment ions, but a single structure cannot be confidently assigned.

Level 4 is assigned when only the molecular formula can be unequivocally determined, using accurate mass and isotopic patterns, whereas level 5, representing the lowest confidence, indicates that only the exact mass of a compound is known with no structural information.

PFAS are a large class (a few thousand) of synthetic compounds. Considering their number, the subsequent lack of standards, their ubiquity, and persistence, specific identification criteria (e.g., mass defects) associated to a dedicated confidence level scale have been elaborated [29, 44].

More recently, a machine learning approach to classify identification reliability was proposed [45]. This identification point system, designed for the NTS by LC-HRMS, is compatible with the five-level system and could be automatized.

8 Concluding Remarks [2, 25] [28–30] [2, 25, 33]

In recent years, LC-HRMS has undergone significant technological advancements, becoming a crucial tool for environmental monitoring. Nevertheless, HRMS alone cannot tackle the increasing number of contaminants and TPs. To address this challenge, ion mobility data, in silico approaches, and machine learning algorithms are increasingly employed to support the detection and characterization of unknown compounds by NTS approach,

Simultaneously, substantial efforts are being made to allow the scientific community to adopt harmonized and standardized protocols, workflows, and terminology, thereby enabling the reliable exchange of data from non-targeted analyses.

Questions

- What are the main advantages and drawbacks of LC-HRMS vs GC-EI-MS for the analysis of contaminants of emerging concern and their transformation products in environmental matrices?
- How do suspect screening and non-target screening differ in terms of prior knowledge, analytical goals, data acquisition strategies, and achievable levels of identification confidence?
- Why are the main challenges in detecting contaminants of emerging concern, and their transformation products?
- Can in silico tools improve compound identification confidence in non-target screening?
- How can data processing workflows reduce LC-HRMS data complexity?
- What is meant with "feature" in non-target analysis? How do identification confidence levels distinguish between annotation and confirmed identification?

Glossary

- AIF: all-ion fragmentation
- APCI: atmospheric pressure chemical ionization
- APPI: atmospheric pressure photoionization
- CEC: contaminants of emerging concern
- DIA: data independent analysis
- DDA: data dependent analysis
- ESI: electrospray ionization
- FWHM: full width at half maximum
- HILIC: hydrophilic interaction liquid chromatography
- HLB: hydrophilic-lipophilic balance
- HR: high-resolution
- IMS: ion mobility spectrometry
- LR: low-resolution
- NOM: natural organic matter
- NTS: non-target screening
- PAHs: polycyclic aromatic hydrocarbons
- PFAS: perfluoroalkyl substances
- POPs: persistent organic pollutants
- PPCPs: pharmaceuticals and personal care products

(continued)

- QA: quality assurance
- QC: quality control
- QSRR: quantitative structure-retention relationships
- RPLC: reversed-phase LC
- SS: suspect screening
- SWATH: sequential window acquisition of all theoretical fragment-ion spectra
- TPs: transformation products
- WWTP: wastewater treatment plant

Further Reading

- Dulio, V.; Alygizakis, N.; Ng, K.; Schymanski, E. L.; Andres, S.; Vorkamp, K.; Hollender, J.; Finckh, S.; Aalizadeh, R.; Ahrens, L.; Bouhoulle, E.; Čirka, Ľ.; Derksen, A.; Deviller, G.; Duffek, A.; Esperanza, M.; Fischer, S.; Fu, Q.; Gago-Ferrero, P.; Haglund, P.; Junghans, M.; Kools, S. A. E.; Koschorreck, J.; Lopez, B.; Lopez de Alda, M.; Mascolo, G.; Miège, C.; Osté, L.; O'Toole, S.; Rostkowski, P.; Schulze, T.; Sims, K.; Six, L.; Slobodnik, J.; Staub, P. F.; Stroomberg, G.; Thomaidis, N. S.; Togola, A.; Tomasi, G.; von der Ohe, P. C. Beyond Target Chemicals: Updating the NORMAN Prioritisation Scheme to Support the EU Chemicals Strategy with Semi-Quantitative Suspect/Non-Target Screening Data. *Environmental Sciences Europe* 2024, *36* (1), 113. https://doi.org/10.1186/S12302-024-00936-3.
- Baskaran, S.; Chirsir, P.; Joudan, S.; Wolf, R.; Bolton, E. E.; Thiessen, P. A.; Schymanski, E. L. Reporting Chemical Data in the Environmental Sciences. *ACS Environmental Au* 2025, *5* (5), 444–456. https://doi.org/10.1021/ACSENVIRONAU.5C00034.
- Malm, L.; Liigand, J.; Aalizadeh, R.; Alygizakis, N.; Ng, K.; Frøkjær, E. E.; Nanusha, M. Y.; Hansen, M.; Plassmann, M.; Bieber, S.; Letzel, T.; Balest, L.; Abis, P. P.; Mazzetti, M.; Kasprzyk-Hordern, B.; Ceolotto, N.; Kumari, S.; Hann, S.; Kochmann, S.; Steininger-Mairinger, T.; Soulier, C.; Mascolo, G.; Murgolo, S.; Garcia-Vara, M.; López de Alda, M.; Hollender, J.; Arturi, K.; Coppola, G.; Peruzzo, M.; Joerss, H.; van der Neut-Marchand, C.; Pieke, E. N.; Gago-Ferrero, P.; Gil-Solsona, R.; Licul-Kucera, V.; Roscioli, C.; Valsecchi, S.; Luckute, A.; Christensen, J. H.; Tisler, S.; Vughs, D.; Meekel, N.; Talavera Andújar, B.; Aurich, D.; Schymanski, E. L.; Frigerio, G.; Macherius, A.; Kunkel, U.; Bader, T.; Rostkowski, P.; Gundersen, H.; Valdecanas, B.; Davis, W. C.; Schulze, B.; Kaserzon, S.; Pijnappels, M.; Esperanza, M.; Fildier, A.; Vulliet, E.; Wiest, L.; Covaci, A.; Macan Schönleben, A.; Belova, L.; Celma, A.; Bijlsma, L.; Caupos, E.; Mebold, E.; Le Roux, J.; Troia, E.; de Rijke, E.; Helmus, R.; Leroy, G.; Haelewyck, N.; Chrastina, D.; Verwoert, M.; Thomaidis, N. S.; Kruve,

(continued)

A. Quantification Approaches in Non-Target LC/ESI/HRMS Analysis: An Interlaboratory Comparison. *Analytical Chemistry* 2024, *96* (41), 16215–16226. https://doi.org/10.1021/ACS.ANALCHEM.4C02902.

- Lennon, S.; Chaker, J.; Price, E. J.; Hollender, J.; Huber, C.; Schulze, T.; Ahrens, L.; Béen, F.; Creusot, N.; Debrauwer, L.; Dervilly, G.; Gabriel, C.; Guérin, T.; Habchi, B.; Jamin, E. L.; Klánová, J.; Kosjek, T.; Le Bizec, B.; Meijer, J.; Mol, H.; Nijssen, R.; Oberacher, H.; Papaioannou, N.; Parinet, J.; Sarigiannis, D.; Stravs, M. A.; Tkalec, Ž.; Schymanski, E. L.; Lamoree, M.; Antignac, J. P.; David, A. Harmonized Quality Assurance/Quality Control Provisions to Assess Completeness and Robustness of MS1 Data Preprocessing for LC-HRMS-Based Suspect Screening and Non-Targeted Analysis. *TrAC Trends in Analytical Chemistry* 2024, *174*, 117674. https://doi.org/10.1016/J.TRAC.2024.117674.
- Schulze, B.; Jeon, Y.; Kaserzon, S.; Heffernan, A. L.; Dewapriya, P.; O'Brien, J.; Gomez Ramos, M. J.; Ghorbani Gorji, S.; Mueller, J. F.; Thomas, K. V.; Samanipour, S. An Assessment of Quality Assurance/Quality Control Efforts in High Resolution Mass Spectrometry Non-Target Workflows for Analysis of Environmental Samples. *TrAC Trends in Analytical Chemistry* 2020, *133*, 116063. https://doi.org/10.1016/J.TRAC.2020.116063.
- Zahn, D.; Arp, H. P. H.; Fenner, K.; Georgi, A.; Hafner, J.; Hale, S. E.; Hollender, J.; Letzel, T.; Schymanski, E. L.; Sigmund, G.; Reemtsma, T. Should Transformation Products Change the Way We Manage Chemicals? *Environmental Science & Technology* 2024, *58* (18), 7710–7718. https://doi.org/10.1021/ACS.EST.4C00125.
- Yang, Y.; Yang, L.; Zheng, M.; Cao, D.; Liu, G. Data Acquisition Methods for Non-Targeted Screening in Environmental Analysis. *TrAC Trends in Analytical Chemistry* 2023, *160*, 116966. https://doi.org/10.1016/J.TRAC.2023.116966.
- Qin, H.; Liu, H.; Liu, Y.; Di, S.; Bao, Y.; Zhai, Y.; Zhu, S. Recent Advances in Sample Preparation and Chromatographic Analysis of Pharmaceuticals and Personal Care Products in Environment. *TrAC Trends in Analytical Chemistry* 2023, *164*, 117112. https://doi.org/10.1016/J.TRAC.2023.117112.
- Wang, H.; Zhong, L.; Su, W.; Ruan, T.; Jiang, G. Machine Learning-Assisted Identification of Environmental Pollutants by Liquid Chromatography Coupled with High-Resolution Mass Spectrometry. *TrAC Trends in Analytical Chemistry* 2024, *180*, 117988. https://doi.org/10.1016/J.TRAC.2024.117988.
- Zweigle, J.; Tisler, S.; Bevilacqua, M.; Tomasi, G.; Nielsen, N. J.; Gawlitta, N.; Lübeck, J. S.; Smilde, A. K.; Christensen, J. H. Prioritization Strategies for Non-Target Screening in Environmental Samples by Chromatography – High-Resolution Mass Spectrometry: A Tutorial. *Journal of Chromatography A* 2025, *1751*, 465944. https://doi.org/10.1016/J.CHROMA.2025.465944.

Bibliography

1. Escher BI, Stapleton HM, Schymanski EL (2020) Tracking complex mixtures of chemicals in our changing environment. Science 367(6476):388–392. https://doi.org/10.1126/science.aay6636
2. Hollender J, Schymanski EL, Ahrens L, Alygizakis N, Béen F, Bijlsma L, Brunner AM, Celma A, Fildier A, Fu Q, Gago-Ferrero P, Gil-Solsona R, Haglund P, Hansen M, Kaserzon S, Kruve A, Lamoree M, Margoum C, Meijer J, Merel S, Rauert C, Rostkowski P, Samanipour S, Schulze B, Schulze T, Singh RR, Slobodnik J, Steininger-Mairinger T, Thomaidis NS, Togola A, Vorkamp K, Vulliet E, Zhu L, Krauss M (2023) NORMAN guidance on suspect and non-target screening in environmental monitoring. Environ Sci Europe 35(1):75. https://doi.org/10.1186/s12302-023-00779-4
3. Ibáñez M, Bijlsma L, Pitarch E, López FJ, Hernández F (2021) Occurrence of pharmaceutical metabolites and transformation products in the aquatic environment of the Mediterranean area. Trend Environ Analy Chem 29:e00118. https://doi.org/10.1016/J.TEAC.2021.E00118
4. Montone CM, Giannelli Moneta B, Laganà A, Piovesana S, Taglioni E, Cavaliere C (2024) Transformation products of antibacterial drugs in environmental water: identification approaches based on liquid chromatography-high resolution mass spectrometry. J Pharm Biomed Anal 238:115818. https://doi.org/10.1016/J.JPBA.2023.115818
5. Hettiyadura APS, Laskin A (2022) Quantitative analysis of polycyclic aromatic hydrocarbons using high-performance liquid chromatography-photodiode Array-high-resolution mass spectrometric detection platform coupled to electrospray and atmospheric pressure photoionization sources. J Mass Spectrom 57(2):e4804. https://doi.org/10.1002/jms.4804
6. Singh RR, Chao A, Phillips KA, Xia XR, Shea D, Sobus JR, Schymanski EL, Ulrich EM (2020) Expanded coverage of non-targeted LC-HRMS using atmospheric pressure chemical ionization: a case study with ENTACT mixtures. Anal Bioanal Chem 412(20):4931–4939. https://doi.org/10.1007/S00216-020-02716-3
7. Kondyli A, Schrader W (2020) Evaluation of the combination of different atmospheric pressure ionization sources for the analysis of extremely complex mixtures. Rapid Commun Mass Spectrom 34(8):e8676. https://doi.org/10.1002/rcm.8676
8. Cappiello A, Famiglini G, Palma P, Pierini E, Termopoli V, Trufelli H (2011) Direct-EI in LC-MS: towards a universal detector for small-molecule applications. Mass Spectrom Rev 30(6):1242–1255. https://doi.org/10.1002/mas.20329
9. Famiglini G, Palma P, Termopoli V, Cappiello A (2021) The history of electron ionization in LC-MS, from the early days to modern technologies: a review. Anal Chim Acta 1167(5):338350. https://doi.org/10.1016/j.aca.2021.338350
10. Palma P, Famiglini G, Trufelli H, Pierini E, Termopoli V, Cappiello A (2011) Electron ionization in LC-MS: recent developments and applications of the direct-EI LC-MS Interface. Anal Bioanal Chem 399(8):2683–2693. https://doi.org/10.1007/s00216-010-4637-0
11. Famiglini G, Palma P, Siviero A, Rezai MA, Cappiello A (2005) Determination of endocrine disrupting compounds in marine water by nanoliquid chromatography/direct-electron ionization mass spectrometry. Anal Chem 77(23):7654–7661. https://doi.org/10.1021/ac050842m
12. Famiglini G, Palma P, Pierini E, Trufelli H, Cappiello A (2008) Organochlorine pesticides by LC–MS. Anal Chem 80(9):3445–3449. https://doi.org/10.1021/ac8000435
13. Brack W, Hollender J, de Alda ML, Müller C, Schulze T, Schymanski E, Slobodnik J, Krauss M (2019) High-resolution mass spectrometry to complement monitoring and track emerging chemicals and pollution trends in European water resources. Environ Sci Eur 31(1):62. https://doi.org/10.1186/S12302-019-0230-0
14. Cavaliere C, Antonelli M, Capriotti AL, La Barbera G, Montone CM, Piovesana S, Laganà A (2019) A triple quadrupole and a hybrid quadrupole orbitrap mass spectrometer in comparison for polyphenol quantitation. J Agric Food Chem 67(17):4885–4896. https://doi.org/10.1021/acs.jafc.8b07163
15. Fontaine F, Morettoni L, Anderson K, Choi B, Zamora I, Bateman KP (2025) Molecular structure and mass spectral data quality–driven processing of high-resolution mass spectrometry for quantitative analysis. Rapid Commun Mass Spectrom 0:e10000. https://doi.org/10.1002/RCM.10000

16. Vergara-Luis I, Righetti L, Olivares M, Padilla-Gonzalez F, Prieto A, Arrizabalaga-Larrañaga A, Ayala-Cabrera JF (2026) A review of suspect and non-target screening of agriculture-related organic contaminants. Anal Chim Acta 1383:344774. https://doi.org/10.1016/J.ACA.2025.344774
17. Mandal V, Ajabiya J, Khan N, Tekade RK, Sengupta P (2024) Advances and challenges in non-targeted analysis: an insight into sample preparation and detection by liquid chromatography-mass spectrometry. J Chromatogr A 1737:465459. https://doi.org/10.1016/J.CHROMA.2024.465459
18. Place BJ, Ulrich EM, Challis JK, Chao A, Du B, Favela K, Feng YL, Fisher CM, Gardinali P, Hood A, Knolhoff AM, McEachran AD, Nason SL, Newton SR, Ng B, Nuñez J, Peter KT, Phillips AL, Quinete N, Renslow R, Sobus JR, Sussman EM, Warth B, Wickramasekara S, Williams AJ (2021) An introduction to the benchmarking and publications for non-targeted analysis working group. Anal Chem 93(49):16289–16296. https://doi.org/10.1021/ACS.ANALCHEM.1C02660
19. Pérez-Fernández V, Mainero Rocca L, Tomai P, Fanali S, Gentili A (2017) Recent advancements and future trends in environmental analysis: sample preparation, liquid chromatography and mass spectrometry. Anal Chim Acta 983(5):9–41. https://doi.org/10.1016/j.aca.2017.06.029
20. Jia W, Liu H, Ma Y, Huang G, Liu Y, Zhao B, Xie D, Huang K, Wang R (2024) Reproducibility in nontarget screening (NTS) of environmental emerging contaminants: assessing different HLB SPE cartridges and instruments. Sci Total Environ 912:168971. https://doi.org/10.1016/J.SCITOTENV.2023.168971
21. Brandsma SH, De Boer J, Leonards PEG, Cofino WP, Covaci A, Leonards PEG (2021) Recent advances in data-mining techniques for measuring transformation products by high-resolution mass spectrometry. TrAC Trends Anal Chem 143:116409. https://doi.org/10.1016/j.trac.2012.12.004
22. Hajeb P, Zhu L, Bossi R, Vorkamp K (2022) Sample preparation techniques for suspect and non-target screening of emerging contaminants. Chemosphere 287:132306. https://doi.org/10.1016/J.CHEMOSPHERE.2021.132306
23. Quero-Pastor MJ, Garrido-Perez MC, Acevedo A, Quiroga JM (2024) Advances in monitoring pharmaceuticals in an aquatic environment: greenness assessment of analytical procedures. TrAC Trends Anal Chem 180:117921. https://doi.org/10.1016/j.scitotenv.2013.07.067
24. Zubarev RA, Makarov A (2013) Orbitrap mass spectrometry. Anal Chem 85(11):5288–5296. https://doi.org/10.1021/AC4001223
25. Celma A, Alygizakis N, Belova L, Bijlsma L, Fabregat-Safont D, Menger F, Gil-Solsona R (2024) Ion mobility separation coupled to high-resolution mass spectrometry in environmental analysis—current state and future potential. Trend Environ Analy Chem 43:e00239. https://doi.org/10.1016/J.TEAC.2024.E00239
26. Dodds JN, Baker ES (2019) Ion mobility spectrometry: fundamental concepts, instrumentation, applications, and the road ahead. J Am Soc Mass Spectrom 30(11):2185–2195. https://doi.org/10.1007/s13361-019-02288-2
27. Baker ES, Hoang C, Uritboonthai W, Heyman HM, Pratt B, MacCoss M, MacLean B, Plumb R, Aisporna A, Siuzdak G (2023) METLIN-CCS: an ion mobility spectrometry collision cross section database. Nat Method 20(12):1836–1837. https://doi.org/10.1038/s41592-023-02078-5
28. Teri D, Aly NA, Dodds JN, Zhang J, Thiessen PA, Bolton EE, Joseph KM, Williams AJ, Schymanski EL, Rusyn I, Baker ES (2025) Reference library for suspect screening of environmental toxicants using ion mobility spectrometry-mass spectrometry. Commun Chem 8(1):224. https://doi.org/10.1038/s42004-025-01619-7
29. Boatman AK, Chappel JR, Kirkwood-Donelson KI, Fleming JF, Reif DM, Schymanski EL, Rager JE, Baker ES (2025) Updated guidance for communicating PFAS identification confidence with ion mobility spectrometry. Environ Sci Technol 59(33):17711–17721. https://doi.org/10.1021/ACS.EST.5C01354
30. Elapavalore A, Ross DH, Grouès V, Aurich D, Krinsky AM, Kim S, Thiessen PA, Zhang J, Dodds JN, Baker ES, Bolton EE, Xu L, Schymanski EL (2025) PubChemLite plus collision cross section (CCS) values for enhanced interpretation of nontarget environmental data. Environ Sci Technol Lett 12(2):166–174. https://doi.org/10.1021/ACS.ESTLETT.4C01003

31. Celma A, Sancho JV, Schymanski EL, Fabregat-Safont D, Ibáñez M, Goshawk J, Barknowitz G, Hernández F, Bijlsma L (2020) Improving target and suspect screening high-resolution mass spectrometry workflows in environmental analysis by ion mobility separation. Environ Sci Technol 54(23):15120–15131. https://doi.org/10.1021/acs.est.0c05713
32. Ahrens L, Daneshvar A, Lau AE, Kreuger J (2022) Enhancing spectral quality in complex environmental matrices: supporting suspect and non-target screening in zebra mussels with ion mobility. Environ Int 170:107585. https://doi.org/10.1016/j.envint.2022.107585
33. Poltash ML, McCabe JW, Shirzadeh M, Laganowsky A, Russell DH (2020) Native IM-Orbitrap MS: resolving what was hidden. TrAC Trends Anal Chem 124:115533. https://doi.org/10.1016/J.TRAC.2019.05.035
34. Hollerbach AL, Ibrahim YM, Meras V, Norheim RV, Huntley AP, Anderson GA, Metz TO, Ewing RG, Smith RD (2023) A dual-gated structures for lossless ion manipulations-ion mobility Orbitrap mass spectrometry platform for combined ultra-high-resolution molecular analysis. Anal Chem 95(25):9531–9538. https://doi.org/10.1021/acs.analchem.3c00881
35. Merel S (2023) Critical assessment of the Kendrick mass defect analysis as an innovative approach to process high resolution mass spectrometry data for environmental applications. Chemosphere 313:137443. https://doi.org/10.1016/j.chemosphere.2022.137443
36. Drees F, Vosough M, Schmidt TC (2025) Unraveling non-target screening variability for LC-HRMS data: a chemometric comparative analysis of river water samples impacted by treated wastewater. Anal Bioanal Chem 417(27):6075–6088. https://doi.org/10.1007/S00216-025-05966-1
37. Zedda M, Zwiener C (2012) Is nontarget screening of emerging contaminants by LC-HRMS successful? A plea for compound libraries and computer tools. Anal Bioanal Chem 403(9):2493–2502. https://doi.org/10.1007/S00216-012-5893-Y
38. Ciccarelli D, Samanipour S, Rapp-Wright H, Bieber S, Letzel T, O'Brien JW, Marczylo T, Gant TW, Vineis P, Barron LP (2025) Bridging knowledge gaps in human chemical exposure via drinking water with non-target screening. Crit Rev Environ Sci Technol 55(3):190–214. https://doi.org/10.1080/10643389.2024.2396690
39. Wicker J, Lorsbach T, Gütlein M, Schmid E, Latino D, Kramer S, Fenner K (2016) EnviPath—the environmental contaminant biotransformation pathway resource. Nucleic Acids Res 44(D1):D502–D508. https://doi.org/10.1093/nar/gkv1229
40. Djoumbou-Feunang Y, Fiamoncini J, Gil-de-la-Fuente A, Greiner R, Manach C, Wishart DS (2019) BioTransformer: a comprehensive computational tool for small molecule metabolism prediction and metabolite identification. J Chem 11(1):2. https://doi.org/10.1186/s13321-018-0324-5
41. Zhang Y, Liu F, Li XQ, Gao Y, Li KC, Zhang QH (2024) Generic and accurate prediction of retention times in liquid chromatography by post-projection calibration. Commun Chem 7(1). https://doi.org/10.1038/S42004-024-01135-0
42. Beck AG, Muhoberac M, Randolph CE, Beveridge CH, Wijewardhane PR, Kenttämaa HI, Chopra G (2024) Recent developments in machine learning for mass spectrometry. ACS Measure Sci Au 4(3):233–246. https://doi.org/10.1021/ACSMEASURESCIAU.3C00060
43. Schymanski EL, Jeon J, Gulde R, Fenner K, Ruff M, Singer HP, Hollender J (2014) Identifying small molecules via high resolution mass spectrometry: communicating confidence. Environ Sci Technol 48(4):2097–2098. https://doi.org/10.1021/es5002105
44. Charbonnet JA, Mcdonough CA, Xiao F, Schwichtenberg T, Cao D, Kaserzon S, Thomas KV, Dewapriya P, Place BJ, Schymanski EL, Field JA, Helbling DE, Higgins CP (2022) Communicating confidence of per-and Polyfluoroalkyl substance identification via high-resolution mass spectrometry. Environ Sci Technol Lett 29(6):473–481. https://doi.org/10.1021/acs.estlett.2c00206
45. Alygizakis N, Lestremau F, Gago-Ferrero P, Gil-Solsona R, Arturi K, Hollender J, Schymanski EL, Dulio V, Slobodnik J, Thomaidis NS (2023) Towards a harmonized identification scoring system in LC-HRMS/MS based non-target screening (NTS) of emerging contaminants. TrAC Trends Anal Chem 159:116944. https://doi.org/10.1016/J.TRAC.2023.116944

Comprehensive Two-Dimensional Gas Chromatography in Environmental Analysis: Principles, Instrumental Configurations, and Data Exploitation

Andrea Caratti, Cecilia Cagliero, and Chiara Cordero

Abstract

Comprehensive two-dimensional gas chromatography (GC×GC) has become a powerful separation technique for addressing the chemical complexity that characterizes many environmental samples. By expanding the chromatographic separation space from a single dimension to a two-dimensional retention space, GC×GC dramatically increases peak capacity, reduces co-elution, and reveals structured chromatographic patterns associated with chemically related compound classes. These features provide analytical information that is fundamentally inaccessible using conventional one-dimensional gas chromatography.

This chapter presents an integrated overview of the principles, instrumental configurations, and analytical potential of GC×GC in environmental analysis. Key aspects of separation design, including column selection, modulation strategies, and carrier gas management, are discussed in relation to their impact on resolving power, robustness, and interpretability of comprehensive separations. The role of different detection systems is addressed, highlighting how both single-channel and multi-channel detectors can be effectively coupled with GC×GC to support quantitative analysis, class-based characterization, and comprehensive profiling of complex mixtures.

While the primary focus of the chapter is on comprehensive chromatographic separation, the contribution of sample preparation is briefly considered as an enabling component that influences system performance and data quality. Emphasis is placed on the need for compatibility between extraction strategies and the stringent requirements imposed by modulation and high-speed second-dimension separations, rather than on exhaustive treatment of preparative methodologies.

A. Caratti · C. Cagliero (✉) · C. Cordero
Department of Drug Science and Technology, University of Turin, Torino, Italy
e-mail: cecilia.cagliero@unito.it

M. Minella (ed.), *Complexity in Environmental Chemistry*,
https://doi.org/10.1007/978-3-032-31774-2_5

Representative environmental applications illustrate how GC×GC enables class-resolved, isomer-sensitive, and pattern-based characterization of complex samples, including petroleum-derived mixtures, halogenated contaminants, and chemically unresolved environmental fractions. Collectively, the chapter positions GC×GC as a problem-solving analytical platform whose unique separation capabilities support advanced environmental investigations beyond the limits of one-dimensional gas chromatography.

Keywords

Comprehensive two-dimensional gas chromatography (GC×GC) · Environmental analysis · Sample preparation · Non-target and suspect screening · Class-based and group-type characterization · Chemical fingerprinting

Chapter Learning Goals

- Understand the fundamental principles and advantages of comprehensive two-dimensional gas chromatography compared to one-dimensional GC
- Recognize the strategic role of sample preparation in defining the chemical information space accessible by GC×GC–MS
- Compare conventional, green, and miniaturized sample preparation techniques and their suitability for comprehensive separations
- Evaluate modulation strategies, column configurations, and detection systems used in GC×GC–MS
- Interpret structured GC×GC chromatograms and understand their value for class-based and non-target environmental analysis
- Appreciate how GC×GC–MS is positioned as a problem-solving tool for the characterization of complex environmental mixtures

1 Introduction

Gas chromatography (GC) remains a cornerstone technique for the separation and characterization of volatile and semi-volatile organic compounds. In its conventional one-dimensional implementation, chromatographic performance is governed by two interdependent parameters: separation efficiency, commonly expressed as peak capacity, and selectivity, which is determined primarily by the chemical nature of the stationary phase. While selectivity can be tailored through stationary-phase chemistry, the peak capacity of one-dimensional (1D) GC is intrinsically constrained by the finite separation space available along a single column operated under a temperature program. Even under optimized conditions, typical peak capacity values achieved with standard capillary columns are insufficient to resolve the complexity of many real-world samples.

The conceptual impact of GC×GC compared with 1D-GC lies in the expansion of chromatographic separation space from a single time axis to a two-dimensional

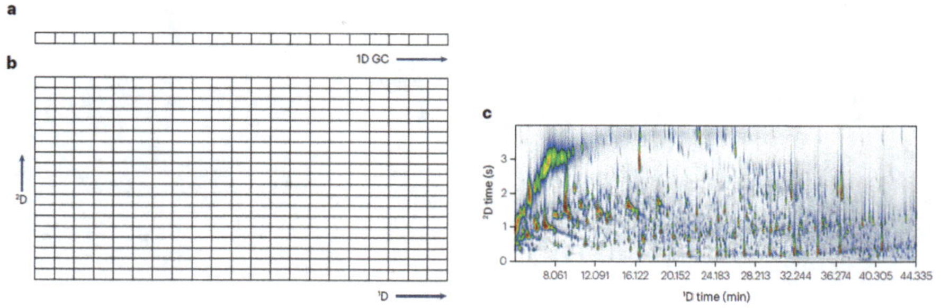

Fig. 1 Conceptual comparison between 1D GC (a) and GC×GC (b) separation capacity. The expansion of the separation space from a single time axis to a two-dimensional retention plane leads to a marked increase in peak capacity and a substantial reduction in co-elution for complex samples as illustrated in the contour plot of sub-figure (c) corresponding to human milk volatiles. Reproduced from: https://doi.org/10.1038/s43586-024-00379-3 [1]

retention plane. This transformation results in a substantial increase in peak capacity and dramatically reduces the probability of co-elution, even in highly complex samples [1]. The difference between 1D and comprehensive 2D separations, and the associated gain in resolving power, is illustrated schematically in Fig. 1.

The concept of expanding chromatographic resolving power by introducing additional separation dimensions emerged shortly after the advent of GC. Early multidimensional approaches were based on so-called heart-cutting (HC) techniques, in which discrete time segments of the effluent from a first column are selectively transferred to a second column characterized by a different stationary phase. This configuration enables an increase in selectivity and effective peak capacity for targeted fractions of interest, and represented a major conceptual advance over single-column separations. However, heart-cutting multidimensional GC inherently provides only partial information on sample composition, as the vast majority of the first-dimension effluent bypasses the secondary separation.

GC×GC was developed to overcome this intrinsic limitation by subjecting the entire first-dimension (^{1}D) effluent to a second, independent separation. This comprehensive operation is achieved through the introduction of a modulation interface positioned between the two columns. The modulator periodically collects narrow time portions of the ^{1}D effluent, refocuses them, and rapidly reinjects them into a short second-dimension (^{2}D) column. As a consequence, GC×GC transforms a single chromatographic separation into a sequence of highly resolved, fast secondary separations that collectively span the full analysis time.

From an instrumental perspective, GC×GC–MS extends the architecture of conventional GC–MS through the introduction of a modulation interface positioned between the ^{1}D and ^{2}D columns. The modulator periodically segments the entire first-dimension effluent (or a systematic and representative sampling of the entire ^{1}D effluent, preserving the ^{1}D separation) and transfers it to a short, high-speed second-dimension column, where rapid secondary separations take place. A schematic overview of a typical GC×GC–MS system, including the principal

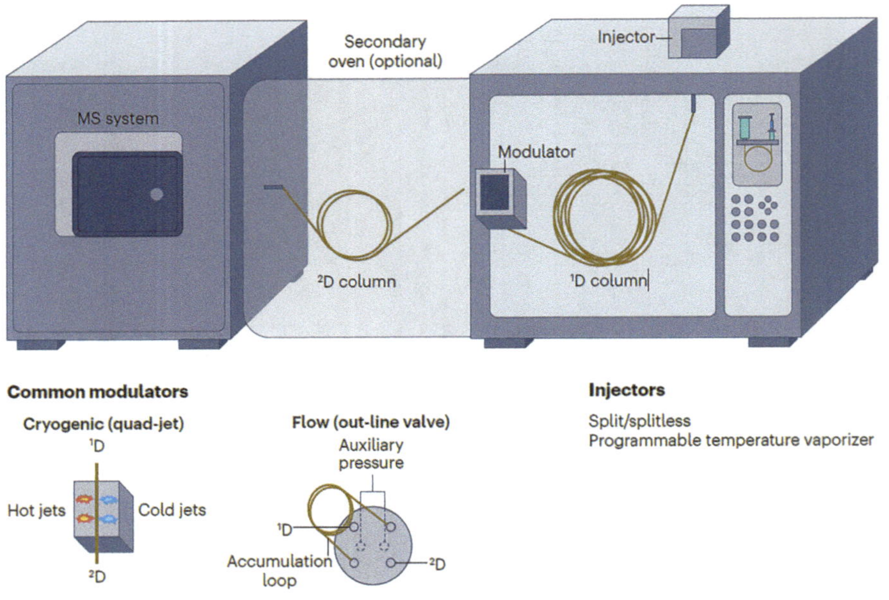

Fig. 2 Schematic representation of a GC×GC–MS system. The configuration includes a conventional injector, a 1D column, a modulation interface (thermal/cryogenic or flow-based), a short high-speed 2D column, and an MS system. Common modulator designs and alternative instrumental configurations are indicated. Reproduced from https://doi.org/10.1038/s43586-024-00379-3 [1]

instrumental components and commonly employed modulation strategies, is shown in Fig. 2.

The introduction of modulation represents the defining technological innovation of GC×GC. Depending on the modulator design, analyte bands may undergo active re-concentration prior to reinjection or may simply be segmented and redirected without significant focusing. In both cases, modulation imposes strict temporal constraints on the ^{2}D separation, which typically occurs on the time scale of seconds. The ^{1}D separation, by contrast, proceeds on a time scale comparable to that of conventional GC. This strong asymmetry between the two dimensions necessitates the use of short, high-speed second-dimension columns capable of generating extremely narrow peaks.

The dramatic increase in separation space afforded by GC×GC results in peak capacities that exceed those achievable with HC multidimensional (MD)GC by a large margin and surpass conventional 1D-GC by approximately an order of magnitude. Beyond increased resolving power, comprehensive operation yields highly structured chromatographic patterns, particularly when homologous or chemically related compound series are analyzed. Such ordered distributions provide an additional, visually intuitive layer of chemical information that is inaccessible in one-dimensional separations. The formation of structured retention patterns in GC×GC is conceptually illustrated in Fig. 3.

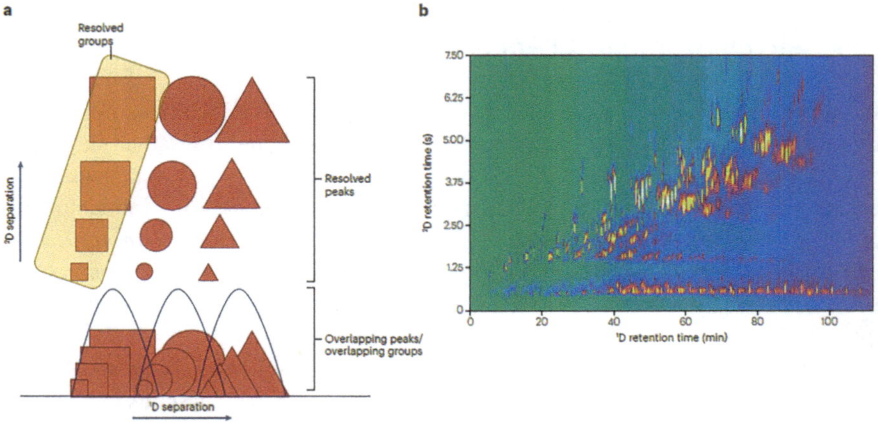

Fig. 3 Schematic illustration of structured chromatographic patterns in GC×GC from a mineral oil sample (i.e., light cycle oil) (a) and actual pattern illustrated by a contour plot from a non polar × semi-polar column combination (b). Primary separation in the 1D combined with rapid secondary separation in the 2D results in organized bands and regions associated with chemically related compound classes, providing additional interpretative information beyond simple peak resolution. From https://doi.org/10.1038/s43586-024-00379-3 [1]

The extremely narrow peak widths generated in the ^{2}D impose stringent requirements on detection systems. Detectors must combine high acquisition frequencies with minimal contribution to band broadening in order to preserve chromatographic resolution and ensure reliable quantitative performance. In this context, MS has emerged as the detector of choice for GC×GC, as it provides both the speed required to accurately sample narrow peaks and the molecular-level information necessary for confident analyte identification.

The coupling of GC×GC with MS results in a multidimensional analytical platform in which separation occurs across two chromatographic dimensions and a third, spectrometric dimension based on mass-to-charge ratio (*m/z*). Although the three dimensions are not fully orthogonal—owing in part to the influence of volatility on both chromatographic retention and ion formation—the combined system delivers unparalleled information density. Comprehensive GC×GC–MS chromatograms commonly contain hundreds to thousands of resolved peaks, rendering MS detection effectively indispensable for data interpretation.

The objective of this chapter is to present a critical and integrated overview of GC×GC applied to environmental analysis, focusing on the strategic role of sample preparation, the fundamental principles of comprehensive separations, instrumental configurations, modulation strategies, detection technologies, and data handling approaches. Emphasis is placed on understanding GC×GC as a coherent analytical platform, in which sample preparation, chromatographic design, modulation, detection, and data processing must be jointly optimized to fully exploit the technique's capabilities.

2 Sample Preparation: Strategic Role in Environmental Analysis

In comprehensive two-dimensional gas chromatography (GC×GC), sample preparation constitutes a defining component of the analytical platform rather than a preliminary or auxiliary operation. The substantial increase in separation space, peak capacity, and detectability afforded by comprehensive separations fundamentally changes how upstream processing influences analytical outcomes. In contrast to one-dimensional GC, where sample preparation is often designed to compensate for limited resolving power, in GC×GC it delineates the chemical domain that can be meaningfully separated, detected, and interpreted.

Environmental samples represent some of the most chemically heterogeneous matrices encountered in analytical chemistry. They encompass gaseous media (e.g., ambient and indoor air), liquid matrices (e.g., surface water, groundwater, seawater, wastewater), and solid or semisolid materials (e.g., soil, sediment, sludge, particulate matter). These matrices are characterized by pronounced spatial and temporal variability, wide concentration ranges, and the coexistence of compounds spanning broad distributions of volatility, polarity, molecular mass, and thermal stability. Frequently, analytes of interest are present at trace or ultra-trace levels within a background dominated by naturally occurring organic matter and anthropogenic interferents [2].

The principal challenge in coupling sample preparation with GC×GC lies in reconciling this matrix complexity with the intrinsic requirements of comprehensive gas chromatographic separation. GC×GC is inherently restricted to volatile and semi-volatile compounds that are sufficiently stable to withstand high-temperature separation and modulation. Sample preparation must therefore isolate or transform analytes that are compatible with gas-phase separation while simultaneously minimizing the transfer of non-volatile or thermally labile constituents that could compromise chromatographic performance or detector stability.

Additional constraints arise from the modulation process and the high-speed second-dimension separation. The extremely narrow peak widths places stringent demands on sample cleanliness, injection reproducibility, and background suppression. Excessive matrix transfer can lead to elevated baseline noise, amplification of column bleed, modulator fouling, and detector saturation. These effects are exacerbated in comprehensive separations by the repeated reinjection inherent to modulation, rendering sample preparation strategies acceptable in one-dimensional GC potentially unsuitable for GC×GC.

At the same time, GC×GC is increasingly applied to analytical objectives that extend beyond narrowly targeted determinations. Class-based characterization of chemically related contaminants, non-target screening, and chemical fingerprinting are central application domains. In these contexts, sample preparation must balance selectivity and inclusivity. Overly selective extraction risks eliminating diagnostically relevant compound classes or distorting relative abundance relationships, whereas insufficient selectivity may overwhelm the chromatographic system and obscure chemically meaningful information.

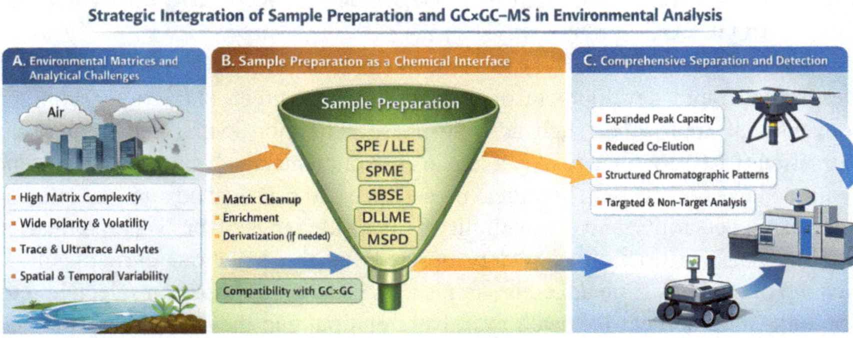

Fig. 4 Scheme reporting the integration sample preparation and GC×GC with an emphasis on the challenges of environmental matrices and potential in obtaining different information-rich chemically structured datasets

Chemical derivatization represents a further dimension of this coupling. Many environmentally relevant compounds, particularly polar species such as organic acids, alcohols, amines, and multifunctional molecules, are not directly amenable to gas chromatographic analysis. Derivatization is therefore often required, not merely to enable volatility, but also to shape chromatographic selectivity and mass spectral characteristics. In GC×GC, derivatization can influence retention behavior in both dimensions and affect the degree of orthogonality between them, thereby contributing directly to the structure of the two-dimensional chromatogram.

An extreme yet illustrative example of the interdependence between sample preparation and GC×GC is provided by pyrolysis-based approaches [3]. In pyrolysis–GC×GC workflows, analyte generation and chromatographic analysis become inseparable, as non-volatile polymeric or macromolecular material is thermally converted into volatile fragments prior to comprehensive separation. In such cases, sample preparation determines not only which compounds are analyzed, but also which compounds are present within the analytical system.

Early GC×GC studies often employed relatively simple sample preparation protocols, reflecting a primary focus on demonstrating the conceptual novelty and resolving power of comprehensive separations. The perception that GC×GC could compensate for suboptimal upstream processing through enhanced selectivity and sensitivity contributed to this tendency. However, accumulated experience has clearly shown that comprehensive separations magnify both the strengths and weaknesses of sample preparation [4]. While GC×GC can separate analytes from interferences more effectively than one-dimensional GC, it cannot recover compounds that were not extracted, nor can it fully mitigate the consequences of excessive matrix transfer. Properly optimized sample preparation is therefore indispensable for generating authentic, reproducible, and chemically interpretable GC×GC–MS data (Fig. 4).

2.1 Conventional Extraction Approaches: Role and Limitations in GC×GC

Despite the rapid development of miniaturized and green extraction techniques, conventional approaches such as solid-phase extraction (SPE) and liquid–liquid extraction (LLE) continue to play a role in environmental analysis by GC×GC. Their continued use reflects both historical precedence and practical advantages related to robustness, versatility, and compatibility with a wide range of analytes and matrices.

Solid-phase extraction is widely applied for the isolation, enrichment, and cleanup of analytes from aqueous samples. Its principle is based on selective retention on a solid sorbent followed by solvent elution, and the availability of diverse sorbent chemistries allows adaptation to a broad spectrum of compound classes. From a GC×GC perspective, SPE offers high enrichment factors from large sample volumes and effective removal of non-volatile matrix components, thereby protecting chromatographic hardware and improving detectability.

However, SPE is inherently an exhaustive extraction technique that often requires substantial solvent consumption and multiple processing steps. More importantly, the selectivity imposed by sorbent chemistry and elution conditions can bias the extracted chemical space, limiting the suitability of SPE for non-targeted or class-based GC×GC applications. Weakly retained compounds may be lost, while strongly retained species may not be fully recovered, leading to distorted chromatographic patterns and altered relative abundances.

Liquid–liquid extraction represents another classical approach that has been extensively used in environmental analysis and combined with GC×GC in both targeted and exploratory workflows. LLE relies on partitioning between immiscible liquid phases and can be effective for moderately nonpolar to semipolar compounds. Its simplicity and minimal equipment requirements are advantageous, but these benefits are offset by high solvent consumption, emulsion formation, and limited selectivity.

In GC×GC workflows, LLE often leads to substantial co-extraction of matrix components, thereby increasing the risk of column contamination, and baseline noise. These drawbacks are amplified by comprehensive operation, where repeated modulation cycles expose the system to cumulative matrix load. Consequently, while SPE and LLE remain useful for targeted or pre-targeted applications, their limitations have motivated the transition toward more selective, sustainable, and versatile alternatives.

2.2 Transition Toward Green and Miniaturized Sample Preparation

Over the past two decades, sample preparation has undergone a marked transformation driven by the principles of Green Analytical Chemistry and the need for more sustainable, efficient, and information-rich methodologies. This evolution has been particularly pronounced in environmental analysis, where traditional workflows often involved large sample volumes, extensive solvent use, and energy-intensive procedures.

Green Analytical Chemistry emphasizes reduction of solvent consumption, minimization of waste, energy efficiency, operator safety, and method simplification. Applied to sample preparation, these principles have encouraged a shift away from exhaustive extraction toward miniaturized, non-exhaustive approaches that integrate extraction, enrichment, and cleanup into a single step, as postulated by the Green Sample Preparation principles. This transition aligns closely with the requirements of GC×GC, which benefits from reduced matrix load, and simplified workflows.

Early efforts focused on accelerating conventional extraction and reducing solvent use through techniques such as Ultrasound-Assisted Extraction (UAE), Microwave-Assisted Extraction (MAE), Pressurized Liquid Extraction (PLE), and Supercritical Fluid Extraction (SFE). While these approaches improved efficiency, they often retained reliance on organic solvents and additional cleanup steps. More transformative developments arose from microextraction techniques that operate with minimal sorbent volumes and negligible solvent use, enabling direct coupling with GC-based systems.

Miniaturization has also facilitated new sampling paradigms, including passive sampling, in-field extraction, and remote deployment. These approaches address challenges related to sample preservation, transport, and representativeness, while enabling spatially and temporally resolved environmental analysis.

Among the diverse approaches developed in the context of Green Sample Preparation, sorptive microextraction techniques occupy a central position. Their conceptual simplicity, operational flexibility, and intrinsic compatibility with gas chromatographic systems have made them foundational tools for modern environmental analysis. Importantly, these techniques represent a shift from exhaustive isolation toward equilibrium or kinetically controlled enrichment processes that preserve chemically meaningful relationships within complex samples, an aspect that is particularly relevant for GC×GC-based fingerprinting and class-based analysis.

2.2.1 Solid-Phase Microextraction: Versatility and Conceptual Alignment with GC×GC

Solid-phase microextraction (SPME) represents the most versatile and extensively adopted sustainable sample preparation technique for GC×GC–based environmental analysis. Since its introduction, SPME has transformed analytical workflows for volatile and semi-volatile organic compounds by integrating extraction, enrichment, and cleanup into a single solvent-free step. Its conceptual alignment with GC×GC is noteworthy: both techniques rely on selective partitioning, tunable interactions, and miniaturized formats to manage chemical complexity rather than suppress it.

SPME is based on the partitioning of analytes between the sample matrix and a thin sorptive coating immobilized on a solid support. Because the extraction phase volume is small relative to the sample, SPME operates under non-exhaustive conditions. Nevertheless, quantitative analysis is achievable when equilibrium or well-defined kinetic regimes are established. From a sustainability standpoint, the elimination of organic solvents, minimal waste generation, and reusability of extraction devices place SPME among the most environmentally benign sample preparation approaches currently available.

A key strength of SPME lies in its structural and chemical versatility. While fiber-based formats remain widely used, alternative geometries such as thin films, blades, and planar devices have been developed to increase surface area, improve extraction kinetics, and enhance mechanical robustness. These configurations extend the applicability of SPME to diverse environmental matrices and sampling scenarios while maintaining compatibility with GC×GC injection systems.

Equally important is the tunability of SPME coatings. Sorbent phases can be engineered to exploit hydrophobic interactions, π–π interactions, hydrogen bonding, or polar interactions, enabling selective enrichment of specific compound classes or broad chemical coverage for non-targeted analysis. In GC×GC, such selectivity tuning directly influences analyte distribution across the two-dimensional retention plane and can enhance the emergence of structured chromatographic patterns.

SPME also offers substantial flexibility in sampling mode. Headspace SPME consists in a three-phase equilibrium among the sample, its headspace and the sorbent SPME coating, limiting the transfer of non-volatile components and reducing the risk of chromatographic contamination. Direct immersion SPME, in contrast, facilitates the extraction of less volatile or more polar compounds but requires careful control to mitigate matrix effects. The ability to select and optimize sampling mode provides analysts with a powerful means of balancing sensitivity, selectivity, and robustness.

Pressure-assisted variants, such as vacuum-assisted SPME (both in fiber-based or thin film formats), further expand the applicability of the technique by accelerating mass transfer and improving recovery of semi-volatile compounds. While these approaches can significantly reduce extraction times, they may also increase background contributions from system components. In GC×GC workflows, where background signals can be amplified by band compression (in time or in space) generated by the modulator, such trade-offs must be carefully evaluated during method development.

Overall, SPME functions not merely as a green alternative to conventional extraction, but as a conceptual bridge between sample preparation and multidimensional separation.

2.2.2 Stir-Bar Sorptive Extraction: Enhanced Capacity and Associated Trade-Offs

Stir-bar sorptive extraction (SBSE) can be viewed as a direct extension of the SPME concept, distinguished primarily by the substantially larger volume of extraction phase. In SBSE, analytes partition into a sorptive coating, typically based on nonpolar or moderately polar polymers, immobilized on a magnetic stir bar immersed directly in the sample. The increased sorbent volume provides markedly higher extraction capacity compared to conventional SPME fibers, resulting in enhanced sensitivity and lower detection limits.

From a sustainability perspective, SBSE retains many advantages of SPME, including solvent-free operation and reusability. However, the increased sorbent capacity also introduces additional considerations in GC×GC workflows. The likelihood of co-extracting matrix components is higher, particularly in complex environmental samples. When thermal desorption is employed, bleeding from the sorbent coating may contribute background signals that complicate non-targeted analysis and spectral interpretation.

In comprehensive separations, these effects are magnified by modulation, which repeatedly introduces extracted material into the second dimension. Consequently, SBSE–GC×GC–MS workflows often rely on the selectivity of mass spectrometric detection and advanced data-processing strategies to discriminate analyte signals from background contributions. This reliance underscores the importance of aligning extraction capacity with analytical objectives, particularly in exploratory or fingerprinting studies.

Conceptually, SBSE exemplifies the balance between sensitivity and chemical inclusivity. Its high enrichment capability is advantageous for trace-level analysis, but excessive capacity may obscure structured chromatographic patterns if matrix contributions are not adequately controlled. Careful optimization is therefore essential to fully exploit SBSE within GC×GC–MS frameworks.

2.2.3 Miniaturized Liquid-Phase Microextraction Techniques

In parallel with sorptive microextraction approaches, miniaturized liquid-phase extraction techniques have gained prominence as more sustainable alternatives to traditional liquid–liquid extraction. These methods, including single-drop microextraction and dispersive liquid–liquid microextraction (DLLME), employ microliter-scale solvent volumes and rapid extraction kinetics.

DLLME, in particular, is characterized by the formation of a fine dispersion of extraction solvent within the sample, generating a large interfacial area that promotes efficient analyte transfer. After phase separation, the analyte-enriched extract can be introduced into a GC system with minimal additional processing.

From a GC×GC standpoint, DLLME offers both advantages and limitations. High enrichment factors can be achieved, which is beneficial for trace-level detection. However, the introduction of liquid extracts increases the risk of transferring residual solvents and co-extracted matrix components into the chromatographic system. Careful selection of solvents and extraction conditions is therefore required to ensure compatibility with comprehensive separations.

2.2.4 Sustainable Approaches for Solid and Semisolid Samples

Solid and semisolid environmental samples pose distinct challenges due to matrix heterogeneity, strong analyte–matrix interactions, and uneven analyte distribution. Sustainable sample preparation strategies for these matrices must therefore address extraction efficiency and representativeness while minimizing solvent use and preprocessing.

Matrix solid-phase dispersion (MSPD) represents one of the most established sustainable approaches for solid samples. In MSPD, the sample is mechanically dispersed in the presence of a solid support, disrupting matrix–analyte interactions and enabling subsequent elution with relatively small solvent volumes. By integrating homogenization, extraction, and cleanup into a single step, MSPD reduces processing time and solvent consumption compared to traditional solid–liquid extraction methods.

SP(M)E-based approaches have also been adapted for solid matrices through assisted extraction modalities, including heating-assisted, ultrasound-assisted, cooling-assisted, and vacuum-assisted configurations. These strategies enhance analyte release from the matrix and improve mass transfer to the extraction phase without extensive solvent use. In GC×GC workflows, such approaches allow access to volatile and semivolatile fractions from solid samples while preserving compatibility with comprehensive separations.

Taken together, sustainable and versatile sample preparation techniques fundamentally reshape how environmental samples are introduced into GC×GC systems. Rather than serving solely as preparatory steps, these approaches actively define the chemical information space accessible to comprehensive separation and detection. Their selectivity, capacity, and operational characteristics directly influence chromatographic structure, modulation behavior, and data interpretability.

Within this framework, SPME emerges not simply as a green alternative to conventional extraction, but as a unifying platform that conceptually mirrors the analytical principles of GC×GC. Extensions such as SBSE, along with complementary techniques including DLLME and MSPD, expand the available toolkit, enabling tailored solutions for diverse environmental challenges while adhering to sustainability principles.

2.3 In-Situ Sampling, Automation, and Remote Sample Preparation Platforms

The evolution of sample preparation strategies for environmental analysis increasingly extends beyond laboratory-based workflows toward *in-situ*, automated, and remotely deployable approaches. This transition reflects both practical imperatives, such as safety, accessibility, and representativeness, and conceptual shifts toward spatially and temporally resolved chemical assessment.

Sorptive microextraction approaches, particularly SPME and related formats, are especially well suited to remote deployment due to their solvent-free operation, mechanical simplicity, and compatibility with portable analytical systems. Although comprehensive two-dimensional gas chromatography remains largely laboratory-bound, recent developments demonstrate how remote sampling platforms can generate stable, transportable extracts suitable for subsequent GC×GC analysis.

Conventional environmental monitoring typically involves manual sample collection followed by transport to centralized laboratories. While effective in many contexts, this approach introduces limitations related to analyte loss, contamination,

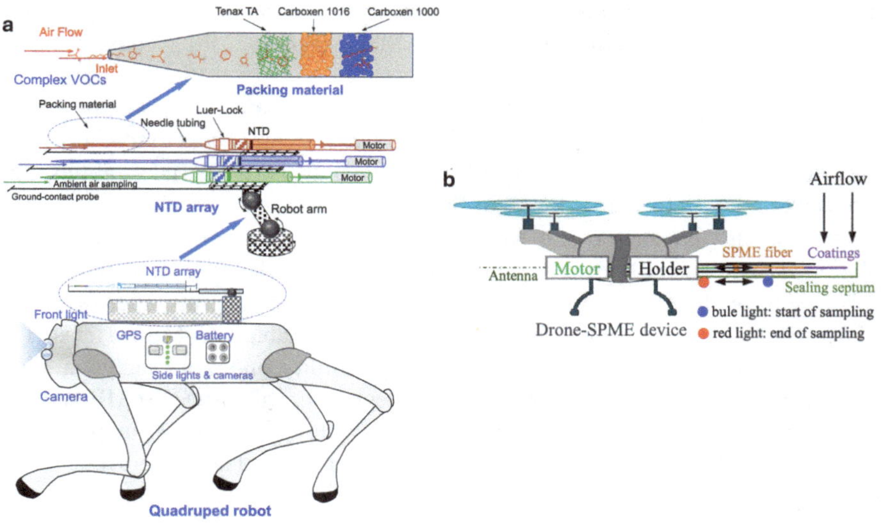

Fig. 5 Examples of robotic sampling platforms. a) Robot-MS system for the on-site detection of hazardous VOCs based on needle trap devices. Adapted from Liu et al. https://doi.org/10.1021/acs.analchem.4c01555 [5], b) Schematic diagram of drone–SPME device. Adapted from Chen et al., https://doi.org/10.1021/acs.est.2c05259 [6]

and temporal resolution. Volatile and reactive compounds are particularly susceptible to transformation during storage and transport, and intermittent sampling may fail to capture transient events or spatial gradients.

In-situ sampling addresses these challenges by enabling analyte collection directly at the point of interest. When combined with automation or remote control, such approaches also reduce human exposure to hazardous environments and permit sampling in locations that are otherwise inaccessible. For GC×GC, the key requirement is that *in-situ* sample preparation generates stable analyte-enriched phases compatible with high-resolution laboratory analysis.

Robotic sampling platforms equipped with miniaturized extraction devices (Fig. 5) represent a promising solution for hazardous or inaccessible environments.

A representative case study is discussed in the work of Liu and co-workers [5] that developed a quadruped robotic system equipped with needle trap devices (NTDs) for the collection of hazardous volatile organic compounds. The robot was remotely operated to navigate complex environments and position the sampling device in targeted locations. Air was actively drawn through the sorbent-packed needle, capturing analytes that were subsequently analyzed using portable mass spectrometric instrumentation.

This approach illustrates several features relevant to future GC×GC workflows. Sorptive sampling enables solvent-free, preconcentrative extraction directly at the sampling site, while immobilization of analytes on a stable sorbent allows safe transport to centralized laboratories. Although the reported implementation relied on portable GC–MS for rapid identification, the same sampling strategy is fully compatible with subsequent GC×GC–MS analysis for comprehensive characterization.

Drone-based platforms represent one of the most versatile solutions for remote environmental sampling. Unmanned aerial vehicles provide access to spatially complex environments, including elevated or confined areas, industrial facilities, and atmospheric layers that are difficult to sample manually.

In a notable case study by Chen et al. [6], SPME devices were mounted on drone platforms to collect airborne contaminants during controlled flights. Sampling was achieved through passive exposure or airflow-assisted mechanisms, after which the SPME devices were retrieved and analyzed using portable GC–MS systems. The approach demonstrated good sensitivity, reproducibility, and rapid analytical turnaround. From the perspective of GC×GC integration, this drone–SPME strategy offers several advantages. Analytes are collected without solvents or pumps, immobilized on a stable sorptive phase, and preserved for delayed analysis. While portable GC–MS enables rapid screening, laboratory-based GC×GC–MS can be employed for detailed non-targeted or class-based characterization, effectively decoupling sampling from comprehensive analysis.

Automation is central to these developments. Automated control of sampling duration, exposure conditions, and device handling improves reproducibility and reduces operator-dependent variability, an important consideration for GC×GC, where subtle differences in sample preparation can significantly influence chromatographic structure.

In-situ sampling, automation, and remote platforms represent a natural evolution of sample preparation strategies in environmental analysis. Case studies involving robotic samplers and drone–based systems demonstrate that these approaches are already capable of generating high-quality analytical data under challenging conditions. As these technologies mature, their integration with GC×GC(–MS) is expected to deepen, enabling chemically comprehensive, spatially resolved, and safer environmental monitoring strategies.

3 Experimental Considerations in GC×GC

The implementation of GC×GC–MS requires an instrumental configuration that extends conventional GC–MS while introducing additional constraints related to modulation, high-speed separations, and multidimensional data acquisition. Although injection strategies largely overlap with those employed in 1D-GC, the presence of a second chromatographic dimension and a modulator fundamentally alters the requirements for column selection, carrier gas management, detector performance, and data handling.

A typical GC×GC–MS system consists of a standard GC inlet, a ^{1}D column, a modulation interface, a ^{2}D column, and an MS positioned at the outlet of the ^{2}D. The overall architecture, shown in Fig. 2, may be realized within a single oven or using independently controlled ovens for the two chromatographic dimensions. The latter configuration provides superior flexibility in method optimization and is increasingly regarded as best practice for demanding applications.

3.1 Column Configuration and Separation Design

Column selection in GC×GC is a central determinant of separation performance and information content. The guiding principle is to maximize the complementarity between the two chromatographic dimensions while maintaining compatibility with modulation and detection constraints. In practice, this often involves combining a relatively long ^{1}D column with a short, high-speed ^{2}D column.

The ^{1}D is typically responsible for the primary separation according to volatility-related properties, whereas the ^{2}D introduces an additional selectivity based on polarity or specific intermolecular interactions. While combinations of low-polarity and medium- to high-polarity stationary phases are frequently adopted, alternative configurations may provide superior performance for specific sample classes. Importantly, the stationary phase offering the highest resolving power for the sample under investigation should generally be assigned to the ^{1}D in order to minimize unresolved overlap at the modulator interface.

Column dimensions must be selected with particular care. Narrow-bore second-dimension columns generate extremely sharp peaks that enhance detectability but also increase susceptibility to overload and wrap-around phenomena. Using columns with identical internal diameters in both dimensions can alleviate carrier gas velocity mismatches and improve robustness, especially when MS detection is employed under reduced-pressure conditions.

Wrap-around occurs when the retention time of an analyte in the second dimension exceeds the modulation period, causing the signal to appear in a subsequent modulation cycle. While excessive wrap-around complicates chromatographic interpretation, moderate wrap-around may be tolerated or even exploited when it does not interfere with structured retention patterns or quantitative accuracy. Adjustments to modulation period and ^{2}D temperature are generally sufficient to control this phenomenon.

3.2 Modulation Technologies

The modulator is the defining element of GC×GC, governing analyte transfer, peak focusing, and temporal synchronization between the two chromatographic dimensions. Over the past decades, modulation technologies have evolved substantially, giving rise to a range of commercially available solutions.

Thermal modulation remains the reference approach due to its superior focusing efficiency and sensitivity enhancement. Cryogenic modulators operate by rapidly cooling short segments of the column to trap analytes, followed by rapid heating to release them as narrow bands. Modern designs employ multi-jet or loop-based configurations that enable dual-stage modulation, improving focusing efficiency and minimizing breakthrough of highly volatile compounds. While liquid cryogens provide the most effective cooling, cryogen-free systems based on mechanical refrigeration have gained relevance by reducing operational complexity and safety concerns.

Flow modulation offers an alternative strategy in which pressure differentials are used to periodically divert portions of the ^{1}D effluent into the ^{2}D. These systems eliminate the need for cryogens and reduce operating costs, but generally provide lower sensitivity due to the absence of active refocusing in space. Flow modulators may operate in high or low duty-cycle modes, with corresponding trade-offs between analyte transfer efficiency, detectability, and compatibility with mass spectrometric pumping capacities. A schematic diagram of thermal quad-jet and differential-flow modulators is reported in Fig. 2.

The choice between thermal and flow modulation should be guided by analytical priorities, including sensitivity requirements, matrix complexity, operational constraints, and sustainability considerations.

3.3 Detection and Data Acquisition

The extremely narrow peak widths generated in the second dimension impose strict demands on detection systems. Adequate sampling requires acquisition frequencies that are significantly higher than those used in conventional GC–MS in order to preserve peak shape and quantitative accuracy.

MS is the dominant detection technique in GC×GC due to its ability to combine high acquisition rates with molecular-level identification. Modern quadrupole mass spectrometers (QMS) have evolved to provide sufficient scan speeds for comprehensive separations, supporting both full-scan and targeted acquisition modes within a single analysis. Time-of-flight mass spectrometers (TOFMS) offer even higher acquisition frequencies and consistent spectral quality across narrow peaks, making them particularly well suited for non-targeted analyses and advanced deconvolution workflows.

The selection of detector and acquisition mode must be aligned with the objectives of the analysis, balancing spectral richness, quantitative performance, and data processing complexity.

3.3.1 Detection Systems in GC×GC–MS

The effectiveness of GC×GC critically depends on the performance of the detection system. The very narrow chromatographic bands generated in the ^{2}D—particularly when thermal modulation is employed—place stringent requirements on detector speed, sensitivity, and signal fidelity. Inadequate detector performance directly compromises the benefits of the enhanced separation space provided by GC×GC, resulting in distorted peak shapes, loss of quantitative accuracy, and reduced information content.

Detectors used in GC×GC can be broadly categorized according to the dimensionality of the data they generate. Single-channel detectors provide a single intensity value per acquisition point, producing a time-ordered signal analogous to that of conventional GC. Multi-channel detectors, by contrast, acquire an array of values at each time point, thereby introducing an additional analytical dimension that is essential for resolving co-eluting compounds in highly complex separations.

3.3.2 Single-Channel Detectors

Among single-channel detectors, the flame ionization detector (FID) remains the most widely employed in GC×GC due to its robustness, broad dynamic range, and highly reproducible response. Its near-universal sensitivity to organic compounds and excellent linearity make it particularly attractive for quantitative applications. Moreover, the predictable relationship between FID response and molecular structure enables the use of response factors, reducing reliance on extensive external calibration.

In GC×GC configurations employing flow modulation with high second-dimension carrier gas flows, the FID is often operated in parallel with mass spectrometry. In such arrangements, part of the effluent is directed to the FID to accommodate flow rates that would otherwise exceed the pumping capacity of the mass spectrometer. This dual-detection strategy combines the quantitative reliability of the FID with the qualitative power of mass spectrometry. However, it is important to remark that possible trade-offs of post-column flow splitting include reduced MS sensitivity and potential variability in the split ratio during the run when analysis is not carried out under constant pressure conditions.

Within the framework of single-channel detectors, the micro-electron capture detector (μECD) represents a highly selective, concentration-sensitive device providing a single integrated response as a function of time. Unlike spectroscopic or multi-signal detectors, the μECD delivers a univariate electrical signal governed exclusively by electron capture events, with no intrinsic spectral or mass-resolved information. Its operating principle—based on the attenuation of a steady-state electron current by electronegative analytes—makes the μECD intrinsically selective toward halogenated compounds, while remaining chemically blind to the majority of hydrocarbons and polar species. The miniaturized detector architecture ensures the fast response and reduced dead volume required for GC×GC, positioning the μECD as a paradigmatic example of a single-channel, highly selective detector for targeted environmental applications, particularly in the trace-level analysis of persistent halogenated contaminants.

Other selective single-channel detectors, including sulfur- and nitrogen-specific devices, have been applied in GC×GC for targeted analyses, particularly in petrochemical and environmental contexts. However, the increasing availability and performance of mass spectrometric detectors, especially those capable of high acquisition rates, have limited the widespread adoption of these selective detectors in comprehensive separations.

3.3.3 Multi-Channel Detectors

Multi-channel detectors play a pivotal role in exploiting the full information content of GC×GC separations. By providing compound-specific signatures at each chromatographic point, they enable reliable identification and deconvolution in regions characterized by partial or complete chromatographic overlap.

Vacuum ultraviolet (VUV) spectroscopy represents a notable alternative to mass spectrometry, offering absorption-based spectral fingerprints that are highly specific to molecular structure. When coupled with GC×GC, VUV detection allows class-based identification and supports the analysis of complex mixtures through spectral deconvolution in the time domain. This approach has proven particularly useful in the characterization of fuels and other hydrocarbon-rich matrices, where compositional classes can be resolved even in the presence of co-elution.

MS, however, remains the dominant multi-channel detection technique in GC×GC. The first demonstrations of GC×GC–MS highlighted the limitations of early QMS, whose acquisition speeds were insufficient to sample ^{2}D peaks accurately. Subsequent technological developments have largely overcome these constraints. Modern quadrupole instruments are capable of rapid scanning while maintaining consistent spectral profiles, enabling their routine use in comprehensive separations for both targeted and untargeted analyses.

TOFMS is particularly well suited to GC×GC due to its intrinsically high acquisition frequencies and uniform spectral acquisition across the entire mass range. Unlike scanning analyzers, TOFMS acquires full mass spectra for packets of ions generated at high repetition rates, ensuring minimal spectral distortion even for sub-second chromatographic peaks. This characteristic, combined with its strong deconvolution performance, has established low-resolution TOFMS as the reference detector for many GC×GC applications.

Although high-resolution (HR)MS and tandem MS configurations have been explored in GC×GC, their added analytical value must be weighed against increased cost, data complexity, and processing requirements. In practice, the combination of comprehensive chromatographic separation with low-resolution mass spectrometry is sufficient for many routine and class-based applications. However, HRMS and MS/MS become particularly valuable in non-target and suspect screening workflows, where accurate mass measurements and enhanced selectivity support the discrimination of isobaric species and increase confidence in structural elucidation. Moreover, when thermal modulation is employed, cryogenic focusing produces sharper second-dimension peaks and improved signal-to-noise ratios, enhancing detectability and reinforcing identification reliability through better spectral purity and reduced co-elution.

3.4 Results and Data Handling in GC×GC–MS

The exceptional separation power of comprehensive two-dimensional gas chromatography is accompanied by the generation of large, information-rich datasets whose interpretation cannot rely on manual inspection alone. As GC×GC transforms chemically complex samples into multidimensional raw data arrays, data handling tools become an integral component of the analytical workflow, effectively constituting a further dimension of the technique. The quality, reliability, and usability of GC×GC–MS results are therefore strongly dependent on appropriate strategies for data acquisition, visualization, processing, and interpretation.

Modern GC×GC–MS software platforms are designed to manage the full analytical pipeline, from raw signal acquisition to automated reporting. These systems support routine, high-throughput operation through batch processing and standardized workflows, while also allowing interactive data exploration for investigational studies. The balance between automation and user intervention is a defining feature of GC×GC data handling, reflecting the diverse analytical contexts in which the technique is applied.

3.4.1 Data Visualization and Representation

Visualization represents the first and arguably most critical step in the interpretation of GC×GC data. By rendering complex multidimensional datasets into intuitive graphical formats, visualization tools enable rapid assessment of chromatographic performance, identification of structured retention patterns, and preliminary qualitative comparisons between samples.

Although detection occurs as a one-dimensional time-ordered signal at the outlet of the second-dimension column, GC×GC data are reconstructed into a two-dimensional retention space based on modulation timing. Each modulation cycle corresponds to a short second-dimension separation, and successive cycles collectively span the entire first-dimension analysis. Data are commonly displayed using a Cartesian coordinate system, where first-dimension retention time defines the horizontal axis and second-dimension retention time defines the vertical axis. The raw data interpretation is illustrated in Fig. 6.

Intensity values associated with each data point are typically encoded using grayscale or pseudocolor scales, allowing rapid identification of high- and low-abundance features. In the case of MS or spectroscopic detection, visualization can be based on total signal intensity or on selectively extracted channels. This flexibility enables the generation of targeted images that highlight specific compound classes, homologous series, or diagnostic fragments, thereby enhancing chemical interpretability beyond what is achievable with non-selective representations.

Three-dimensional visualization tools further extend this concept by incorporating signal intensity as a third axis, providing insight into quantitative relationships across wide dynamic ranges. While such representations are less commonly used for routine analysis, they are valuable for method development, troubleshooting, and educational purposes. An example of 3D visualization is shown in Fig. 7 for a complex pattern of volatile organic compounds (VOCs) sampled from ambient air by headspace solid phase microextraction (HS-SPME) in a cheese ageing cave.

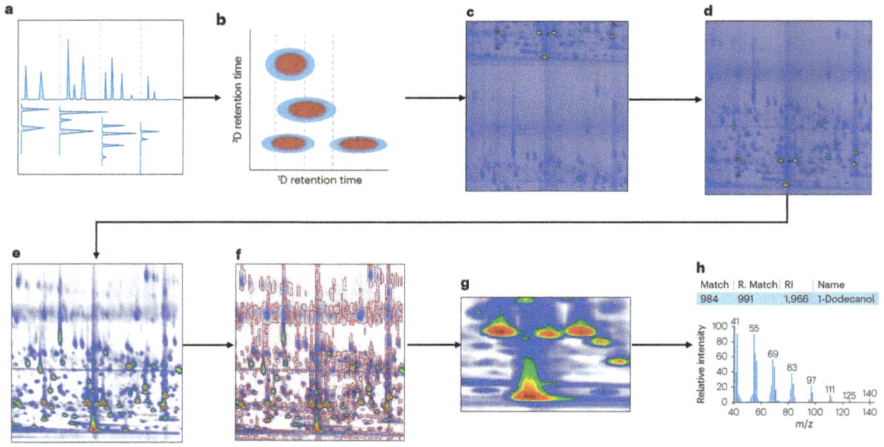

Fig. 6 Representative GC×GC–MS chromatogram displayed as a raw signal and corresponding 2D chromatograms (a), as schematic contour plot (b), pseudocolorization for visualizations before (c) and after (d-e) shift operations. Peak detection (f) and analytes identification (g) by MS spectral signature and retention index (h) completes the process. First-dimension retention time is shown on the x-axis, ^{2}D retention time on the y-axis, and signal intensity is encoded by color, highlighting the high information density and structured nature of comprehensive separations. Adapted from https://doi.org/10.1038/s43586-024-00379-3 [1]

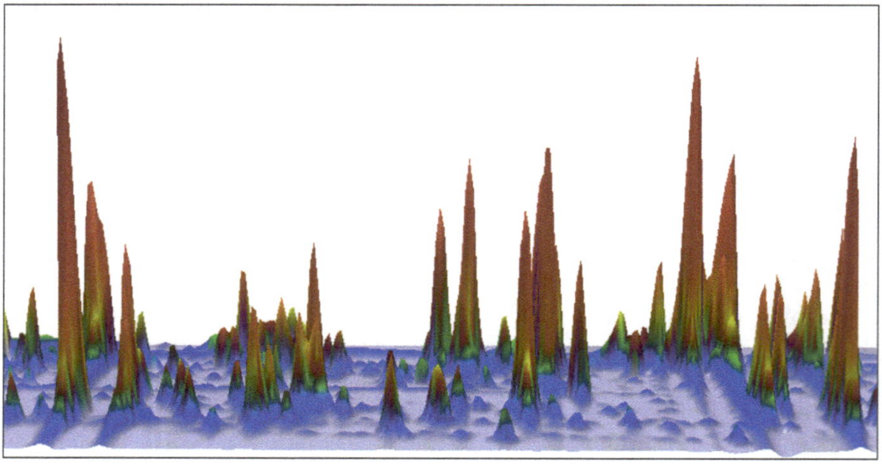

Fig. 7 Three-dimensional view of a GC×GC-TOFMS chromatogram—Total Ion Current TIC—representing the VOCs samples by HS-SPME in a cheese ripening cave. Separation is by a polar ^{1}D (HeavyWax) coupled to a semi-polar ^{2}D (OV1701) and thermal modulation by loop-type thermal modulator (Zoex Corporation)

3.4.2 Pre-processing of GC×GC Data

Prior to peak detection and identification, raw GC×GC–MS data typically undergo a series of pre-processing steps aimed at correcting systematic artifacts and enhancing signal quality. One critical operation is modulation-phase alignment, which ensures that the temporal boundaries of each second-dimension separation are correctly synchronized with the modulation cycle. Minor misalignments can arise from timing offsets between the modulator and detector and, if uncorrected, may degrade peak shape and retention-time consistency. The phase shift process is illustrated in Fig. 6 at steps **c** and **d**.

Baseline correction represents another essential pre-processing step. In GC×GC, baseline contributions may arise from detector noise, column bleed, or unresolved chemical background. Depending on the detection mode, baseline estimation may be performed locally around individual features or globally across the entire chromatographic plane. For multi-channel detectors, baseline correction is often integrated with spectral centroiding to ensure consistent treatment across all channels.

Pre-processing strategies must be carefully optimized to avoid the introduction of artifacts or loss of low-abundance features. Excessive smoothing or aggressive background subtraction can compromise the detection of trace components, particularly in non-targeted analyses.

3.4.3 Two-Dimensional Peak Detection and Deconvolution

Peak detection in GC×GC presents challenges that are fundamentally different from those encountered in 1D-GC. Features in GC×GC data occupy finite areas in the 2D retention plane rather than simple time intervals, and may be distorted by chromatographic effects, background fluctuations, or co-elution.

Two general approaches to peak detection can be distinguished. In the first, conventional ^{1}D algorithms are applied sequentially to each ^{2}D separation, followed by aggregation of related peaks across modulation cycles. In this paradigm, spectral information is typically exploited after chromatographic peak detection, mainly to merge modulated peaks belonging to the same compound and to support deconvolution in cases of partial overlap. In the second, true ^{2}D algorithms treat the chromatogram as an image and identify contiguous regions of elevated signal intensity. Here, spectral similarity can be incorporated directly into the detection step, allowing pixels or voxels to be grouped not only by spatial continuity in the retention plane but also by coherence of mass spectral patterns. These regions, often referred to as blobs, may correspond to individual analytes or to unresolved combinations of co-eluting compounds. The distinction between chromatographic peaks and detected blobs is non-trivial. Over-segmentation may occur when a single analyte is divided into multiple detected features, while under-segmentation may result in multiple analytes being merged into a single feature. To address these issues, peak detection is frequently coupled with deconvolution strategies that exploit spectral information.

In GC×GC–MS, spectral deconvolution plays a central role in resolving co-eluting components. Algorithms may operate by identifying characteristic mass spectral patterns with minimal interference and then reconstructing individual component spectra through mathematical unmixing. However, deconvolution performance is strongly influenced by spectral distinctiveness and signal quality. Compounds exhibiting highly similar or collinear mass spectra—such as homologous series or structurally related isomers—pose intrinsic challenges, as mathematical separation becomes underdetermined when spectral profiles lack orthogonal features. In addition, low signal-to-noise ratios, detector saturation, or truncated mass spectra can reduce spectral purity and compromise the stability of deconvolution algorithms. High acquisition rates and consistent spectral quality, particularly in TOFMS, significantly enhance the effectiveness of these approaches by improving peak definition and preserving representative ion ratios across narrow second-dimension peaks.

3.4.4 Analyte Identification and Quantification

Analyte identification in GC×GC–MS is based on the combined use of retention information and detector-specific signatures. The 2D retention coordinates provide an additional level of confidence relative to 1D-GC, as the probability of coincidental co-elution across both dimensions is substantially reduced.

Retention indices calculated along the first dimension can be used to further constrain identification, particularly when combined with MS matching against reference libraries. Modern software platforms integrate retention information directly into spectral search algorithms, improving identification robustness and reducing false positives.

Quantification in GC×GC–MS may be performed using total signal intensities or selectively extracted channels. The latter approach is particularly advantageous in complex regions of the chromatogram, where selective ions or spectral features can be used to isolate the contribution of individual analytes. Careful consideration of modulation effects, peak integration boundaries, and detector response characteristics is essential to ensure quantitative accuracy.

3.4.5 Profiling, Fingerprinting, and Comparative Analysis

The high resolving power of GC×GC–MS enables both targeted profiling of predefined analytes and non-targeted fingerprinting of complex samples. In targeted approaches, known compounds are tracked across samples to assess compositional differences, while non-targeted approaches aim to capture the full chemical signature of a sample without prior assumptions.

Feature-based representations of GC×GC data form the basis for comparative analysis. Features may be defined at different levels of abstraction, including individual data points, peak regions, or larger chromatographic tiles. Each representation involves trade-offs between chemical specificity, robustness to retention-time variability, and computational complexity.

Advanced workflows combine chromatographic alignment, feature extraction, and multivariate analysis to support classification, discrimination, and pattern recognition across large sample sets. In this context, GC×GC–MS data provide a rich substrate for chemometric and machine-learning approaches, enabling the extraction of subtle yet chemically meaningful differences between samples.

3.5 Reproducibility, Robustness, and Data Availability

As with any advanced analytical platform, reproducibility represents a fundamental prerequisite for the reliability and scientific validity of GC×GC results. In comprehensive separations, reproducibility must be evaluated not only in terms of conventional chromatographic parameters—such as retention time stability, peak area precision, and detector response—but also with respect to modulation performance and second-dimension behavior.

Many factors influencing reproducibility in GC×GC–MS are shared with conventional GC–MS, including sample preparation consistency, injection conditions, carrier gas purity, and column aging. However, the use of a modulation interface and a second chromatographic dimension introduce additional sources of variability that must be carefully controlled. In particular, precise synchronization between the modulator, the chromatographic system, and the mass spectrometer is essential. Minor timing discrepancies, when propagated across hundreds of modulation cycles, can lead to cumulative retention-time shifts that compromise chromatographic alignment and downstream data analysis.

Modern GC×GC systems employ high-precision electronic controllers with switching times on the order of milliseconds, effectively mitigating these risks under routine operating conditions. When properly configured and maintained, modulation does not constitute a limiting factor for reproducibility and can achieve levels of temporal stability comparable to those of conventional GC–MS.

Data availability and traceability are increasingly important considerations in the context of complex, data-intensive techniques such as GC×GC–MS. Comprehensive datasets must be accompanied by detailed metadata describing experimental conditions, instrumental configurations, modulation parameters, and data-processing workflows. Without such contextual information, the interpretability and reusability of GC×GC–MS data are severely compromised.

The principles governing data storage and sharing in GC×GC–MS largely mirror those established for GC–MS. Standardized data formats facilitate long-term storage, interoperability between software platforms, and data exchange within the scientific community. The inclusion of both raw data and processed outputs, together with transparent documentation of processing steps, is essential to ensure reproducibility and support independent reanalysis.

3.6 Applications

The application space of GC×GC is particularly well aligned with the analytical challenges encountered in environmental chemistry, where samples are frequently characterized by extreme chemical complexity, wide concentration ranges, and the presence of both known and unknown constituents. In such contexts, GC×GC is rarely adopted for incremental improvements over 1D-GC, but rather when conventional approaches reach intrinsic limits in terms of separation power, detectability, or interpretability.

Environmental applications of GC×GC can be broadly grouped into three partially overlapping categories: (i) trace-level target analysis in highly complex matrices, (ii) class-based or group-type characterization of chemically related contaminants, and (iii) comprehensive profiling of unresolved or poorly characterized mixtures for source apportionment, fate assessment, or treatment evaluation. The following sections highlight representative examples where GC×GC has demonstrated clear analytical advantages, illustrating how its unique attributes translate into environmentally relevant information.

3.6.1 Group-Type and Isomer-Resolved Characterization of Naphthenic Acids in Produced Waters

Produced waters generated during petroleum extraction and processing represent one of the most chemically complex aqueous effluents encountered in environmental analysis. Among their most environmentally relevant constituents are naphthenic acids (NAs), a broad and structurally diverse class of alicyclic and aromatic carboxylic acids associated with aquatic toxicity and long-term environmental persistence. The analytical challenge posed by NAs derives not only from their low concentrations and severe matrix interferences, but also from extensive structural isomerism, with numerous species sharing identical nominal masses and closely related physicochemical properties.

GC×GC has proven particularly effective in addressing these challenges by enabling isomer-resolved and class-based characterization of NA mixtures following appropriate derivatization and prior selective extraction of this class of compounds by DI-SPME [7]. A fully automated DI-SPME method was developed to combine extraction, fiber rinsing and drying, and on-fiber derivatization with N-methyl-N-tert-butyldimethylsilyltrifluoroacetamide. When coupled with high-resolution mass spectrometry (GC×GC–HRMS), NA derivatives are organized into structured chromatographic patterns reflecting systematic variations in carbon number, ring content, and degree of alkylation. This organization supports both targeted quantification of defined NA subclasses and non-targeted evaluation of compositional trends across samples.

In produced water analysis, GC×GC-based workflows have been successfully applied to discriminate NA profiles associated with different extraction processes, to track transformation pathways during treatment or environmental weathering, and to support risk-oriented assessments based on compositional fingerprints rather than single-compound metrics. The ability of GC×GC–HRMS to translate highly complex NA mixtures into chemically interpretable retention–structure relationships exemplifies how structured chromatograms extend the analytical scope of gas chromatography in environmental systems.

As illustrated in Fig. 8, GC×GC–HRMS enables class-resolved interpretation of NA mixtures in produced water by converting an extremely congested chemical space into ordered two-dimensional retention patterns. Following on-fiber derivatization, NA-related features distribute across the 2D separation plane as a coherent band consistent with systematic variations in molecular size and structural motifs. Automated data-processing scripts were applied to visualize NAs in

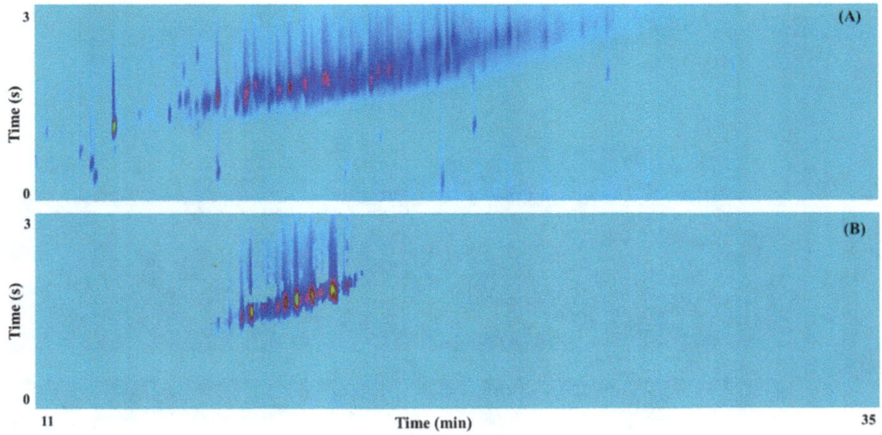

Fig. 8 Representative GC×GC–HRMS two-dimensional chromatogram obtained from produced-water analysis after derivatization (a), highlighting the structured distribution of naphthenic-acid–related features across the retention plane (b) and enabling class-resolved profiling of a highly complex environmental mixture. Adapted from Crucello et al., https://doi.org/10.1016/j.chroma.2023.463844 [7]

produced-water samples by selectively filtering GC×GC–HRMS data according to diagnostic fragmentation behavior of *tert*-butyldimethylsilyl (t-BDMS) derivatives. NA-related features were identified based on the concurrent presence of three characteristic even-electron ions, namely the pseudo-molecular ion $[M–57]^+$ arising from loss of the tert-butyl group, the adduct-related $[M–39]^+$ ion, and the fragment at m/z 75.0261 corresponding to the $[Si(CH_3)_2OH]^+$ moiety. Accurate-mass constraints were combined with an ordinal-ranking approach, whereby these diagnostic ions were required to appear among the most intense fragments of each chromatographic peak. Implementation within GC Image software (GC Image, Lincoln, Nebraska USA) enabled the generation of class-specific filtered chromatograms (Fig. 8b) by suppressing isobaric interferences through tight mass tolerances. The resulting visualizations isolate NA subclasses within the two-dimensional retention space, yielding structured chromatographic patterns that support class-resolved qualitative interpretation.

3.6.2 Determination of Chlorinated Paraffins in Environmental Matrices

Short-chain chlorinated paraffins (SCCPs) are persistent organic pollutants of regulatory concern due to their bioaccumulative behavior and potential toxicity. From an analytical perspective, SCCPs constitute a particularly challenging target, as they comprise thousands of congeners differing in carbon chain length and degree of chlorination. This extreme compositional complexity results in extensive co-elution and pronounced matrix interferences when using conventional one-dimensional GC approaches.

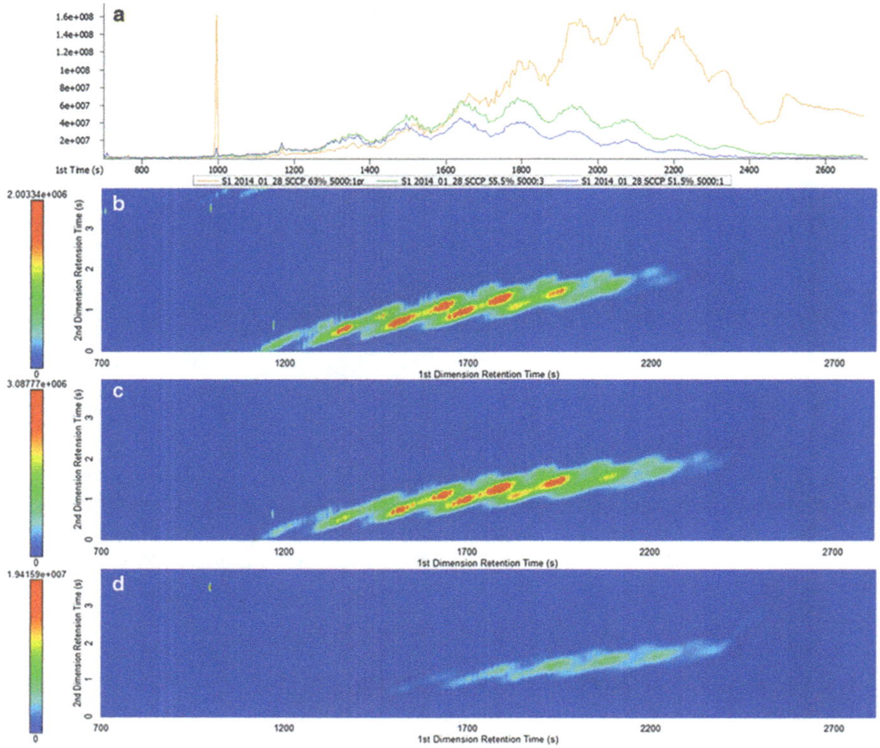

Fig. 9 Example GC×GC–μECD two-dimensional chromatogram for SCCPs in a sediment extract, illustrating the appearance of distinct ordered bands associated with congener-group organization and enabling group-type quantification with improved resistance to matrix interferences. Adapted from Muscalu et al., https://doi.org/10.1007/s00216-016-0153-1 [8]

GC×GC has emerged as an effective solution for SCCP analysis in complex environmental matrices, including sediments, soils, and biota [8]. By exploiting orthogonal selectivity between the two chromatographic dimensions, GC×GC enables partial resolution of SCCP congeners according to chain length and chlorination pattern, yielding structured 2D chromatograms that reflect congener class distributions. When coupled with highly selective detection systems, such as μECD detection (GC×GC–μECD) or MS, this approach substantially improves both qualitative discrimination and quantitative reliability.

In sediment analysis, GC×GC-based methods demonstrate enhanced robustness against matrix-related interferences and increased confidence in SCCP determination compared to one-dimensional GC. Notably, SCCP-related signals in GC×GC–μECD form ordered bands across the two-dimensional separation space, consistent with systematic variations in carbon number and chlorination level as illustrated in

Fig. 9. This structured chromatographic response enhances selectivity toward SCCPs over co-extracted sediment constituents and supports class- or group-based quantification strategies aligned with regulatory frameworks and risk-oriented monitoring requirements.

Overall, this application illustrates the role of GC×GC not merely as a higher-resolution separation technique, but as an enabling methodology for analytically defensible, regulation-relevant quantification of highly complex pollutant classes such as chlorinated paraffins.

3.6.3 Comprehensive Characterization of Petroleum-Derived Environmental Mixtures

Environmental contamination by petroleum-derived materials frequently involves highly complex mixtures that cannot be adequately described using a limited set of target analytes. Representative examples include oily sludges, weathered hydrocarbons, and water-soluble fractions generated during accidental spills or waste handling operations. In such contexts, the analytical objective extends beyond the identification of individual compounds toward comprehensive characterization of mixture composition, biodegradability, and transformation behavior.

GC×GC–TOFMS has been widely applied to the analysis of these petroleum-derived mixtures, enabling the detection of hundreds to thousands of constituents spanning multiple chemical classes. The structured nature of GC×GC chromatograms allows components to be organized into chemically meaningful patterns based on homologous series, functionalization degree, or aromaticity. These organized distributions provide robust descriptors of mixture composition that are not accessible through one-dimensional separations [9].

In studies focused on the environmental fate or biodegradability of oily sludge–derived materials, GC×GC fingerprints have been effectively used to monitor compositional changes during treatment or degradation processes. For petroleum-derived mixtures such as oily-sludge water-soluble fractions, GC×GC–TOFMS generates information-dense chromatographic signatures in which thousands of features are distributed across interpretable regions of the two-dimensional separation space as illustrated in Fig. 10, in particular when coupled to SBSE that has the potential for detection of a wide range of compounds present in the aqueous phase at low concentration levels. These signatures can be leveraged to track treatment- or biodegradation-driven shifts in composition and to link chromatographic pattern evolution to removal efficiency and transformation pathways.

Overall, this application underscores one of the most distinctive contributions of GC×GC to environmental analysis: the ability to deliver comprehensive yet interpretable chemical fingerprints that support process-oriented and mechanistic investigations of complex contamination scenarios, rather than reliance on isolated compound measurements.

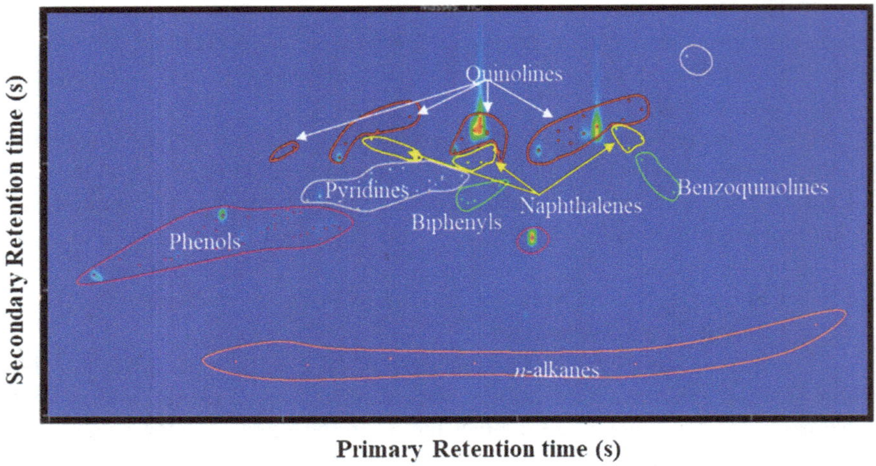

Fig. 10 GC×GC–TOFMS fingerprint of the water-soluble fraction of oily sludge, illustrating the high feature density and structured distribution of mixture constituents across the two-dimensional retention space; the resulting signature supports environmental mixture characterization and transformation tracking. Adapted from Chand et al., https://doi.org/10.1016/j.envpol.2022.119177 [9]

3.7 Positioning of GC×GC Within Environmental Analytical Strategies

The environmental applications discussed in this chapter illustrate that GC×GC should not be regarded as a universal replacement for 1D-GC, but rather as a strategically deployed analytical platform whose strengths emerge most clearly under conditions of extreme chemical complexity. In environmental analysis, such conditions are not exceptional but rather commonplace, arising from heterogeneous matrices, broad concentration ranges, extensive isomerism, and the frequent coexistence of known and unknown constituents.

Across diverse environmental scenarios—including produced waters, contaminated sediments, and petroleum-derived mixtures—GC×GC demonstrates a consistent ability to overcome analytical bottlenecks associated with co-elution, matrix interference, and limited interpretability of conventional chromatograms. The combination of increased peak capacity, enhanced detectability, and structured chromatographic output enables analytical strategies that extend beyond compound-by-compound determination toward chemically meaningful representations of complex mixtures. This shift in perspective is particularly relevant for environmental chemistry, where regulatory frameworks, risk assessments, and remediation strategies increasingly emphasize classes of compounds and mixture effects rather than isolated target analytes.

From a practical standpoint, GC×GC offers its greatest value when the analytical question requires either comprehensive screening or robust discrimination within chemically complex fractions. In such cases, the technique provides a level of confidence that cannot be achieved through incremental optimization of 1D methods. At the same time, its complexity, operational demands, and data-processing requirements impose a clear responsibility on the analyst to justify its use in relation to the specific objectives of the study. GC×GC should therefore be positioned as a problem-solving tool rather than a routine default, deployed when the cost of incomplete or ambiguous chemical information outweighs the additional analytical effort.

Moreover, sample preparation and GC×GC–MS must be viewed as a tightly integrated analytical alliance rather than as independent methodological components. The evolution toward green, miniaturized, and versatile sample preparation techniques has not diminished the importance of upstream processing; instead, it has elevated its strategic role in defining the quality, scope, and interpretability of comprehensive chromatographic data. When carefully aligned, modern sample preparation approaches and GC×GC–MS together provide an exceptionally powerful framework for addressing the analytical challenges posed by complex environmental systems.

Looking forward, the future impact of GC×GC in environmental analysis is expected to be shaped less by radical innovations in chromatographic hardware and more by advances in detection, data processing, and information extraction. Continued improvements in MS acquisition speed, sensitivity, and resolving power will further enhance the utility of GC×GC for non-target and suspect screening workflows. In parallel, the integration of advanced chemometric, machine-learning, and artificial intelligence–based approaches is poised to transform how comprehensive chromatographic datasets are interpreted, enabling automated pattern recognition, class-based quantification, and cross-sample comparison at scales that are currently impractical.

In this evolving scenario, GC×GC, combined with *in-situ* sampling, automation, and remote platforms, is likely to play a central role in addressing emerging environmental challenges related to complex mixtures, transformation products, and unknown contaminants. Its ability to generate information-rich yet chemically structured datasets uniquely positions the technique to support next-generation environmental monitoring, mixture-oriented risk assessment, and mechanistic studies of contaminant fate and behavior. As such, GC×GC should be viewed not only as an advanced separation technique, but as a foundational analytical platform for future developments in environmental analytical chemistry.

Questions

1. How does the expansion of separation space from one to two chromatographic dimensions fundamentally change peak capacity and the probability of co-elution in complex environmental samples?
2. What is the role of modulation in GC×GC, and how do modulation period and modulator design influence second-dimension separation performance and data interpretability?

(continued)

3. Why is column configuration (stationary-phase selection, column length, and internal diameter) critical in GC×GC, and how does it affect orthogonality and structured chromatographic patterns?
4. In what ways do structured retention patterns in GC×GC provide chemically meaningful information beyond simple peak resolution, and how can this be exploited in environmental analysis?
5. How do the requirements imposed by fast second-dimension separations influence the choice and performance of detection systems in GC×GC?
6. Under which analytical circumstances does GC×GC provide clear advantages over one-dimensional GC, and how should its use be justified within an environmental analytical strategy?

Glossary

- **Comprehensive two-dimensional gas chromatography** (GC×GC): A chromatographic technique in which the entire effluent from a first-dimension column is continuously transferred to a second column with a different selectivity via a modulator. providing a two-dimensional separation with greatly increased peak capacity.
- **Modulation**: The periodic process in GC×GC whereby small, sequential fractions of the first-dimension effluent are trapped, focused, and rapidly reinjected into the second-dimension column, while preserving the first-dimension separation in the form of discrete modulation bands.
- **First dimension (^{1}D):** The primary chromatographic separation in GC×GC, typically governed by volatility-related interactions and conducted on a conventional-length column.
- **Second dimension (^{2}D):** A short, high-speed chromatographic separation performed on each modulated fraction of the first-dimension effluent, providing complementary selectivity.
- **Peak capacity:** A measure of the maximum number of peaks that can theoretically be resolved within a chromatographic separation space; substantially increased in GC×GC relative to one dimensional GC.
- **Single-channel detector:** A detector that provides a single intensity signal as a function of time (e.g., FID, μECD), commonly used for quantitative or selective GC×GC applications.
- **Multi-channel detector:** A detector that records compound-specific signatures at each acquisition point (e.g., mass spectrometry, VUV spectroscopy), enabling enhanced identification and deconvolution in GC×GC.

(continued)

- **Structured chromatographic patterns**: Organized distributions of analytes in the two-dimensional retention plane of GC×GC that reflect chemical similarity, such as homologous series or functional classes.
- **Solid-phase microextraction (SPME)**: A solvent-free, non-exhaustive sample preparation technique based on analyte partitioning into a sorptive coating, widely used in GC×GC workflows.
- **Stir-bar sorptive extraction (SBSE)**: A sorptive extraction technique employing a polymer-coated stir bar with higher sorbent capacity than SPME, offering enhanced sensitivity but increased matrix transfer risk.
- **Thermal modulation**: A modulation strategy in GC×GC that uses rapid cooling and heating to actively trap and refocus analytes prior to second-dimension separation.
- **Flow modulation**: A modulation approach based on pressure or flow switching without cryogenic trapping, generally offering lower sensitivity but simpler operation when compared with thermal modulation.
- **Non-target screening**: An analytical strategy aimed at detecting and characterizing unknown or unexpected compounds without prior selection of target analyte
- **Chemical fingerprinting:** The use of comprehensive chromatographic patterns to characterize and compare complex mixtures based on overall composition rather than individual compounds.

Further Reading

- Martínez-Pérez-Cejuela, H.; Gionfriddo, E. Evolution of Green Sample Preparation: Fostering a Sustainable Tomorrow in Analytical Sciences. *Analytical Chemistry* 2024, *96*, 7840–7863. https://doi.org/10.1021/acs.analchem.4c01328
- Blomberg, J.; Schoenmakers, P. J.; Beens, J.; Tijssen, R. Comprehensive two-dimensional gas chromatography (GC × GC) and its applicability to the characterization of complex (petrochemical) mixtures. *Journal of High Resolution Chromatography*. 1997, *20*, 539–544. https://doi.org/10.1002/jhrc.1240201005
- Trinklein, T. J.; Cain, C. N.; Ochoa, G. S.; Schöneich, S.; Mikaliunaite, L.; Synovec, R.E. Recent advances in GC × GC and chemometrics to address emerging challenges in nontargeted analysis. *Analytical Chemistry* 2023, *95*, 264–286. https://doi.org/10.1021/acs.analchem.2c04235

Bibliography

1. Mondello L, Cordero C, Janssen HG, Synovec RE, Zoccali M, Tranchida PQ (2025) Comprehensive two-dimensional gas chromatography–mass spectrometry. Nat Rev Method Prim 5:7. https://doi.org/10.1038/s43586-024-00379-3
2. Martínez-Pérez-Cejuela H, Gionfriddo E (2025) Innovative sample preparation strategies for emerging pollutants in environmental samples. Ann Rev Analy Chem 18:73–95. https://doi.org/10.1146/annurev-anchem-071224-093709
3. Dumont G, Rodrigues A, Velimirovic M, Lievens S, Jordens J, Focant JF, Stefanuto PH (2025) Microplastic and nanoplastic analysis: from pyrolysis gas chromatography-mass spectrometry to pyrolysis two-dimensional gas chromatography-mass spectrometry—a critical review. J Separat Sci 48(10):e70287. https://doi.org/10.1002/jssc.70287
4. Tranchida PQ, Maimone M, Purcaro G, Dugo P, Mondello L (2015) The penetration of green sample-preparation techniques in comprehensive two-dimensional gas chromatography. TrAC Trend Analy Chem 71:74–84. https://doi.org/10.1016/j.trac.2015.03.011
5. Liu X, Huang Q, Deng J, Liu X, Hu B (2024) Portable mass spectrometry for on-site detection of hazardous volatile organic compounds via robotic extractive sampling. Analy Chem 96(23):9325–9331. https://doi.org/10.1021/acs.analchem.4c01555
6. Chen W, Zou Y, Mo W, Di D, Wang B, Wu M, Huang Z, Hu B (2022) Onsite identification and spatial distribution of air pollutants using a drone-based solid-phase microextraction array coupled with portable gas chromatography-mass spectrometry via continuous-airflow sampling. Environ Sci Technol 56(23):17100–17107. https://doi.org/10.1021/acs.est.2c05259
7. Crucello J, Sampaio NMFN, Junior IM, Carvalho RM, Gionfriddo E, Marriott PJ, Hantao LW (2023) Automated method using direct-immersion solid-phase microextraction and on-fiber derivatization coupled with comprehensive two-dimensional gas chromatography high-resolution mass spectrometry for profiling naphthenic acids in produced water. J Chromatog A 1692:463844. https://doi.org/10.1016/j.chroma.2023.463844
8. Muscalu AM, Morse D, Reiner EJ, Górecki T (2017) The quantification of short-chain chlorinated paraffins in sediment samples using comprehensive two-dimensional gas chromatography with μECD detection. Analy Bioanaly Chem 409:2065–2074. https://doi.org/10.1007/s00216-016-0153-1
9. Chand P, Suryendu D, Mukherji S (2022) Characterization and biodegradability assessment of water-soluble fraction of oily sludge using stir bar sorptive extraction and GCxGC-TOF MS. Environ Pollut 304:119177. https://doi.org/10.1016/j.envpol.2022.119177

The 5 W of Chemometrics: Why Multivariate Thinking Matters

Cristina Malegori and Paolo Oliveri

Abstract

Chemometrics is the discipline that enables chemists to extract meaningful information from complex, multivariate data. Environmental systems inherently generate such a type of data, in particular when working with modern analytical techniques; in this context chemometrics provides the conceptual and practical tools required to interpret structure, reduce dimensionality, reveal patterns, build predictive models, and ensure scientifically robust decision-making. The present chapter introduces chemometrics through the accessible and intuitive journalistic framework of the 5 W—What, Who, Why, When, and Where—clarifying the nature of the discipline, identifying the professionals best suited to perform it, highlighting its scientific importance, and defining the context in which it should be applied across analytical chemistry and related fields. Through this approach, readers will understand that multivariate thinking is essential for modern chemical research and practice, and how chemometrics transforms raw data into chemical knowledge.

Keywords

Chemometrics · Multivariate Analysis · Data Interpretation · Chemical Modelling · Pattern Recognition · Environmental science

C. Malegori (✉) · P. Oliveri
Department of Pharmacy (DIFAR), University of Genova, Genoa, Italy
e-mail: cristina.malegori@unige.it

M. Minella (ed.), *Complexity in Environmental Chemistry*,
https://doi.org/10.1007/978-3-032-31774-2_6

Chapter Learning Goals

- Understand chemometrics through the 5 W journalistic framework.
- Recognize the importance of multivariate thinking in chemistry.
- Appreciate the role of the chemometrician compared to general data scientists.
- Learn when and why chemometrics is essential for both exploring and controlling variability.
- Identify the contexts and platforms where chemometrics can be effectively applied.

1 Introduction: Why the 5 W Approach?

Chemometrics is a rich, powerful, and often misunderstood discipline. Originally created to solve complex chemical problems, it is today an essential component of modern analytical chemistry, yet, it is still perceived as abstract or inaccessible by many practitioners. For this reason, the present chapter adopts the 5 W structure used in scientific journalism—What, Who, Why, When, and Where. This narrative style introduces chemometrics in a clear, intuitive way, allowing readers to grasp both its conceptual foundations and its practical value. In light of these considerations, it is worth noting that how chemometrics can play a crucial role in addressing the complexity of environmental data (Fig. 1).

2 What Is Chemometrics?

Chemometrics emerged as a formal discipline in the early 1970s, at a time when chemistry faced an unprecedented transformation driven by the rapid expansion of instrumental technologies. Modern spectrometers, chromatographs, and process analyzers were producing more data than chemists had ever handled before. It was in this context that Svante Wold first introduced the word "kemometri" in a 1971 research proposal submitted at the University of Umeå [1]. Although the term appeared modest, it marked the beginning of a new way of thinking about chemical data.

Wold's intellectual background may have made this step natural: he was the son of the statistician Herman Wold, a key figure in econometrics, and the grandson of Svante Arrhenius, Nobel laureate in chemistry (1903). Inspired by the naming of fields such as econometrics and psychometrics—where mathematical tools are used to extract meaningful information from complex datasets—he proposed an analogous approach for chemistry. According to his early definition, chemometrics is "the art of extracting chemically relevant information from data produced in chemical experiments" [1].

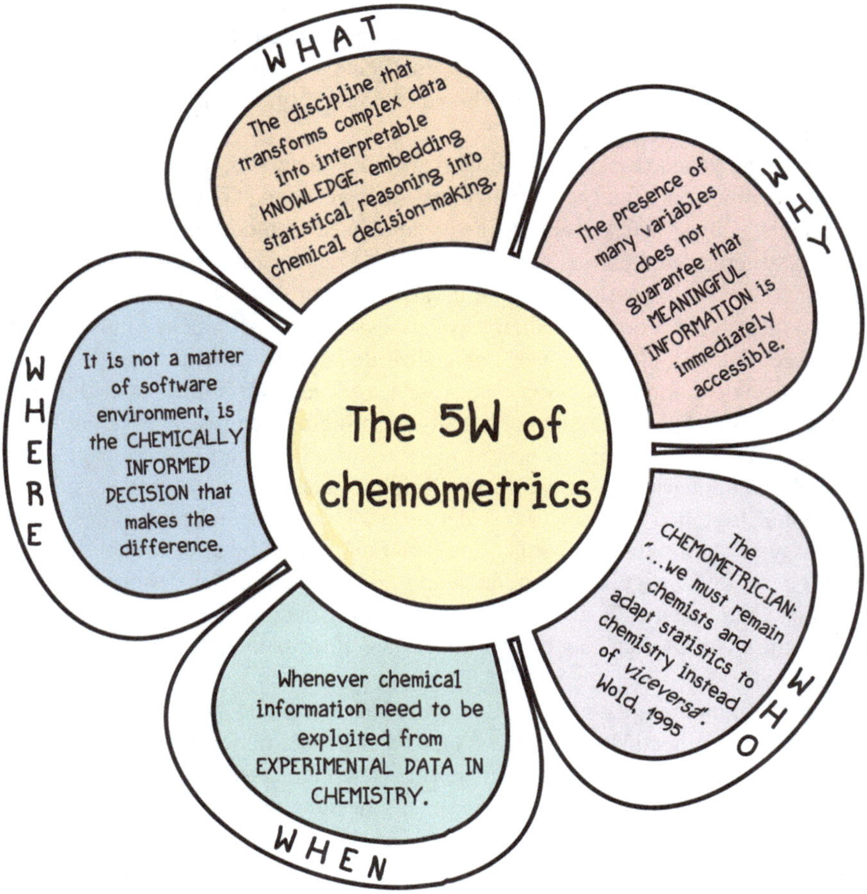

Fig. 1 The 5 W of chemometrics: a conceptual map

This definition resonated widely. Analytical chemistry was becoming intrinsically multivariate: a gas chromatogram may contain dozens of peaks; a vibrational spectrum hundreds or thousands of wavelength-specific intensities; a metabolomics profiling experiment tens of thousands of mass features. Even environmental chemistry—when describing an environmental matrix by means of target and untargeted analysis of contaminants—is generating heterogeneous yet complementary data. In all the branches of chemistry, the chemists have needed, and still need, methods capable of handling this complexity [2].

Early pioneers such as Bruce Kowalski at the University of Washington recognized this need and played a key role in establishing chemometrics as a global discipline. In 1974, Kowalski and Wold founded the International Chemometrics Society (ICS), which rapidly became a reference point for the community.

What distinguishes chemometrics from general data science is not the algorithms themselves—but the insistence on chemical interpretability. Most chemometric methods allow analysts to extract model coefficients or latent variables that reflect underlying chemical phenomena. This interpretability has always been a core requirement of the field.

During the years, several methodological pillars emerged, adapting various statistical learning tools for specific chemical interpretation. These include principal component analysis (PCA), which is an unsupervised methodology aimed at exploring data structure, and partial least squares (PLS) regression, which is a supervised strategy aimed at estimating a response variable from a block of predictors [3, 4]. Beyond their exploratory or predictive role, these methods offer a key interpretative advantage: they allow the analyst to visualize underlying structure, emphasize the most informative variables, reduce noise and irrelevant variability, and ultimately connect statistical patterns with chemically meaningful features.

A widely cited definition appeared in 1997, when Massart and colleagues formalized chemometrics as "a chemical discipline that uses mathematics, statistics and formal logic (a) to design or select optimal experimental procedures; (b) to provide maximum relevant chemical information by analysing chemical data; and (c) to obtain knowledge about chemical systems." This broader definition emphasizes that chemometrics is not only about modelling data: it also guides experimental design, ensuring that laboratory studies are informative, efficient, and fit for purpose [5].

The chemometric perspective also embraces a critical philosophical notion: models are simplifications of reality. As Box and Draper famously stated, "all models are wrong, but some are useful" [6]. This statement carries particular weight in chemometrics, where models are intentionally designed to describe a specific portion of the chemical domain, not the entirety of chemical behaviour. The goal is not generality but practical utility—to answer a real analytical question reliably.

Achieving such usefulness from experimental data requires careful validation, an area in which chemometrics has arguably contributed some of its most influential ideas. Cross-validation strategies, test sets, and permutation testing all help assess whether a model captures genuine chemical structure or merely noise. Without proper validation, even the most sophisticated model risks overfitting, producing excellent results on the training set but failing dramatically on new samples [7, 8]. Chemometrics therefore places validation at the centre of its scientific workflow.

In summary, to answer the question 'What is chemometrics?', chemometrics is the discipline that allow chemists to fully exploit the information contained in complex data. It transforms raw data into interpretable knowledge, ensures experimental soundness, supports prediction, and embeds statistical reasoning into the core of chemical decision-making. Far from being a mere accessory, chemometrics is now an essential component of modern environmental science.

3 Why Does Chemometrics Matter?

Chemometrics matters especially because chemistry is inherently multivariate. Every chemical sample, every instrumental signal, and every analytical measurement contains far more information than can be captured—or perceived—by univariate thinking. Traditional one-variable-at-a-time approaches ignore the complex interplay among variables, while chemometrics—which typically implements multivariate strategies—reveals patterns of joint variation that better reflect underlying chemical structure.

In many analytical contexts, a single measurement already represents a high-dimensional data structure. A chromatogram of environmental contaminants, for example, provides an entire compositional profile in a single analytical run, with each chromatographic peak representing a quantitative variable that contributes to the chemical fingerprint of the sample. Spectroscopic techniques push this even further: in vibrational spectroscopy, in fact, each spectrum contains hundreds or thousands of absorbance values recorded at sequential wavelengths, forming a data vector where the order of variables carries physical significance, but not all chemical data are homogeneous. In environmental monitoring, for example, samples are commonly characterized by a combination of chemical variables, physicochemical parameters, spatial and temporal descriptors, and climate conditions. These heterogeneous datasets are still multivariate: each sample corresponds to a vector of measurements, although the variables may differ widely in scale, magnitude, and noise behaviour. Without proper data pre-treatment and multivariate analysis, the most informative parts of such datasets remain hidden [12].

The presence of many variables does not guarantee that meaningful information is immediately accessible. Data must be explored, structured, and interpreted. This is where chemometrics provides unparalleled tools. Multivariate methods reveal patterns—clustering, trends, latent gradients—that univariate methods cannot detect. They describe how variables co-vary, which is essential because chemical phenomena rarely affect one variable at a time. Actually, patterns of covariance often reflect meaningful chemical structure.

Yet, despite their power, multivariate approaches are still underused. Some analysts apply multivariate methods only when all simpler strategies have failed, hoping that chemometrics can extract value from data that may not contain any informative structure at all. This attitude is risky. Applying complex models to poor-quality or uninformative data can produce superficially impressive but misleading results, driven not by real chemical information but by statistical artefacts [13].

One of the most common dangers is overfitting, where a predictive model learns noise instead of signal [7]. Overfitted models perform extremely well on the training set but generalize poorly to new samples. The risk grows with the complexity of the modelling technique and with the dimensionality of the measured variables. Chemometrics therefore places enormous emphasis on rigorous validation: cross-validation, test sets, and permutation tests are embedded into the workflow to ensure that the model captures genuine chemical structure.

In conclusion, chemometrics matters because it transforms variability into knowledge. It reveals structure where univariate approaches do not succeed; it highlights relationships hidden within high-dimensional space; it enables prediction, optimization, and informed decision-making processes. Thus it ensures that decisions in environmental chemistry are grounded in statistically sound and chemically meaningful reasoning.

4 Who Does Chemometrics?

Chemometrics is often mistakenly perceived as a branch of statistics applied to chemistry. This misconception has led many to assume that the ideal professional to perform chemometric analysis is a statistician or data scientist. In reality, the opposite is true: chemometrics must be done by chemists—chemists with a strong mathematical mindset, yes, but above all, chemists who deeply understand the chemical system under study. The distinction is not semantic: it is foundational. In particular when chemistry is applied to complex systems, such as environmental science, it is fundamental that the scientist in charge of the extraction of useful information from the data knows the environmental problem: which is the question to be answered, what is the nature of the samples and the sampling strategy employed, which analytical approaches were carried out, and, only as one step of the full process, which data processing tools are available for the specific context. Just in light of the knowledge of the system under study, results can be interpreted.

Chemometric methods, in fact—whether exploratory or predictive approaches, or advanced pre-processing strategies—are powerful only when used by someone who understands the chemical nature of the data. This is because chemometric techniques are inherently interpretive: they allow users to identify the contribution of original variables in transferring the physical–chemical meaning to the developed model. Without domain expertise, these elements risk being misunderstood or, worse, misused.

This interpretive requirement is what distinguishes chemometricians from general data scientists. A data scientist may be highly skilled in algorithmic optimization, machine learning, or computational modelling, but may lack familiarity with sampling variability, matrix effects, spectroscopic behavior, scattering phenomena, or chromatographic artefacts. Chemometric decisions—such as choosing an appropriate pre-processing method, selecting relevant variables, designing a calibration strategy, or evaluating model reliability—cannot be made in isolation from chemical knowledge.

A paradigmatic example is data pre-processing, a stage that is often underestimated by newcomers. Software packages offer dozens of mathematical transformations—standard normal variate (SNV), derivatives, autoscaling, baseline correction, detrending, multiplicative scatter correction (MSC), logarithmic transforms—to be applied individually or in combination. Each transformation has a specific mathematical effect, but whether it is chemically appropriate depends entirely on the system being analyzed. For instance, SNV may be ideal for mitigating scattering effects in near-infrared or visible spectroscopy when quantifying analytes, but the same transformation would eliminate meaningful structural features if the objective were to study particle size distributions [9].

The need for chemically informed interpretation has been emphasized repeatedly by the founders of the field. In his 1995 reflection on the first 20 years of chemometrics, Svante Wold wrote: "In essence, we must remain chemists and adapt statistics to chemistry instead of vice versa" [10]. This principle remains a cornerstone of the discipline. Chemometrics is not meant to replace chemical reasoning but to strengthen it.

The history demonstrates, in fact, how chemical insight can drive methodological innovation. A particularly illustrative case is that of William Sealy Gosset, the analytical chemist at Guinness brewery who, under the pseudonym Student, developed the *t* distribution to solve practical brewing problems [11]. Although often cited in statistics textbooks, Gosset's work was fundamentally chemometric: he adapted statistical concepts to real chemical questions, not the other way around.

Thus, the true chemometrician is not primarily a statistician, nor a programmer, nor a machine learning expert—although chemometricians may possess these skills. Chemometricians are first and foremost chemists who know their data intimately: how they were collected, what experimental factors influence them, what artefacts may distort them, and what physical–chemical mechanisms underpin the observed variability.

This deep awareness of the analytical process—from sampling to instrumental acquisition, from signal behavior to matrix effects—enables chemometricians to make informed choices at every step of the modelling workflow. It is this integration of chemical thinking with quantitative tools that defines the profession and ensures the reliability, interpretability, and practical usefulness of chemometric models.

To summarize, what truly determines the quality of chemometric work is the professional's ability to—understand how the data were generated—anticipate which artefacts or sources of variability may be present—select appropriate preprocessing strategies—choose and validate the most suitable models—and interpret results in a chemically meaningful way.

5 When Is It Useful to Apply Chemometrics?

Chemometrics is useful whenever chemical information is embedded in data. Unlike traditional approaches, chemometrics should not be reserved for particularly complex datasets or for situations where everything else has failed. In fact, it is always valuable in at least two broad and complementary scenarios: when variability is present and must be understood, and when variability is unwanted and must be controlled.

In many chemical systems, natural or experimental variability is not a nuisance but a source of insight. Differences among samples encode patterns reflecting composition, structure, origin, processing conditions, or environmental influences. Moreover, current analytical methods produce data that are often too rich to be interpreted visually or with traditional statistics, making chemometrics an indispensable companion to extract meaningful information [14].

In other contexts—such as quality control, batch consistency, product release, and stability studies—the goal is not to characterize diversity but to ensure similarity. Multivariate statistical process control (MSPC), for example, defines a "normal operating space" for a process or product. Once this multivariate fingerprint is established, chemometrics can detect deviations and anomalies far earlier than univariate control charts. Even when variability is minimal, chemometric models can capture the tiny fluctuations that determine whether a process remains stable or begins to drift [15].

Rather than being a last resort, chemometrics should be considered an integral component of sound analytical practice, guiding both exploration and control.

6 Where Is Chemometrics Applied?

The "where" of chemometrics is not a physical location, nor a specific software environment. In fact, nowadays, many data-processing platforms are available, from the most modern ones—open source, online and freely available—to the professional dedicated tools, developed by specialized companies. Whether models are built in MATLAB, R, Python, SIMCA, PLS_Toolbox, The Unscrambler, Orange, KNIME, or any other platform, the underlying algorithms should be mathematically equivalent. Different software packages may vary in terms of interface, speed, visualization, or specialized features, but none of these aspects replaces the chemist's knowledge of the system.

It is important to notice that some of the platforms available for multivariate data processing and machine learning are developed outside the chemometrics context and, therefore, they may lack key aspects related to the chemometric workflow, such as data pre-processing, with dedicated transforms, and validation strategies, to monitor overfitting.

Algorithms and software will continue to evolve, but the ability to pose the right questions, recognize meaningful patterns, and make chemically informed decisions will always lie at the core of the discipline.

Speaking about evolution, artificial intelligence (AI) and the related terminology have become deeply embedded in modern science and society. Deep learning (DL), currently one of the most prominent families within AI, implements multilayered and non-linear artificial neural networks inspired by biological learning processes. These techniques have demonstrated extraordinary performance in fields such as image analysis, language processing, and large-scale pattern recognition.

However, DL also introduces challenges. One major limitation is the requirement for extremely large and well-designed training sets: as has been noted in numerous applications, careful design of representative training samples is essential—but often impractical in chemical studies, where data acquisition may be costly, time-consuming, or experimentally constrained. As a consequence, many DL studies risk being fundamentally limited not by the algorithm, but by the dataset available.

A second and equally important challenge lies in interpretability. The coefficients of deep neural networks are notoriously opaque (hidden layers): while they may encode useful representations, they rarely offer chemically meaningful insights. For chemists who must justify decisions, understand mechanisms, or validate analytical conclusions, this "black box" limitation is nontrivial.

This is where chemometrics maintains a crucial role. AI and chemometrics should not be viewed as competing paradigms. Instead, they form a continuum of data-driven chemistry: for large datasets where prediction is the primary goal, DL may bring unprecedented capabilities; for smaller experimental studies—where interpretability, validation, and chemical meaning are essential—classical chemometric approaches remain indispensable [16].

Questions

- What distinguishes chemometrics from traditional statistical approaches?
- Why is multivariate thinking essential in chemical data interpretation?
- How does the chemist's domain knowledge influence chemometric modelling?
- When does variability represent information, and when is it unwanted noise?
- Why is software choice secondary compared to chemical understanding?

Glossary

- Artificial intelligence (AI): a broad field of computational methods that enable machines to perform tasks that usually require human intelligence, through pattern recognition and data-driven modelling approaches.
- Black box: a predictive model whose internal structure or parameters do not provide direct interpretative insight into the underlying physical–chemical mechanisms.
- Calibration: a mathematical relationship developed to predict a quantitative response variable from a set of measured predictor variables.
- Cross-validation: a model validation strategy in which data are repeatedly partitioned into training and test subsets, according to a predefined split scheme.
- Data pre-processing: mathematical transformations applied to raw data to reduce unwanted systematic effects, improve comparability, and enhance relevant chemical information before modelling.

(continued)

- Deep learning (DL): a family of artificial intelligence methods based on multilayered neural networks capable of modelling complex non-linear relationships, typically requiring large datasets.
- Test set: an independent dataset not used during model development, employed to evaluate predictive performance and robustness.
- Training set: a dataset used to develop a model.
- Latent variables: variables obtained from multivariate data as linear combinations of the measured variables; they summarize structured variation and represent underlying patterns in the dataset.
- Overfitting: a modelling weakness that occurs when a model captures noise rather than true structure, resulting in excellent performance on the training set but poor prediction on new data (test set or unknown samples).
- Permutation test: a diagnostic validation strategy in which class labels or response variables are randomly reassigned to evaluate whether model performance significantly exceeds what would be expected by chance.
- Supervised method: a modelling approach that uses known response variables to build predictive models.
- Unsupervised method: a modelling approach that explores data structure without using additional information about the samples.

Further Reading

Bro, R. AI? You Mean Chemometrics? The Analytical Scientist, September 2024.

P Oliveri et al., "the impact of signal pre-processing on the final interpretation of analytical outcomes – a tutorial," Anal Chim Acta, 1058, 9–17 (2019). https://doi.org/10.1016/j.aca.2018.10.055

Bibliography

1. Wold S (2001) Personal memories of the early PLS development. Chemometr Intell Lab Syst 58:83–84. https://doi.org/10.1016/S0169-7439(01)00152-6
2. Dupont MF, Elbourne A, Cozzolino D, Chapman J, Truong VK, Crawford RJ, Latham K (2020) Chemometrics for environmental monitoring: a review. Anal Methods 12(38):4597–4620
3. Bro R, Smilde AK (2014) Principal component analysis. Analy Method 6(9):2812–2831
4. Wold S, Sjöström M, Eriksson L (2001) PLS-regression: a basic tool of chemometrics. Chemom Intell Lab Syst 58(2):109–130
5. Massart DL, Vandeginste BGM, Buydens LMC, De Jong S, Lewi PJ, Smeyers-Verbeke J (1997) Chemometrics: a textbook. Elsevier
6. Box GEP, Draper NR (1987) Empirical model-building and response surfaces. Wiley
7. Hawkins DM (2004) The problem of overfitting. J Chem Inf Comput Sci

8. Szymańska E, Saccenti E, Smilde A, Westerhuis J (2012) Double-checking PCA: validation and interpretation. Metabolomics
9. Rinnan Å, Van den Berg F, Engelsen SB (2009) Review of normalization and scatter correction methods. TrAC Trends Anal Chem
10. Wold S (1995) Chemometrics; what do we mean with it, and what do we want from it? Chemom Intell Lab Syst 30:109–115. https://doi.org/10.1016/0169-7439(95)00042-9
11. Gosset WS (1908) The probable error of a mean. Biometrika
12. Mas S, de Juan A, Tauler R, Olivieri AC, Escandar GM (2010) Application of chemometric methods to environmental analysis of organic pollutants: a review. Talanta 80(3):1052–1067
13. Kjeldahl K, Bro R (2010) Some common misunderstandings in chemometrics. J Chemom 24(7–8):558–564
14. Pomerantsev AL, Rodionova OY (2012) Process analytical technology: a critical view of the chemometricians. J Chemom 26(6):299–310
15. MacGregor JF, Kourti T (1995) Statistical process control of multivariate processes. Control Eng Pract 3(3):403–414
16. Oliveri P (2025) Chemometrics: a bridge to the AI age. The analytical scientist 2025. https://url.us.m.mimecastprotect.com/s/IzpjCJ6R2Ri0vB8ktkH5CyZoE5?domain=urlsand.esvalabs.com. Accessed 6 Nov 2025

Practical Strategies for the Analysis of Chemical Data: From Traditional Chemometric Tools to Artificial Intelligence

Eugenio Alladio

Abstract

Environmental chemistry generates datasets of remarkable complexity, characterized by numerous interrelated variables measured across spatial and temporal dimensions. This chapter provides a comprehensive methodological guide to the principal chemometric and machine learning techniques employed for analyzing such multivariate environmental data. The chemometric toolkit is organized into two complementary domains: experimental design strategies that optimize data collection, and multivariate data analysis methods that extract information from complex datasets.

Design of Experiments (DoE) provides a systematic framework for efficient experimental planning, enabling detection of factor interactions and simultaneous optimization of multiple responses while minimizing resource expenditure. Principal Component Analysis (PCA) serves as the cornerstone exploratory technique, revealing data structure, identifying patterns and outliers, and reducing dimensionality while preserving essential information. Multivariate Curve Resolution (MCR) and related methods recover physically meaningful source profiles from mixed environmental signals, enabling quantitative source apportionment of atmospheric particulates, dissolved organic matter, and other complex mixtures.

Supervised learning approaches are examined for both classification and regression applications. Linear Discriminant Analysis (LDA), Partial Least Squares Discriminant Analysis (PLS-DA), Support Vector Machines (SVM), Random Forests, and Soft Independent Modelling of Class Analogy (SIMCA) build predictive models for categorical outcomes including source identification, quality classification, and anomaly detection. Regression methods (Multiple

E. Alladio (✉)
Department of Chemistry, University of Turin, Torino, Italy
e-mail: eugenio.alladio@unito.it

M. Minella (ed.), *Complexity in Environmental Chemistry*,
https://doi.org/10.1007/978-3-032-31774-2_7

Linear Regression (MLR), Principal Component Regression (PCR), and Partial Least Squares Regression (PLS-R)) extend prediction to continuous quantities such as pollutant concentrations and environmental quality parameters. The chapter also introduces deep learning architectures, including Convolutional Neural Networks for spatial and spectral data, Recurrent Neural Networks for temporal sequences, and autoencoders for anomaly detection, highlighting both their transformative potential and current limitations.

Throughout all applications, rigorous validation emerges as the critical determinant of model reliability. Cross-validation strategies, external validation with independent test sets, and temporal validation protocols are essential for ensuring that models generalize beyond development data. The chapter emphasizes that the chemometrician's expertise lies not in mechanical application of favored methods but in thoughtful selection and integration of techniques matched to specific research objectives, data characteristics, and application requirements, ultimately transforming environmental measurements into actionable knowledge for effective environmental stewardship.

Keywords

Chemometrics · Multivariate analysis · Source apportionment · Machine learning · Environmental monitoring · Model validation

Chapter Learning Goals
What you will learn in this chapter:

- The principles of Design of Experiments (DoE) for efficient experimental planning in environmental chemistry.
- The fundamentals of Principal Component Analysis (PCA) for exploratory analysis, dimensionality reduction, and pattern recognition.
- The distinction between PCA and Multivariate Curve Resolution (MCR) methods, and when physically meaningful source profiles are required for quantitative source apportionment.
- The capabilities and limitations of cluster analysis for sample grouping and pattern discovery.
- The characteristics of supervised classification methods (LDA, PLS-DA, SVM, Random Forests, SIMCA) and criteria for their selection.
- The principles of regression techniques (MLR, PCR, PLS-R) for quantitative prediction of environmental parameters.
- An introduction to deep learning approaches and their potential for complex environmental data.
- The critical importance of rigorous validation protocols to ensure model reliability and generalizability.

1 Introduction

This chapter presents a comprehensive overview of the main chemometric techniques used in the analysis of environmental data. Environmental chemistry produces highly complex datasets characterized by many interrelated variables measured across spatial and temporal dimensions. The multivariate nature of many environmental applications, including air quality assessment, water pollution monitoring, and soil contamination evaluation, requires analytical approaches capable of extracting meaningful patterns from high-dimensional data. The techniques described here form the methodological foundation of environmental chemometrics, enabling researchers to transform raw analytical measurements into actionable knowledge [1].

The chemometric toolkit for environmental applications can be broadly organized into two complementary domains: experimental design strategies that optimize data collection, and multivariate data analysis methods that extract information from complex datasets. Within the latter category, techniques are further classified according to their learning paradigm, i.e., unsupervised methods that discover inherent data structure without prior class information, and supervised approaches that evaluate known classifications or quantitative responses to build predictive models. Within this framework, each strategy offers specific advantages for particular environmental applications, and their thoughtful combination often yields the most comprehensive understanding of environmental systems [2]. The evolution of environmental monitoring has significantly transformed the data available for analysis. Traditional monitoring programs typically collected a limited number of parameters at weekly or monthly intervals from a small set of stations. Modern monitoring increasingly relies on continuous automated measurements across extensive sensor networks, hyperspectral imaging producing thousands of spectral channels per spatial pixel, and comprehensive chemical profiling that quantifies hundreds of analytes simultaneously. This dramatic increase in data dimensionality and volume has made chemometric approaches not only useful but essential for interpreting environmental data [3, 4]. At the same time, it has increased the risk of extracting spurious patterns from complex datasets, making rigorous validation an indispensable component of any chemometric analysis [3, 4].

The following sections are organized to reflect a logical progression from data generation to data interpretation. The chapter first introduces Design of Experiments as a strategy for efficient experimental planning. It then presents unsupervised multivariate analysis methods for exploratory data analysis and source resolution, followed by supervised classification and regression techniques for predictive modeling. Advanced machine learning and deep learning approaches are discussed as extensions of classical chemometric methods for handling increasingly large and complex environmental datasets. The chapter concludes with remarks on methodological integration and future perspectives. An essential preliminary step in any chemometric workflow is the thoughtful design of experiments and data collection strategies. The way data are generated strongly influences the type and reliability of the information that can subsequently be extracted through multivariate analysis.

For this reason, the chapter begins by introducing Design of Experiments as a methodological framework for efficient and informative data generation, before moving to data analysis techniques.

2 Design of Experiments (DoE)

Design of Experiments (DoE) is a systematic, statistically grounded approach to planning experimental investigations. In environmental chemistry, where experiments often involve expensive analyses, limited sample availability, or time-sensitive sampling campaigns, the efficient allocation of experimental resources becomes essential [5]. The DoE methodology provides a rational framework for determining which experiments to perform, how many replicates are necessary, and how factors (i.e., variables) should be varied across runs to maximize the information extracted from each experimental run [6, 7].

Traditional experimental optimization in chemistry has long relied on the One-Variable-at-a-Time (OVAT) approach, in which a single experimental factor is varied while all others are held constant. While intuitively appealing due to its simplicity and apparent logical structure, this strategy suffers from fundamental limitations that become particularly problematic in complex environmental systems where synergistic and antagonistic interactions between variables are common. It is important to note that, although OVAT is often informally referred to as an "optimization" strategy, it cannot achieve true optimization in a multidimensional experimental space, as it inherently ignores interactions between factors. The OVAT approach implicitly assumes that experimental factors act independently, an assumption rarely satisfied in real environmental and analytical systems [1, 2].

Consider optimizing an analytical method for extracting pollutants from soil samples [3]. Extraction efficiency depends on multiple factors, including temperature, extraction time, solvent composition, solid-to-liquid ratio, and mixing intensity. An OVAT optimization might identify local optima for each factor individually but fail to discover that temperature and time interact synergistically; a moderate increase in both yields far better extraction than maximizing either alone. The researcher following the OVAT approach would systematically optimize temperature while holding time constant, then optimize time at the newly found optimal temperature, never realizing that the true optimum lies at a combination of moderate temperature and moderate time that was never tested. This ability to identify optima at points not actually sampled during the experimental campaign represents a crucial advantage of DoE over OVAT [4].

Furthermore, OVAT approaches are inherently inefficient, requiring numerous experiments to explore a multidimensional experimental space and often converging on suboptimal conditions. For example, if five factors are to be optimized and each factor is tested at five levels, a complete factorial exploration would require 5^5 = 3125 experiments, which is clearly impractical for most analytical applications. OVAT might reduce this to about 25 experiments (five levels for each of five factors tested sequentially), but this comes at the cost of missing all interaction effects and potentially converging on a local rather than a global optimum. Therefore, the inefficiency of the OVAT approach becomes increasingly severe as the number of factors increases, leading to a rapidly growing number of experiments with limited information gain [5].

On the other hand, DoE addresses these limitations by varying multiple factors simultaneously, using predefined experimental matrices derived from statistical principles. The key insight underlying DoE is that information about main effects and interactions can be extracted efficiently through strategic factor combinations rather than exhaustive enumeration. A crucial consequence of this approach is that DoE enables the identification of a true optimum within the experimental domain, which may lie at a combination of factor levels that has not been explicitly tested during the experimental campaign. This feature represents a fundamental advantage of DoE over OVAT strategies, which can only identify optima among the limited set of conditions that are directly explored. This approach offers several critical advantages over traditional OVAT methodology, including [5]:

- *Detection of factor interactions*: DoE enables the identification and quantification of synergistic and antagonistic effects between experimental variables revealing interactions invisible to OVAT approaches. A synergistic interaction occurs when the combined effect of two factors exceeds the sum of their individual effects; an antagonistic interaction shows the opposite pattern. Such interactions are pervasive in environmental and analytical chemistry, making their detection essential for true optimization.
- *Experimental efficiency*: by strategically distributing experiments across the experimental domain, DoE extracts maximum information with minimal experimental effort which is particularly important when samples are limited or analyses are expensive. A well-designed fractional factorial experiment can screen ten or more factors for significance in as few as 16 to 32 experiments, identifying the critical few factors that warrant detailed optimization while eliminating the trivial many from further consideration.
- *Simultaneous optimization of multiple responses*: environmental analyses often require balancing competing objectives such as sensitivity versus selectivity, analysis time versus precision, or cost versus throughput. DoE facilitates multi-response optimization using response surface methodology and desirability functions, identifying conditions that provide an acceptable compromise across multiple performance criteria rather than optimizing each criterion in isolation.
- *Statistical rigor*: DoE offers built-in replication and randomization strategies that enable proper estimation of experimental error and statistical significance of observed effects. Randomization protects against confounding by uncontrolled factors that may change during the experimental campaign, while replication provides an error estimate against which factor effects are judged significant.
- *Mathematical modeling*: experimental data from DoE studies can be used to construct predictive mathematical models of system behavior, enabling interpolation and optimization across the experimental domain. These response surface models express the relationship between factors and responses as polynomial equa-

tions, providing a quantitative map of the experimental landscape that guides identification of optimal operating conditions [6, 7].

The DoE framework includes a variety of experimental designs tailored to different research objectives and constraints. Understanding the purpose and characteristics of each design type allows for appropriate selection to meet specific environmental research needs.

Screening designs are used when many potential factors may affect the response, but resources allow only limited experimentation. Fractional factorial designs and Plackett-Burman designs efficiently identify which factors among many candidates have significant effects on the response. These designs sacrifice the ability to estimate some interaction effects in exchange for substantial reductions in the number of required experimental runs. For example, a Plackett-Burman design can screen up to 11 factors in just 12 experimental runs, helping to identify the most important factors for further detailed study [9].

Optimization designs focus on a smaller set of factors already identified as significant, enabling the construction of response surface models that map the relationship between factors and responses across the experimental domain. Central composite designs and Box-Behnken designs add axial and center points to factorial cores, allowing estimation of quadratic (curved) response surfaces that can identify true optima within the experimental domain. These designs support response surface methodology, which guides the identification of optimal operating conditions.

Finally, mixture designs address the special case in which factors represent proportions of mixture components must sum up to a fixed total. In environmental applications, mixture designs are valuable for optimizing solvent compositions for extraction, investigating the effects of pollutant mixtures on biological endpoints, or studying the effects of sediment composition on contaminant bioavailability. The constraint that component proportions sum to unity requires specialized design structures, such as simplex-lattice or simplex-centroid designs [10].

In environmental chemistry, DoE finds extensive application across method development, environmental system investigation, and process optimization. Method development represents a classic and widespread application of DoE. The development of extraction procedures for emerging contaminants in environmental matrices exemplifies this use case. Factors such as extraction solvent, extraction temperature, extraction time, sample mass, solvent volume, and pH must be optimized to maximize recovery of target analytes while minimizing co-extraction of interfering matrix components. A screening design is typically employed first to identify which factors significantly affect recovery, followed by an optimization design to locate optimal conditions for the critical factors [11–13].

Chromatographic method optimization similarly benefits from DoE approaches: e.g., the mobile phase composition, gradient profile, column temperature, and flow rate interact in complex ways to determine separation quality. DoE enables systematic exploration of these interactions, often revealing that adequate separation can be achieved more quickly or with less expensive solvents than OVAT optimization would suggest. The efficiency gains are particularly valuable in high-throughput environmental laboratories where even small reductions in analysis time can translate into substantial gains in throughput when applied across thousands of samples, yielding substantial cost savings and increased throughput [8].

Beyond analytical method development, DoE supports the investigation of environmental processes themselves. Studies of contaminant degradation in natural waters may use factorial designs to examine how temperature, pH, dissolved organic matter, and sunlight intensity interact to determine degradation rates. Remediation process optimization uses DoE to identify treatment conditions that maximize contaminant removal while minimizing cost and secondary environmental impacts. Even the design of environmental sampling strategies can benefit from DoE principles, balancing spatial coverage, temporal resolution, and resource constraints to maximize the information obtained from monitoring investments [14].

Recent advances have integrated DoE with computational approaches and machine learning, enabling more sophisticated experimental planning and adaptive designs that modify experimental strategies as results accumulate. Bayesian optimization approaches combine prior knowledge with experimental observations to guide sequential experimentation toward optimal regions with minimal wasted effort. These developments are particularly relevant for environmental monitoring networks too, where sampling strategies must balance spatial coverage, temporal resolution, and resource constraints while adapting to observed patterns and emerging concerns [9, 15, 16].

3 Exploratory Analysis: Principal Component Analysis

Principal Component Analysis represents one of the cornerstone techniques for the exploratory analysis of multivariate environmental data. Developed originally by Karl Pearson in 1901 [10] and formalized by Harold Hotelling in the 1930s [17], PCA has become ubiquitous in environmental chemometrics due to its ability to reveal structure in complex datasets, reduce dimensionality while preserving essential information, and provide intuitive graphical representations of multivariate relationships. No technique has been more widely applied to environmental data, and understanding PCA is a prerequisite to understanding most subsequent multivariate methods.

The fundamental premise of PCA is that environmental datasets, though measured across many variables, often contain redundant information due to correlations between measured parameters. When pollutants share common sources, when chemical species participate in related reactions, or when measurement methods respond to overlapping sample characteristics, the resulting variables become correlated. PCA exploits these correlations by constructing new orthogonal (uncorrelated) variables, i.e., the principal components, which are linear combinations of the original measurements.

These components are ordered by the amount of variance they explain: the first principal component captures the direction of maximum variance in the data, representing the single linear combination of original variables that exhibits the greatest spread across samples. The second principal component captures the maximum remaining variance orthogonal to the first, and so forth. Each successive component extracts progressively less variance, eventually reaching components that capture primarily measurement noise rather than meaningful signal [18].

Mathematically, PCA decomposes the original data matrix X (with samples as rows and variables as columns) into the product of two matrices: scores (T) representing sample coordinates in the new principal component space, and loadings (P) representing the contribution of each original variable to each principal component. This decomposition can be expressed as $X = TP' + E$, where E represents the residual matrix containing variation not explained by the retained components. Dimensionality reduction is achieved by retaining only a subset of the extracted components, referred to as retained components, which explain a substantial portion of the total variance, typically accounting for 80–90% of the cumulative explained variance. However, the appropriate threshold depends on data quality and analytical objectives [19].

Before applying PCA, data preprocessing strategies significantly influence the subsequent models and results. Centering (subtracting the mean of each variable) is standard practice, ensuring that principal components describe variation around the data center rather than distance from zero. Scaling, typically performed by dividing each variable by its standard deviation to give it unit variance, prevents variables with large numerical values from dominating the principal components solely because of their measurement scale. While autoscaling (centering plus unit-variance scaling) is the most common preprocessing method, there are alternatives [14, 20].

PCA serves multiple interconnected objectives in environmental data analysis, each contributing to a comprehensive understanding of complex environmental datasets:

- *Pattern recognition and sample clustering*: visualization of samples in reduced-dimensional space reveals natural groupings, gradients, and outliers. Environmental samples from similar sources or conditions typically cluster together in score plots. Spatial patterns in monitoring data become visible when samples are colored by location, while temporal patterns emerge when samples are marked by sampling date. Unexpected sample groupings may reveal unrecognized similarities or shared influences, while isolated outliers warrant investigation for unusual conditions, contamination events, or potential measurement errors [21].
- *Variable correlation structure*: the evaluation of the loadings identifies groups of correlated variables, suggesting common underlying factors or sources. In air quality studies, for example, correlated pollutants often share emission sources, traffic-related compounds cluster together, as do industrial emissions markers and secondary aerosol components. Understanding which variables are correlated informs interpretation of environmental processes and may suggest measurement redundancies that could be eliminated in future monitoring [22–25].
- *Noise reduction and signal enhancement*: by focusing on principal components that explain substantial variance, PCA effectively filters out measurement noise

concentrated in minor components. Reconstructing data using only significant components produces smoothed estimates that emphasize systematic variation while suppressing random fluctuations. This noise filtering proves particularly valuable for spectroscopic data where detector noise contributes variation unrelated to sample composition [26].

- *Data preprocessing for subsequent analyses*: PCA scores serve as inputs, providing orthogonal variables free of multicollinearity issues that can compromise regression and classification techniques. When original variables are highly correlated (as is common in spectroscopic data), applying methods such as multiple linear regression directly yields unstable results with inflated coefficient variances. Using PCA scores as predictors eliminates this collinearity by construction, as principal components are mathematically orthogonal [27, 28].
- *Outlier detection and quality control*: Samples that fall far from the main data cluster in score space, or that have large residuals after reconstruction from retained components, are potential outliers warranting investigation. The Hotelling's T^2 statistic quantifies distance from the score space center, while Q residuals (i.e., squared prediction error) measure distance from the PCA model plane. Samples that are extreme on either metric may reflect unusual environmental conditions, sampling or analytical errors, or fundamentally different sample types that were inappropriately included in the dataset. Figure 1 provides a schematic representation of these PCA outputs, including the scree plot for component selection, score and loading plots for pattern visualization, and diagnostic statistics (Hotelling's T^2 and Q residuals) for outlier detection [29].

PCA has been widely applied in all areas of environmental chemistry, providing insights not achievable with univariate analysis. In air quality assessment, PCA of atmospheric particulate composition data reveals pollution source signatures combinations of elements and compounds characteristic of specific emission sources. Traffic-related particles show high loadings for elements linked to brake wear (Cu, Sb, Ba), tire wear (Zn), and combustion (elemental carbon), while crustal dust contributions load heavily on lithophile elements (Al, Si, Ti, Fe). Secondary aerosol formation appears as correlated sulfate, nitrate, and ammonium, often anti-correlated with primary emission markers. These source signatures, identified through PCA loading patterns, form the basis for more quantitative source apportionment approaches [30–34]. In this context, source apportionment refers to the quantitative attribution of measured environmental pollution to its contributing sources.

Water quality monitoring employs PCA to identify spatial patterns in river systems, revealing how tributaries with different land use contribute distinct chemical signatures to receiving waters. Seasonal variations in lake chemistry become apparent when samples are plotted by season, often showing separation along components related to productivity, stratification, or runoff intensity. The influence of point source discharges manifests as gradients in score space, with samples downstream of discharge points displaced from upstream conditions along interpretable component directions. Groundwater investigations use PCA to distinguish water masses

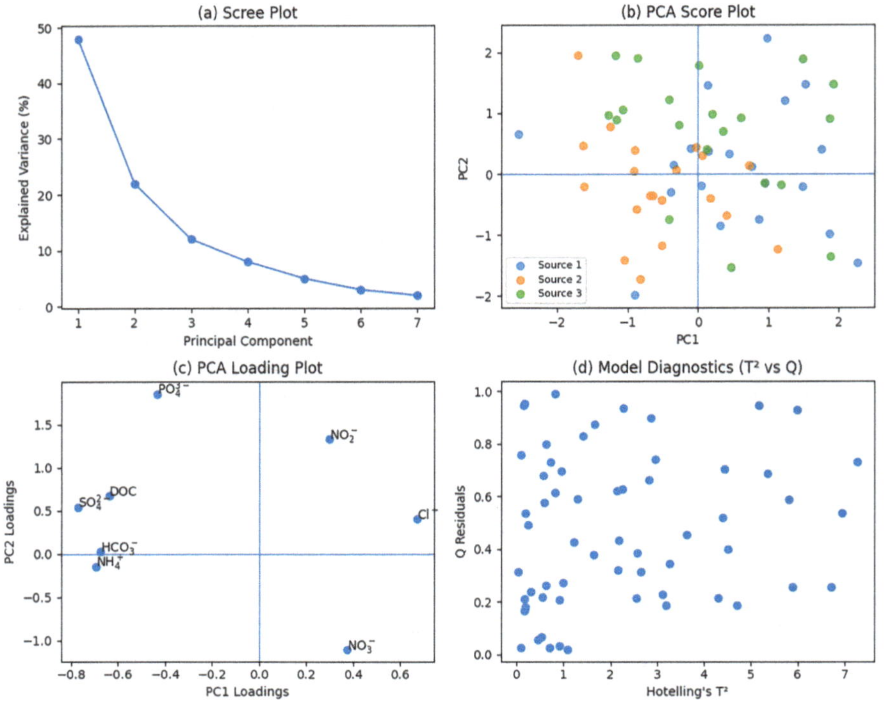

Fig. 1 Schematic representation of typical Principal Component Analysis outputs for simulated environmental water quality data. The figure illustrates (**a**) a scree plot supporting component selection, (**b**) a score plot showing sample distribution and clustering, (**c**) a loading plot highlighting groups of correlated variables, and (**d**) a Hotelling's T^2 versus Q residuals plot for outlier detection

with different recharge sources or flow paths, thereby supporting hydrogeological conceptual models [35–41].

Soil contamination studies use PCA to distinguish natural geochemical backgrounds from anthropogenic pollution and to identify spatial patterns of contamination spreading from source areas. Parent material variations create natural geochemical gradients that PCA can separate from superimposed contamination patterns. Industrial contamination often exhibits distinctive multi-element signatures distinct from those of urban diffuse pollution, enabling source attribution even when multiple contamination types overlap spatially. Depth profiles analyzed by PCA reveal historical contamination layers, with score patterns shifting at horizons corresponding to different deposition periods [30, 31, 42].

Applying PCA to environmental monitoring networks provides insights that inform both scientific understanding and practical management. Analysis across monitoring stations reveals which locations have similar pollution profiles, potentially enabling network rationalization without information loss. Temporal analysis identifies seasonal patterns, long-term trends, and episodic events that might go undetected in univariate time series. Combined spatiotemporal analysis shows how pollution patterns propagate through environmental systems, supporting source identification and impact assessment.

4 Data Deconvolution and Source Apportionment: Multivariate Curve Resolution

While Principal Component Analysis excels at exploratory analysis and dimensionality reduction, the components it generates are mathematical constructs optimized for variance explanation and may not correspond to physically meaningful environmental sources or processes. The orthogonality constraint, requiring each component to be uncorrelated with preceding components, is mathematically convenient but has no physical justification. Real pollution sources are rarely orthogonal: as an example, traffic and industrial emissions may be correlated due to shared spatial patterns or temporal activity cycles [32]. Multivariate Curve Resolution (MCR) addresses this limitation by recovering pure component profiles, the spectral signatures or chemical compositions of individual sources contributing to measured environmental signals, without imposing artificial orthogonality.

MCR techniques decompose a mixed signal into contributions from underlying pure components based on the bilinear model assumption. The measured data matrix D is modeled as the product of a concentration matrix C and a spectral (or compositional) profile matrix S: $D = CS' + E$. Each row of C contains the contributions of all sources to one sample, while each row of S contains the compositional profile of one source across all measured variables. Unlike PCA, which uses mathematical criteria (variance maximization, orthogonality) to determine components, MCR applies physically meaningful constraints that guide the resolution toward interpretable solutions representing real sources or species [33, 34].

The power of MCR lies in the constraints that can be applied during resolution; some of the most exploited are, for instance:

- *Non-negativity*: concentrations and most spectral intensities cannot be negative. This simple constraint dramatically reduces the range of mathematically possible solutions, guiding resolution toward physically realistic profiles. Non-negativity is nearly universally applicable and forms the foundation of most MCR applications [43].
- *Closure*: when data represent compositions or proportions, component contributions must sum to a constant total (e.g., 1 or 100%). Closure constraints ensure mass balance in source apportionment, preventing solutions where apparent source contributions exceed measured totals [34].
- *Selectivity and local rank*: when certain components are known to be absent in specific samples or spectral regions, this knowledge constrains the resolution. For example, if one sampling station is known to be unaffected by a particular industrial source, setting the contribution from that source to 0 in those samples helps resolve the contributions from the remaining sources [44].

- *Known profiles*: when pure-component spectra or source compositions are available from independent characterization, they can be fixed during resolution, reducing degrees of freedom and improving the resolution of unknown components.

MCR-ALS (Multivariate Curve Resolution—Alternating Least Squares) represents the most widely applied MCR variant due to its flexibility in incorporating diverse constraints. The algorithm iteratively optimizes concentration and spectral profiles, alternating between updating C while holding S fixed, then updating S while holding C fixed, applying constraints at each iteration, until convergence. The iterative nature of MCR-ALS provides flexibility in handling diverse data types and constraint combinations, though convergence to global rather than local optima cannot be guaranteed, and results may depend on initialization [33, 45].

In environmental chemometrics, source apportionment refers to the quantitative attribution of measured environmental pollution to its contributing sources, such as emission categories, processes, or material origins.

Source apportionment, i.e., the quantitative attribution of environmental pollution to contributing sources, is a paramount application of MCR methodology and related factor-analytic techniques. Atmospheric particulate matter measured at a receptor site originates from multiple emission sources: traffic exhaust, industrial emissions, construction dust, sea salt, biomass burning, and secondary aerosol formation. Similarly, organic contamination in sediments may derive from petroleum spills, urban runoff, agricultural drainage, and natural organic matter inputs. MCR-based source apportionment resolves these mixed signals into source-specific contributions, answering the critical question of which sources contribute how much to observed pollution levels [46, 47].

The application of MCR and related methods (including, for instance, Positive Matrix Factorization (PMF), which incorporates measurement uncertainty weighting) to source apportionment typically proceeds through several stages. Initial data assembly compiles measurements of multiple chemical species across numerous samples representing different times, locations, or conditions. The chemical species selected should include source-specific markers, compounds, or elements characteristic of particular sources, as well as species that distinguish between candidate sources. Data quality screening removes samples with excessive missing values or obvious measurement problems that would distort resolution [48, 49].

The core MCR-ALS analysis resolves source profiles (characteristic chemical compositions showing the relative abundances of measured species in emissions from each source) and source contributions (the mass each source contributes to each sample). Resolution quality depends on the distinctiveness of source profiles; for example, sources with very similar compositions may not separate cleanly, and on the adequacy of the constraints applied. Multiple solutions from different initializations should be compared to assess solution stability.

While atmospheric particulate source apportionment represents the most mature MCR application in environmental science, the methodology extends to diverse environmental problems. In water quality studies, MCR (as well as other chemometric techniques, such as PARAFAC—PARallel FACtor analysis [50]) decomposes fluorescence excitation-emission matrices (EEMs) into contributions from different classes of dissolved organic matter. Humic-like, protein-like, and anthropogenic fluorophores exhibit distinct spectral signatures that MCR can separate, enabling the tracking of organic matter sources through aquatic systems and the assessment of treatment process performance [51–53].

Spectroscopic analysis of contaminated sites uses MCR to identify and quantify distinct contaminant species that may not be separable by conventional univariate calibration. Petroleum contamination, for example, exhibits complex spectral signatures that change with weathering as volatile components evaporate and residual compounds undergo biodegradation and photodegradation. MCR can resolve fresh and weathered petroleum contributions, track natural attenuation progress, and distinguish petroleum from other organic matter [54–59].

Temporal series of environmental measurements analyzed by MCR and PMF reveal the evolution of pollution composition over time. Source contributions may vary with season (heating emissions peak in winter), day of week (traffic patterns differ on weekdays versus weekends), or meteorological conditions (secondary aerosol formation accelerates under stagnant, sunny conditions). Understanding these temporal patterns supports both source identification (patterns matching expected source behavior provide confirming evidence) and forecasting (known temporal patterns can be extrapolated to predict future conditions) [52, 60].

5 Cluster Analysis: Capabilities and Limitations

Cluster analysis comprises a family of unsupervised techniques that group samples based on their multivariate similarity without requiring prior knowledge of group memberships. The fundamental premise is that samples originating from similar environmental conditions or sources should exhibit similar chemical profiles and therefore cluster together in multivariate space. The application of clustering methods requires a sequence of methodological choices that strongly influence the resulting grouping. These choices can be logically organized into three main steps: the selection of an appropriate distance metric to quantify similarity between samples, the definition of a rule to compute distances between clusters, and the choice of the clustering algorithm itself. Understanding the role of each of these elements is essential for the correct application and interpretation of cluster analysis in environmental studies [61, 62].

Before applying any clustering algorithm, a measure of similarity or dissimilarity between samples must be defined. Distance metrics quantify how similar or different samples are in multivariate space and represent the foundation of all clustering approaches. Commonly used metrics include Euclidean distance, which measures straight-line distance in multivariate space; Manhattan distance, which sums absolute differences across variables; and Mahalanobis distance, which accounts for correlations among variables and differences in scale. Agglomerative (bottom-up) methods begin with each sample as its own cluster and progressively merge the most similar clusters until all samples belong to a single group. At each step, the algorithm identifies the pair of clusters closest by some criterion and merges them, recording the merge distance. Divisive (top-down) methods proceed in reverse, successively splitting clusters, though these are less commonly applied due to computational complexity [63].

Once distances between individual samples have been defined, hierarchical clustering methods require a rule to compute distances between clusters. These linkage criteria determine how clusters are merged and can lead to markedly different clustering structures. Common linkage methods include single linkage, complete linkage, average linkage, and Ward's method, each imposing different assumptions on cluster shape and compactness.

Linkage criteria determine how distances between clusters (rather than individual samples) are calculated, and different criteria can yield dramatically different clusterings from the same data. Single linkage (nearest neighbor) defines inter-cluster distance as the minimum distance between any member of one cluster and any member of another, producing elongated "chain" clusters. Complete linkage (furthest neighbor) uses the maximum inter-member distance, producing compact spherical clusters. Average linkage methods take the means of pairwise distances. Ward's method minimizes within-cluster variance at each merge, producing similarly-sized compact clusters. No single linkage method is universally optimal; the appropriate choice depends on the cluster shapes expected in the data [63, 64].

Clustering algorithms can then be classified according to their underlying approach. Hierarchical methods build a nested structure of clusters and are commonly visualized using dendrograms. Partitional methods, such as k-means clustering, divide samples into a predefined number of clusters by minimizing within-cluster variance. Extensions of these approaches include fuzzy clustering, which allows partial membership of samples to multiple clusters, and density-based methods that identify clusters as regions of high sample density separated by sparse regions. The algorithm begins with k initial cluster centroids (either randomly placed or selected by various initialization schemes), assigns each sample to its nearest centroid, recalculates the centroids as the means of the assigned samples, and iterates assignment and recalculation until assignments stabilize. The objective function minimized is the total within-cluster sum of squares, i.e., the sum across all clusters of the squared distances from each sample to its cluster centroid. K-means offers computational efficiency, which is advantageous for large datasets where hierarchical methods become impractical. However, several limitations warrant consideration. The number of clusters k must be specified a priori, though various indices (silhouette width, gap statistic, within-cluster sum of squares "elbow" plot) can guide selection. Results depend on initial centroid placement; multiple runs with different random initializations should be compared, and sophisticated initialization schemes (k-means++) improve convergence to good solutions. K-means assumes spherical clusters of similar size and performs poorly when natural clusters are elongated, vary in size, or are irregularly shaped [65].

Extensions address some k-means limitations. Fuzzy C-Means allows samples to belong partially to multiple clusters, producing membership degrees rather than hard assignments, thus being useful when cluster boundaries are genuinely ambiguous [66]. Density-based methods like DBSCAN (Density-Based Spatial Clustering of Applications with Noise) and OPTICS (Ordering Points To Identify the Clustering Structure) identify clusters as high-density regions separated by low-density areas, handle irregularly shaped clusters, and automatically determine the number of clusters [67]. Gaussian mixture models fit probabilistic models to cluster distributions, providing formal cluster assignment probabilities [68].

Despite its intuitive appeal, cluster analysis has significant limitations in environmental applications and should be regarded strictly as an exploratory tool. Determining the optimal number of clusters often relies on heuristic criteria or visual inspection of dendrograms, and different analysts examining the same dataset may legitimately reach different conclusions. Moreover, different choices of distance metrics, linkage criteria, or clustering algorithms can produce substantially different solutions from the same data, highlighting the methodological sensitivity of clustering results; when results vary with methodological choices, the natural cluster structure may be weak or ambiguous.

A fundamental problem is that clustering forces samples into discrete groups, potentially obscuring continuous gradients characteristic of many environmental phenomena. For this reason, clustering results should be interpreted in conjunction with independent information, such as spatial patterns, temporal trends, or known source characteristics, and their stability should be assessed through resampling or comparative analyses using multiple methods. A pollution gradient from source to background becomes artificially discretized, and boundary placement between categories is arbitrary [69–71]. Results also depend heavily on variable scaling choices: without appropriate preprocessing, variables with large numerical ranges may dominate clustering, at the expense of variables with smaller ranges. Critically, cluster algorithms will always produce clusters, even from random data with no natural groupings. The existence of a clustering solution does not demonstrate that meaningful clusters exist in the underlying population; validation through resampling, stability assessment, and external criteria is therefore essential. Given these limitations, cluster analysis is best employed as an exploratory tool, with results validated through complementary approaches and domain expertise. Agreement across clustering methods, stability under resampling, and correspondence with independent information, such as spatial patterns or known sources, strengthen confidence in the reality of the clusters [54–58, 72].

6 Supervised Classification Strategies

While unsupervised methods identify inherent data structure without external guidance, supervised classification uses known class memberships to build models that predict the class of unknown samples. The fundamental requirement is a training set of samples with known class assignments, from which the algorithm learns patterns that distinguish classes and can then be applied to classify new samples. Environmental applications include source identification from chemical fingerprints, quality classification according to regulatory standards, authenticity verification, and anomaly detection for pollution events.

Linear Discriminant Analysis (LDA) constructs linear boundaries separating predefined classes by finding the projection direction that maximizes between-class variance relative to within-class variance. This Fisher criterion ensures that classes are separated as much as possible while each class remains as compact as possible. LDA assumes multivariate normality within classes and equal covariance matrices across classes; these assumptions are often violated in environmental data, but are sufficiently robust for many practical applications. The resulting linear discriminant functions provide not only classifications but also insight into which variables best distinguish classes [59, 73].

On the other hand, Quadratic Discriminant Analysis (QDA) relaxes the equal covariance assumption, estimating separate covariance matrices for each class and thereby allowing quadratic (curved) decision boundaries. QDA offers greater flexibility when class distributions genuinely differ in shape or spread, but requires more training data to estimate the additional parameters reliably. With small training sets, QDA may overfit, performing well on training data, but poorly on new samples [74].

Partial Least Squares Discriminant Analysis (PLS-DA) extends PLS regression (described in the subsequent section) to classification by encoding class membership as a binary response matrix, with columns indicating membership in each class. PLS-DA excels when variables outnumber samples, e.g., common in spectroscopic applications where thousands of wavelengths characterize each sample, and when variables exhibit high collinearity. The method simultaneously performs dimensionality reduction, by extracting latent variables that capture class-relevant variation, and class prediction, making it particularly suitable for complex environmental fingerprinting applications. Variable importance measures from PLS-DA identify which features contribute most to classification, supporting interpretation and potential variable selection [75, 76].

Support Vector Machines (SVM) construct optimal separating hyperplanes that maximize the margin, i.e., the distance from the decision boundary to the nearest training samples of each class. This maximum margin principle provides theoretical guarantees of good generalization to new samples. Through kernel functions, SVM can model nonlinear class boundaries by implicitly mapping data to high-dimensional feature spaces where linear separation becomes possible. Common kernels include polynomial (for polynomial decision boundaries) and radial basis function (for highly flexible local boundaries). SVM's mathematical elegance and strong theoretical foundations have become a selected choice for many environmental classification problems [77, 78].

Random Forests aggregate predictions from numerous decision trees, each trained on bootstrap samples (random samples with replacement from training data) using random feature subsets at each split. This ensemble approach provides robust classification that is resistant to overfitting and can handle complex variable interactions without explicit specification. Individual trees may overfit, but averaging across hundreds or thousands of diverse trees yields stable, accurate predictions. Random forests additionally provide variable importance measures indicating which features contribute most to classification-valuable information for understanding environmental discrimination mechanisms and selecting variables for simpler models [79, 80]. Additionally, Soft Independent Modelling of Class Analogy (SIMCA) remains widely used in chemometrics for class modeling and authentication problems, where the goal is to determine whether a sample belongs to a defined class rather than discriminating between multiple classes [28, 74].

Supervised classification is widely applied across environmental disciplines, where samples are assigned to meaningful categories, and the objective is to classify new observations accordingly or to verify membership to predefined classes. One major application is source identification, in which pollution samples are classified based on chemical signatures to determine their origin [81–83]. Examples include petroleum spill forensics that differentiate crude oils from various production regions, refined products from different refinery processes, and weathered residues from fresh spills; water pollution studies that classify contamination sources using characteristic biomarker patterns; and air pollution receptor models that group days by dominant source influence [84–87].

Another important application is quality classification, which involves categorizing environmental samples according to established quality standards to support regulatory compliance and management decisions. This includes classifying water bodies by trophic status (oligotrophic, mesotrophic, or eutrophic) using chemical and biological indicators, assessing soil contamination levels relative to remediation thresholds, and assigning air quality categories based on multi-pollutant measurements [88–90].

Supervised classification is also central to authenticity and origin verification, where the aim is to distinguish natural from anthropogenic contributions, verify environmental claims, or detect adulteration. Typical examples include differentiating biogenic from petrogenic hydrocarbons, identifying the geographic origin of particulate matter using isotopic signatures, and verifying compliance with environmental regulations [91, 92].

Finally, temporal pattern recognition represents another key application, in which time series data are classified to identify pollution episodes, seasonal regimes, or anomalous conditions. Such approaches support both retrospective analyses and real-time monitoring, with event detection systems classifying incoming measurements as normal or anomalous to trigger alerts for unusual conditions that may indicate pollution events, equipment malfunctions, or emerging environmental concerns [93, 94].

Successful classification requires careful attention to model validation, which assesses how well models generalize to independent data rather than merely fitting the training set. Validation therefore represents a methodological prerequisite for all supervised learning approaches discussed in this section, regardless of the specific algorithm employed. In particular, overfitting, i.e., when models capture noise and anomalies in the training data rather than generalizable patterns, produces optimistic training performance that does not translate into reliable classification of new samples. In other words, overfitting refers to a situation in which a model captures noise or dataset-specific features rather than generalizable class-discriminating patterns, resulting in overly optimistic training performance and poor predictive ability on new samples. Therefore, rigorous validation protocols are essential for trustworthy classification models [95, 96].

Cross-validation strategies assess generalization by systematically withholding portions of data for testing. Leave-one-out cross-validation trains on all samples except one, classifies the held-out sample, and repeats for each sample. K-fold cross-validation partitions data into k subsets, training on k-1 and testing on the remaining fold, rotating through all folds. These approaches provide more realistic performance estimates than training set accuracy, though they still use the same data for both training and validation [97, 98].

External validation with truly independent test sets provides the most stringent performance evaluation. Test samples should represent the full range of conditions the model will encounter in application, e.g., different sampling times, locations, and environmental conditions beyond those in the training data. Temporal validation, training on historical data, and testing on subsequent samples prove particularly relevant for environmental monitoring applications where temporal trends and non-stationarity may compromise models trained on limited time windows. Repeated cross-validation and Monte Carlo validation address variability due to particular partition choices by averaging across multiple random partitions. Repeated k-fold validation runs the k-fold process multiple times with different random partitions. Monte Carlo validation repeatedly samples random training/test splits. These approaches provide more stable performance estimates and confidence intervals [99].

Classification metrics should align with application requirements, reflecting the relative cost of misclassification and the specific objectives of the environmental study. The selection of appropriate performance metrics therefore represents an integral component of the overall validation strategy rather than a purely technical detail. Overall accuracy (proportion correctly classified) provides a simple summary but may mislead when classes are imbalanced. Sensitivity (true-positive rate) and specificity (true-negative rate) characterize performance for each class separately. When missing contamination events is costly, sensitivity takes priority; when false alarms trigger expensive responses, specificity matters more. The area under the receiver operating characteristic curve (AUC-ROC) summarizes discrimination ability across all possible classification thresholds, providing a threshold-independent performance measure [100, 101].

7 Supervised Regression Methods

Regression methods extend supervised learning to quantitative prediction by modeling continuous relationships between predictor variables and numerical responses. While classification assigns samples to categories, regression predicts continuous values. Environmental applications include predicting pollutant concentrations from indirect measurements, developing calibration models for spectroscopic sensors, forecasting environmental quality from meteorological and emission data, and estimating difficult-to-measure parameters from easily measured surrogates [102, 103].

Multiple Linear Regression (MLR) models the response as a linear combination of predictor variables, estimating coefficients through ordinary least squares (OLS) to minimize the sum of squared residuals between predicted and observed responses. In multivariate environmental datasets, predictors are often affected by multicollinearity, i.e., strong correlations among explanatory variables, which can severely compromise the stability and interpretability of regression coefficients. While conceptually straightforward and widely understood, MLR struggles with correlated predictors, situations where the number of predictors exceeds the number of samples, and noisy data all common features of environmental datasets. Under these conditions, regression coefficients become unstable and highly sensitive to small variations in the data, limiting both interpretability and predictive reliability. Multicollinearity inflates coefficient standard errors, making individual predictor effects uncertain even when overall prediction remains good. When the number of predictors exceeds the number of samples, OLS has an infinite number of solutions and fails to converge [104–106].

Principal Component Regression (PCR) addresses multicollinearity by first reducing predictor dimensionality through Principal Component Analysis and then regressing the response on a subset of the resulting principal component scores. By construction, these scores are orthogonal, eliminating collinearity among predictors and enabling stable regression even when the original variables are strongly correlated. PCR can handle situations where the number of predictors exceeds the number of samples by retaining only a subset of components. However, PCR selects components solely based on predictor variance, without considering the response. Components capturing most of the predictor variance may be irrelevant to prediction if the response primarily reflects minor variance directions [107–109].

Partial Least Squares Regression (PLS-R) addresses this limitation by simultaneously accounting for both predictor and response variation during latent variable extraction. As a result, the extracted latent variables are explicitly optimized for prediction, making PLS-R particularly well suited to environmental datasets characterized by high dimensionality, collinearity, and limited sample size. PLS-R iteratively constructs latent variables that maximize the covariance between the predictors and the response, ensuring that the retained components are relevant to prediction. From a geometric perspective, PLS-R latent variables represent rotations of principal components toward directions correlated with the response. PLS-R has become the method of choice for calibrating spectroscopic measurements, where thousands of wavelength variables must be related to analyte concentrations, and the collinearity, high dimensionality, and noise of spectral data would defeat simpler approaches [110–112].

The number of latent variables (or components) retained in PCR and PLS-R critically affects model performance and must be selected through appropriate validation procedures rather than based solely on explained variance or goodness-of-fit measures. Too few components underfit, failing to capture response-relevant variation and yielding poor predictions. Too many components overfit, incorporating noise that degrades prediction on new samples. Cross-validation provides a principled framework for component selection: prediction error initially decreases as components capture response-relevant information, reaches a minimum at optimal model complexity, and then increases as additional components incorporate noise rather than signal. The number of components minimizing cross-validated error provides a principled selection criterion [106, 113].

Regression methods underpin a wide range of environmental quantification applications by enabling the prediction of continuous variables from measured data. A prominent example is spectroscopic calibration, in which infrared, Raman, UV-visible, or fluorescence spectra are related to pollutant concentrations, enabling rapid analysis without the labor-intensive wet-chemistry methods [111, 114]. Near-infrared spectroscopy calibrated for soil organic carbon or nutrient content supports precision agriculture, mid-infrared models quantify petroleum hydrocarbons in contaminated soils, and fluorescence signatures are used to estimate dissolved organic carbon and specific pollutant classes in natural waters. In each case, regression transforms complex spectral signals into actionable quantitative environmental information [115–118].

Another important application is the development of surrogate parameters, in which regression models are used to estimate difficult or time-consuming measurements from simpler, faster alternatives. For instance, biochemical oxygen demand, which traditionally requires a 5-day incubation, can be predicted from rapid optical measurements, while total organic carbon often correlates strongly with UV absorbance. Similarly, heavy metal bioavailability, which normally involves complex extraction procedures, can be related to more easily measured extractable fractions. These surrogate relationships reduce analytical costs and enable more frequent and widespread environmental monitoring [119–121].

Model validation in regression follows the same general principles introduced earlier in the chapter and parallels those used in classification. Internal validation through cross-validation provides an estimate of generalization error, while external validation with independent test sets represents the most stringent assessment of predictive reliability. Regression diagnostics, including root mean square error of calibration (RMSEC), cross-validation (RMSECV), and prediction (RMSEP), along with R^2 values and bias assessment, characterize model accuracy. Examination of residuals for systematic patterns, heteroscedasticity, or influential outliers supports model diagnosis and improvement. Prediction intervals quantifying uncertainty around point predictions provide the complete picture needed for environmental decision-making [97, 98].

8 Deep Learning and Advanced Machine Learning Approaches

The emergence of deep learning, including neural network architectures with multiple hidden layers capable of learning hierarchical representations, can be viewed as a natural extension of classical multivariate analysis approaches for environmental data analysis rather than as a separate methodological domain. From a conceptual perspective, many deep learning architectures generalize ideas already familiar in chemometrics. Autoencoders can be interpreted as nonlinear extensions of Principal Component Analysis, where dimensionality reduction is achieved without the constraint of linearity. Similarly, supervised deep learning models share conceptual similarities with Partial Least Squares approaches, as both aim to extract latent representations that are maximally informative with respect to a target response. These approaches, distinguished by their capacity to learn features directly from raw data rather than relying on human-engineered variables, offer unprecedented capabilities for processing complex environmental signals. Satellite imagery spanning numerous spectral bands, continuous sensor streams generating millions of measurements, hyperspectral data with hundreds of wavelength channels, and multimodal datasets integrating diverse data types all become accessible to deep learning analysis [122, 123]. In this sense, deep learning strategies represent a response to the increasing volume, dimensionality, and heterogeneity of environmental data generated in the modern big data era.

Convolutional Neural Networks (CNNs) excel at extracting spatial features from images and can be adapted to one-dimensional signals, such as spectra. Convolutional layers apply learnable filters that detect local patterns, such as edges, textures, and spectral features, while pooling layers provide translation invariance and dimensionality reduction. Deeper layers combine lower-level features into increasingly abstract representations. Environmental applications include analyzing satellite and aerial imagery for land cover classification, change detection, and pollution mapping; processing hyperspectral data for vegetation stress assessment and contamination identification; and extracting features from spectroscopic signals for calibration and classification. The key advantage is automatic feature learning: CNNs discover relevant features from data, eliminating the need for manual feature engineering that limits traditional methods [124–128].

Recurrent Neural Networks (RNNs) and their variants (Long Short-Term Memory, LSTM; Gated Recurrent Unit, GRU) process sequential data by maintaining memory of past inputs. These architectures are valuable for time-series analysis of environmental monitoring data, where current conditions depend on historical evolution. Air quality evolution over days, seasonal water quality dynamics, and ecosystem response trajectories involve temporal dependencies that RNNs can capture. Long-term dependencies, e.g., how conditions months or years ago influence current states, are naturally modeled by LSTM and GRU variants that explicitly manage memory over extended sequences [91, 129–131].

Autoencoders learn compressed representations of input data through an encoder that reduces dimensionality and a decoder that reconstructs the original input. This approach conceptually extends PCA's linear dimensionality reduction to capture nonlinear structure. The compressed representation in the bottleneck layer captures essential data features while discarding noise. Environmental applications include anomaly detection, where unusual pollution events produce large reconstruction errors; dimensionality reduction, an alternative to PCA that can capture nonlinear structure; denoising of sensor measurements by training to reconstruct clean signals from noisy inputs; and missing data imputation by learning to complete incomplete measurements [132–134].

Finally, self-attention models and transformers represent recent advances that enable models to focus on relevant portions of the input when making predictions. Originally developed for language processing, attention mechanisms are increasingly applied to environmental time series and spatial data. Self-attention allows models to weigh the relevance of different time points when predicting current conditions or different spatial locations when classifying satellite image pixels. Transformers, being architectures built entirely on attention, can process long sequences efficiently and capture complex dependencies that challenge RNNs [135–137].

Deep learning provides important advantages for environmental chemometrics by enabling models to learn directly from complex data. Unlike traditional approaches that rely on expert-defined features, deep networks automatically discover relevant patterns, including nonlinear relationships such as thresholds, saturation effects, and interacting influences that are common in environmental systems. Deep learning also supports the integration of diverse data sources, enabling the combination of satellite imagery, ground measurements, meteorological variables, and emission data within unified predictive frameworks. Once trained, these models scale efficiently to very large datasets, making them well-suited for applications such as real-time monitoring and analysis of extensive sensor networks or long-term satellite archives [138, 139].

Despite these strengths, several challenges limit the routine application of deep learning in environmental studies and mirror concerns already encountered with classical chemometric models, albeit on a larger scale. Large training datasets are often required, yet environmental data are frequently scarce, costly, or limited in historical extent, with transfer learning offering only partial relief. Interpretability remains a major concern, as the black-box nature of deep networks can hinder scientific understanding and regulatory acceptance, even though explainable AI methods can provide some insight. In addition, the high capacity of deep models increases the risk of overfitting, necessitating careful regularization and validation, and their training demands substantial computational resources, which may be prohibitive for some research settings [140, 141].

9 Concluding Remarks and Future Perspectives

The chemometric techniques described in this chapter constitute an integrated methodological toolkit for environmental data analysis, with each approach offering specific capabilities suited to different stages of environmental investigation and different analytical objectives. Design of Experiments optimizes data collection, ensuring that limited experimental resources yield the maximum amount of information. Exploratory methods, such as Principal Component Analysis, reveal data structure and inform subsequent analyses by identifying patterns, outliers, and relationships among variables. Multivariate Curve Resolution resolves mixed signals into physically meaningful source contributions, enabling quantitative source apportionment. Classification methods build predictive models for categorical outcomes, such as source identification, quality classification, and anomaly detection. Regression methods extend prediction to continuous quantities, such as concentration estimation, parameter forecasting, and surrogate parameter development (i.e., estimating difficult measurements from simpler alternatives). Emerging deep learning approaches expand capabilities to complex, high-dimensional data streams beyond the reach of traditional methods.

Throughout all applications, rigorous validation ensures that models generalize beyond development data and represent genuine environmental information rather than artefacts of noise, overfitting, or dataset-specific structure. The temptation to trust impressive training performance must yield to honest assessment through cross-validation and, ideally, external validation with truly independent samples. Models that fail validation should be rejected or refined rather than applied, regardless of how elegant their construction or how impressive their calibration statistics are.

The selection of appropriate techniques depends on research objectives, data characteristics, and application requirements. Often, the most comprehensive understanding emerges from the integration of complementary approaches, rather than from the isolated application of any single technique. An initial exploratory analysis with PCA reveals data structure and guides subsequent modeling choices, while thoughtful experimental design upstream determines the quality and informativeness of the data entering all subsequent analyses. Classification for source identification may be followed by regression to quantify source contributions. Validation informs confidence in all results. The chemometrician's expertise therefore lies not in the mechanical application of a single favored method, but in the informed selection, critical evaluation, and integration of techniques matched to specific data characteristics, research objectives, and decision-making contexts.

As environmental monitoring generates increasingly complex data through advanced instrumentation and expanded networks, chemometric methods will remain essential for transforming measurements into environmental knowledge. Sensor networks generating continuous multi-parameter streams, hyperspectral imagers capturing hundreds of wavelengths across landscapes, and mass spectrometers profiling thousands of compounds in each sample all produce data volumes and complexities that demand multivariate analysis. The challenge lies in deciding whether to apply chemometrics and, if so, how best to apply increasingly sophisticated methods to increasingly complex data [142, 143].

The ongoing integration of chemometric principles with artificial intelligence and machine learning promises continued methodological advancement. Deep learning extends pattern recognition to data types and complexities beyond traditional methods. Automated machine learning (AutoML) streamlines method selection and hyperparameter optimization. Explainable AI bridges the gap between black-box predictions and interpretable insights. Integration with environmental process models combines data-driven pattern recognition with mechanistic understanding [144–146].

Yet the fundamental imperative remains unchanged: applying appropriate multivariate tools, with proper validation, to extract reliable information from complex environmental data. This "How?" of environmental chemometrics ultimately determines whether the growing volume of environmental measurements translates into genuine understanding and effective environmental stewardship.

Glossary

- **Chemometrics**: the application of mathematical and statistical methods to extract meaningful information from chemical data, enabling pattern recognition, classification, and prediction from complex multivariate datasets.
- **Design of Experiments (DoE)**: a systematic approach to planning experimental investigations that enables efficient exploration of factor effects and interactions while minimizing the number of required experiments.
- **Principal Component Analysis (PCA)**: an exploratory technique that reduces data dimensionality by transforming correlated variables into a smaller set of uncorrelated components (principal components) that capture the maximum variance in the data.
- **Source apportionment**: The quantitative attribution of environmental pollution to its contributing sources, typically achieved through receptor modeling techniques such as Multivariate Curve Resolution (MCR) or Positive Matrix Factorization (PMF).
- **Supervised learning**: a machine learning paradigm in which models are trained using labeled data with known outcomes, enabling the prediction of class membership (classification) or continuous values (regression) for new samples.
- **Unsupervised learning**: a machine learning paradigm in which models identify inherent patterns and structures in data without prior knowledge of class labels or target values.
- **Machine learning**: a branch of artificial intelligence in which algorithms learn patterns from data to make predictions or decisions, encompassing both traditional statistical learning methods and modern deep learning approaches.

(continued)

- **Overfitting**: a modeling error that occurs when a model captures noise or random fluctuations in training data rather than the underlying pattern, resulting in poor predictive performance on new, unseen samples.
- **Cross-validation**: a resampling procedure used to evaluate model performance by partitioning data into training and validation subsets, ensuring that predictive ability is assessed on data not used during model development.
- **Latent variable**: an unobserved (hidden) variable derived from measured variables that captures underlying structure in the data; principal components and PLS factors are examples of latent variables used in chemometrics.

Further Reading

- Brereton, R.G. (2018). Chemometrics: Data Driven Extraction for Science (2nd ed.). Wiley.
- Varmuza, K., & Filzmoser, P. (2016). Introduction to Multivariate Statistical Analysis in Chemometrics. CRC Press.
- James, G., Witten, D., Hastie, T., & Tibshirani, R. (2021). An Introduction to Statistical Learning with Applications in R (2nd ed.). Springer.
- Prince, S.J.D. (2023). Understanding Deep Learning. MIT Press - freely available at https://udlbook.github.io/udlbook/.

References

1. Fernandes FAN (2024) Experimental design for chemometrics: best practices. In: Chemometrics. Elsevier, pp 39–59. https://doi.org/10.1016/B978-0-443-21493-6.00003-4
2. Deming SN, Morgan SL (1993) Experimental design: a chemometric approach - ISBN 0444891110.
3. Nguyen TTN, Baduel C (2023) Optimization and validation of an extraction method for the analysis of multi-class emerging contaminants in soil and sediment. J Chromatogr A 1710:464287. https://doi.org/10.1016/j.chroma.2023.464287
4. Leardi R (2009) Experimental design in chemistry: a tutorial. Anal Chim Acta 652:161–172. https://doi.org/10.1016/j.aca.2009.06.015
5. Wall BJ, Koeritz MT, Stanley LM, VanVeller B (2025) A practical start-up guide for synthetic chemists to implement Design of Experiments (DoE). ACS Catal 15:8885–8893. https://doi.org/10.1021/acscatal.5c01626
6. Eriksson L, Johansson E, Kettaneh-Wold N, Wikström C, Wold S (2000) Design of experiments, principles and applications - ISBN 9197373001.
7. Brereton RG (2003) Chemometrics: data analysis for the laboratory and chemical plant - ISBN 9780471489771.

8. Jakeman AJ, Letcher RA, Norton JP (2006) Ten iterative steps in development and evaluation of environmental models. Environ Model Softw 21:602–614. https://doi.org/10.1016/j.envsoft.2006.01.004
9. Shahriari B, Swersky K, Wang Z, Adams RP, de Freitas N (2016) Taking the human out of the loop: a review of Bayesian optimization. Proc IEEE 104:148–175. https://doi.org/10.1109/JPROC.2015.2494218
10. Pearson K (1901) LIII. On lines and planes of closest fit to systems of points in space. London, Edinburgh, Dublin Philos Mag J Sci 2:559–572. https://doi.org/10.1080/14786440109462720
11. Dejaegher B, Vander Heyden Y (2011) Experimental designs and their recent advances in set-up, data interpretation, and analytical applications. J Pharm Biomed Anal 56:141–158. https://doi.org/10.1016/j.jpba.2011.04.023
12. Ebrahimi-Najafabadi H, Leardi R, Jalali-Heravi M (2014) Experimental Design in Analytical Chemistry—Part II: applications. J AOAC Int 97:12–18. https://doi.org/10.5740/jaoacint.SGEEbrahimi2
13. Dejaegher B, Vander Heyden Y (2007) Ruggedness and robustness testing. J Chromatogr A 1158:138–157. https://doi.org/10.1016/j.chroma.2007.02.086
14. Mishra P, Biancolillo A, Roger JM, Marini F, Rutledge DN (2020) New data preprocessing trends based on ensemble of multiple preprocessing techniques. TrAC Trends Anal Chem 132:116045. https://doi.org/10.1016/j.trac.2020.116045
15. Joshi PB (2023) Navigating with chemometrics and machine learning in chemistry. Artif Intell Rev 56:9089–9114. https://doi.org/10.1007/s10462-023-10391-w
16. Houhou R, Bocklitz T (2021) Trends in artificial intelligence, machine learning, and chemometrics applied to chemical data. Anal Sci Adv 2:128–141. https://doi.org/10.1002/ansa.202000162
17. Hotelling H (1933) Analysis of a complex of statistical variables into principal components. J Educ Psychol 24:417–441. https://doi.org/10.1037/h0071325
18. Jolliffe IT, Cadima J (2016) Principal component analysis: a review and recent developments. Philos Trans R Soc A Math Phys Eng Sci 374:20150202. https://doi.org/10.1098/rsta.2015.0202
19. Bro R, Smilde AK (2014) Principal component analysis. Anal Methods 6:2812–2831. https://doi.org/10.1039/C3AY41907J
20. van den Berg RA, Hoefsloot HC, Westerhuis JA, Smilde AK, van der Werf MJ (2006) Centering, scaling, and transformations: improving the biological information content of metabolomics data. BMC Genomics 7:142. https://doi.org/10.1186/1471-2164-7-142
21. Vega M, Pardo R, Barrado E, Debán L (1998) Assessment of seasonal and polluting effects on the quality of river water by exploratory data analysis. Water Res 32:3581–3592. https://doi.org/10.1016/S0043-1354(98)00138-9
22. Kanellopoulos PG, Chrysochou E, Koukoulakis K, Vasileiadou E, Kizas C, Savvides C, Bakeas E (2020) Secondary organic aerosol tracers and related polar organic compounds between urban and rural areas in the eastern Mediterranean region: source apportionment and the influence of atmospheric oxidants. Environ Sci Process Impacts. https://doi.org/10.1039/D0EM00238K
23. Popovicheva O, Padoan S, Schnelle- Kreis J, Nguyen D-L, Adam T, Kistler M, Steinkogler T, Kasper-Giebl A, Zimmermann R, Chubarova N (2020) Spring aerosol in urban atmosphere of megacity: analytical and statistical assessment for source impacts. aerosol air Qual Res 20:702–719. https://doi.org/10.4209/aaqr.2019.08.0412
24. Cruz LPS, Santos DF, dos Santos IF, Gomes ÍVS, Santos AVS, Souza KSPP (2020) Exploratory analysis of the atmospheric levels of BTEX, criteria air pollutants and meteorological parameters in a tropical urban area in northeastern Brazil. Microchem J 152:104265. https://doi.org/10.1016/j.microc.2019.104265
25. Chauhan BVS, Berg MJ, Sharma A, Smallbone KL, Wyche KP (2025) Temporal and machine learning-based principal component and clustering analysis of VOCs and their role in urban air pollution and ozone formation. Atmosphere (Basel) 16:724. https://doi.org/10.3390/atmos16060724

26. Perkins JH, Hasenoehrl EJ, Griffiths PR (1992) The use of principal component analysis for the structural interpretation of mid-infrared spectra. Chemom Intell Lab Syst 15:75–86. https://doi.org/10.1016/0169-7439(92)80028-3
27. Alladio E, Martyna A, Salomone A, Pirro V, Vincenti M, Zadora G (2017) Evaluation of direct and indirect ethanol biomarkers using a likelihood ratio approach to identify chronic alcohol abusers for forensic purposes. Forensic Sci Int 271:13–22. https://doi.org/10.1016/j.forsciint.2016.12.019
28. Tominaga Y (1999) Comparative study of class data analysis with PCA-LDA, SIMCA, PLS, ANNs, and k-NN. Chemom Intell Lab Syst 49:105–115. https://doi.org/10.1016/S0169-7439(99)00034-9
29. Gao S, Tian R, Chen P (2022) Principal component analysis for process monitoring in distributed system environment. Concurr Comput Pract Exp 34. https://doi.org/10.1002/cpe.5309
30. Su C, Meng J, Zhou Y, Bi R, Chen Z, Diao J, Huang Z, Kan Z, Wang T (2022) Heavy metals in soils from intense industrial areas in South China: spatial distribution, source apportionment, and risk assessment. Front Environ Sci 10. https://doi.org/10.3389/fenvs.2022.820536
31. Davis HT, Marjorie Aelion C, McDermott S, Lawson AB (2009) Identifying natural and anthropogenic sources of metals in urban and rural soils using GIS-based data, PCA, and spatial interpolation. Environ Pollut 157:2378–2385. https://doi.org/10.1016/j.envpol.2009.03.021
32. Stanimirova I, Tauler R, Walczak B (2011) A comparison of positive matrix factorization and the weighted multivariate curve resolution method. Application to environmental data. Environ Sci Technol 45:10102–10110. https://doi.org/10.1021/es201024m
33. Tauler R (2021) Multivariate curve resolution of multiway data using the multilinearity constraint. J Chemom 35. https://doi.org/10.1002/cem.3279
34. Jaumot J, de Juan A, Tauler R (2015) MCR-ALS GUI 2.0: new features and applications. Chemom Intell Lab Syst 140:1–12. https://doi.org/10.1016/j.chemolab.2014.10.003
35. Gazzaz NM, Yusoff MK, Ramli MF, Aris AZ, Juahir H (2012) Characterization of spatial patterns in river water quality using chemometric pattern recognition techniques. Mar Pollut Bull 64:688–698. https://doi.org/10.1016/j.marpolbul.2012.01.032
36. Yang S, Liang M, Qin Z, Qian Y, Li M, Cao Y (2021) A novel assessment considering spatial and temporal variations of water quality to identify pollution sources in urban rivers. Sci Rep 11:8714. https://doi.org/10.1038/s41598-021-87671-4
37. Zeinalzadeh K, Rezaei E (2017) Determining spatial and temporal changes of surface water quality using principal component analysis. J Hydrol Reg Stud 13:1–10. https://doi.org/10.1016/j.ejrh.2017.07.002
38. Selle B, Schwientek M, Lischeid G (2013) Understanding processes governing water quality in catchments using principal component scores. J Hydrol 486:31–38. https://doi.org/10.1016/j.jhydrol.2013.01.030
39. Ouyang Y (2005) Evaluation of river water quality monitoring stations by principal component analysis. Water Res 39:2621–2635. https://doi.org/10.1016/j.watres.2005.04.024
40. Xu W, Duan L, Wen X, Li H, Li D, Zhang Y, Zhang H (2022) Effects of seasonal variation on water quality parameters and eutrophication in Lake Yangzong. Water 14:2732. https://doi.org/10.3390/w14172732
41. Ghani J, Ullah Z, Nawab J, Iqbal J, Waqas M, Ali A, Almutairi MH, Peluso I, Mohamed HRH, Shah M (2022) Hydrogeochemical characterization, and suitability assessment of drinking groundwater: application of geostatistical approach and geographic information system. Front Environ Sci 10. https://doi.org/10.3389/fenvs.2022.874464
42. Aidoo EN, Appiah SK, Awashie GE, Boateng A, Darko G (2021) Geographically weighted principal component analysis for characterising the spatial heterogeneity and connectivity of soil heavy metals in Kumasi, Ghana. Heliyon 7:e08039. https://doi.org/10.1016/j.heliyon.2021.e08039
43. Bro R, De Jong S (1997) A fast non-negativity-constrained least squares algorithm. J Chemom 11:393–401. https://doi.org/10.1002/(SICI)1099-128X(199709/10)11:5<393::AID-CEM483>3.0.CO;2-L

44. Tauler R, Smilde A, Kowalski B (1995) Selectivity, local rank, three-way data analysis and ambiguity in multivariate curve resolution. J Chemom 9:31–58. https://doi.org/10.1002/cem.1180090105
45. Tauler R (1995) Multivariate curve resolution applied to second order data. Chemom Intell Lab Syst 30:133–146. https://doi.org/10.1016/0169-7439(95)00047-X
46. Alier M, van Drooge BL, Dall'Osto M, Querol X, Grimalt JO, Tauler R (2013) Source apportionment of submicron organic aerosol at an urban background and a road site in Barcelona (Spain) during SAPUSS. Atmos Chem Phys 13:10353–10371. https://doi.org/10.5194/acp-13-10353-2013
47. Tauler R, Viana M, Querol X, Alastuey A, Flight RM, Wentzell PD, Hopke PK (2009) Comparison of the results obtained by four receptor modelling methods in aerosol source apportionment studies. Atmos Environ 43:3989–3997. https://doi.org/10.1016/j.atmosenv.2009.05.018
48. Brown SG, Eberly S, Paatero P, Norris GA (2015) Methods for estimating uncertainty in PMF solutions: examples with ambient air and water quality data and guidance on reporting PMF results. Sci Total Environ 518–519:626–635. https://doi.org/10.1016/j.scitotenv.2015.01.022
49. Paatero P, Tapper U (1994) Positive matrix factorization: a non-negative factor model with optimal utilization of error estimates of data values. Environmetrics 5:111–126. https://doi.org/10.1002/env.3170050203
50. Bro R (1997) PARAFAC. Tutorial and applications. Chemom Intell Lab Syst 38:149–171. https://doi.org/10.1016/S0169-7439(97)00032-4
51. Zhang X, Marcé R, Armengol J, Tauler R (2014) Distribution of dissolved organic matter in freshwaters using excitation emission fluorescence and multivariate curve resolution. Chemosphere 111:120–128. https://doi.org/10.1016/j.chemosphere.2014.03.009
52. Malik A, Tauler R (2013) Extension and application of multivariate curve resolution-alternating least squares to four-way quadrilinear data-obtained in the investigation of pollution patterns on Yamuna River, India—a case study. Anal Chim Acta 794:20–28. https://doi.org/10.1016/j.aca.2013.07.047
53. Kumar K, Mishra AK (2012) Application of 'multivariate curve resolution alternating least square (MCR–ALS)' analysis to extract pure component synchronous fluorescence spectra at various wavelength offsets from total synchronous fluorescence spectroscopy (TSFS) data set of dilute. Chemom Intell Lab Syst 116:78–86. https://doi.org/10.1016/j.chemolab.2012.04.015
54. Lee S, Kim J, Hwang J, Lee E, Lee K-J, Oh J, Park J, Heo T-Y (2020) Clustering of time series water quality data using dynamic time warping: a case study from the Bukhan River water quality monitoring network. Water 12:2411. https://doi.org/10.3390/w12092411
55. Halder S, Bhattacharya S, Roy MB, Roy PK (2023) Application of fuzzy C-means clustering and fuzzy EDAS to assess groundwater irrigation suitability and prioritization for agricultural development in a complex hydrogeological basin. Environ Sci Pollut Res 30:57529–57557. https://doi.org/10.1007/s11356-023-26394-7
56. Güler C, Thyne GD, McCray JE, Turner KA (2002) Evaluation of graphical and multivariate statistical methods for classification of water chemistry data. Hydrogeol J 10:455–474. https://doi.org/10.1007/s10040-002-0196-6
57. Bu J, Liu W, Pan Z, Ling K (2020) Comparative study of Hydrochemical classification based on different hierarchical cluster analysis methods. Int J Environ Res Public Health 17:9515. https://doi.org/10.3390/ijerph17249515
58. Cloutier V, Lefebvre R, Therrien R, Savard MM (2008) Multivariate statistical analysis of geochemical data as indicative of the hydrogeochemical evolution of groundwater in a sedimentary rock aquifer system. J Hydrol 353:294–313. https://doi.org/10.1016/j.jhydrol.2008.02.015
59. FISHER RA (1936) The use of multiple measurements in taxonomic problems. Ann Eugenics 7:179–188. https://doi.org/10.1111/j.1469-1809.1936.tb02137.x

60. Izadmanesh Y, Ghasemi JB, Tauler R (2017) Receptor modeling of environmental aerosol data using MLPCA-MCR-ALS. Chemom Intell Lab Syst 167:50–62. https://doi.org/10.1016/j.chemolab.2017.05.008
61. Drab K, Daszykowski M (2014) Clustering in analytical chemistry. J AOAC Int 97:29–38. https://doi.org/10.5740/jaoacint.SGEDrab
62. Kaufman L, Rousseeuw PJ (1990) Finding groups in data. Wiley. https://doi.org/10.1002/9780470316801
63. Ward JH (1963) Hierarchical grouping to optimize an objective function. J Am Stat Assoc 58:236–244. https://doi.org/10.1080/01621459.1963.10500845
64. Ran X, Xi Y, Lu Y, Wang X, Lu Z (2023) Comprehensive survey on hierarchical clustering algorithms and the recent developments. Artif Intell Rev 56:8219–8264. https://doi.org/10.1007/s10462-022-10366-3
65. Ikotun AM, Ezugwu AE, Abualigah L, Abuhaija B, Heming J (2023) K-means clustering algorithms: a comprehensive review, variants analysis, and advances in the era of big data. Inf Sci (Ny) 622:178–210. https://doi.org/10.1016/j.ins.2022.11.139
66. Nayak J, Naik B, Behera HS (2015) Fuzzy C-Means (FCM) clustering algorithm: a decade review from 2000 to 2014, pp. 133–149. https://doi.org/10.1007/978-81-322-2208-8_14
67. Hajihosseinlou M, Maghsoudi A, Ghezelbash R (2024) A comprehensive evaluation of OPTICS, GMM and K-means clustering methodologies for geochemical anomaly detection connected with sample catchment basins. Geochemistry 84:126094. https://doi.org/10.1016/j.chemer.2024.126094
68. Kartakoullis A, Caporaso N, Whitworth MB, Fisk ID (2025) Gaussian mixture model clustering allows accurate semantic image segmentation of wheat kernels from near-infrared hyperspectral images. Chemom Intell Lab Syst 259:105341. https://doi.org/10.1016/j.chemolab.2025.105341
69. D'Urso P, De Giovanni L, Massari R (2015) Time series clustering by a robust autoregressive metric with application to air pollution. Chemom Intell Lab Syst 141:107–124. https://doi.org/10.1016/j.chemolab.2014.11.003
70. Iizuka A, Shirato S, Mizukoshi A, Noguchi M, Yamasaki A, Yanagisawa Y (2014) A cluster analysis of constant ambient air monitoring data from the Kanto region of Japan. Int J Environ Res Public Health 11:6844–6855. https://doi.org/10.3390/ijerph110706844
71. Gomez MLS, Martin MCR (1987) Application of cluster analysis to identify sources of airborne particles. Atmos Environ 21:1521–1527. https://doi.org/10.1016/0004-6981(87)90314-3
72. Yao R, Yan Y, Wei C, Luo M, Xiao Y, Zhang Y (2022) Hydrochemical characteristics and groundwater quality assessment using an integrated approach of the PCA, SOM, and fuzzy c-means clustering: a case study in the northern Sichuan Basin. Front Environ Sci 10. https://doi.org/10.3389/fenvs.2022.907872
73. Tharwat A, Gaber T, Ibrahim A, Hassanien AE (2017) Linear discriminant analysis: a detailed tutorial. AI Commun 30:169–190. https://doi.org/10.3233/AIC-170729
74. Frank IE, Lanteri S (1989) Classification models: discriminant analysis, SIMCA, CART. Chemom Intell Lab Syst 5:247–256. https://doi.org/10.1016/0169-7439(89)80052-8
75. Ballabio D, Consonni V (2013) Classification tools in chemistry. Part 1: linear models. PLS-DA. Anal Methods 5:3790. https://doi.org/10.1039/c3ay40582f
76. Barker M, Rayens W (2003) Partial least squares for discrimination. J Chemom 17:166–173. https://doi.org/10.1002/cem.785
77. Luts J, Ojeda F, Van de Plas R, De Moor B, Van Huffel S, Suykens JAK (2010) A tutorial on support vector machine-based methods for classification problems in chemometrics. Anal Chim Acta 665:129–145. https://doi.org/10.1016/j.aca.2010.03.030
78. Vapnik VN (1995) The nature of statistical learning theory. Springer New York, New York, NY. https://doi.org/10.1007/978-1-4757-2440-0
79. Breiman L (2001) Random forests. Mach Learn 45:5–32. https://doi.org/10.1023/A:1010933404324

80. Scott IM, Lin W, Liakata M, Wood JE, Vermeer CP, Allaway D, Ward JL, Draper J, Beale MH, Corol DI, Baker JM, King RD (2013) Merits of random forests emerge in evaluation of chemometric classifiers by external validation. Anal Chim Acta 801:22–33. https://doi.org/10.1016/j.aca.2013.09.027
81. Law JC-F, Huang Y, Chow C-H, Lam T-K, Leung KS-Y (2021) Comparative physicochemical properties and toxicity of organic UV filters and their photocatalytic transformation products. Environ Pollut 286:117551. https://doi.org/10.1016/j.envpol.2021.117551
82. Wang Z, Yang C, Parrott JL, Frank RA, Yang Z, Brown CE, Hollebone BP, Landriault M, Fieldhouse B, Liu Y, Zhang G, Hewitt LM (2014) Forensic source differentiation of petrogenic, pyrogenic, and biogenic hydrocarbons in Canadian oil sands environmental samples. J Hazard Mater 271:166–177. https://doi.org/10.1016/j.jhazmat.2014.02.021
83. Wang Z, Yang C, Kelly-Hooper F, Hollebone BP, Peng X, Brown CE, Landriault M, Sun J, Yang Z (2009) Forensic differentiation of biogenic organic compounds from petroleum hydrocarbons in biogenic and petrogenic compounds cross-contaminated soils and sediments. J Chromatogr A 1216:1174–1191. https://doi.org/10.1016/j.chroma.2008.12.036
84. Alexandrino GL, Tomasi G, Kienhuis PGM, Augusto F, Christensen JH (2019) Forensic investigations of diesel oil spills in the environment using comprehensive two-dimensional gas chromatography–high resolution mass spectrometry and chemometrics: new perspectives in the absence of recalcitrant biomarkers. Environ Sci Technol 53:550–559. https://doi.org/10.1021/acs.est.8b05238
85. Ismail A, Toriman ME, Juahir H, Kassim AM, Zain SM, Ahmad WKW, Wong KF, Retnam A, Zali MA, Mokhtar M, Yusri MA (2016) Chemometric techniques in oil classification from oil spill fingerprinting. Mar Pollut Bull 111:339–346. https://doi.org/10.1016/j.marpolbul.2016.06.089
86. Wang Z, Stout SA, Fingas M (2006) Forensic fingerprinting of biomarkers for oil spill characterization and source identification. Environ Forensic 7:105–146. https://doi.org/10.1080/15275920600667104
87. Christensen JH, Tomasi G (2007) Practical aspects of chemometrics for oil spill fingerprinting. J Chromatogr A 1169:1–22. https://doi.org/10.1016/j.chroma.2007.08.077
88. El-Serehy HA, Abdallah HS, Al-Misned FA, Al-Farraj SA, Al-Rasheid KA (2018) Assessing water quality and classifying trophic status for scientifically based managing the water resources of the Lake Timsah, the lake with salinity stratification along the Suez Canal. Saudi J Biol Sci 25:1247–1256. https://doi.org/10.1016/j.sjbs.2018.05.022
89. Uddin MG, Nash S, Mahammad Diganta MT, Rahman A, Olbert AI (2022) Robust machine learning algorithms for predicting coastal water quality index. J Environ Manag 321:115923. https://doi.org/10.1016/j.jenvman.2022.115923
90. Watanabe FSY, Miyoshi GT, Rodrigues TWP, Bernardo NMR, Rotta LHS, Alcântara E, Imai NN (2020) Inland water's trophic status classification based on machine learning and remote sensing data. Remote Sens Appl Soc Environ 19:100326. https://doi.org/10.1016/j.rsase.2020.100326
91. Freeman BS, Taylor G, Gharabaghi B, Thé J (2018) Forecasting air quality time series using deep learning. J Air Waste Manage Assoc 68:866–886. https://doi.org/10.1080/10962247.2018.1459956
92. Castelli M, Clemente FM, Popovič A, Silva S, Vanneschi L (2020) A machine learning approach to predict air quality in California. Complexity 2020:1–23. https://doi.org/10.1155/2020/8049504
93. Houdou A, El Badisy I, Khomsi K, Abdala SA, Abdulla F, Najmi H, Obtel M, Belyamani L, Ibrahimi A, Khalis M (2024) Interpretable machine learning approaches for forecasting and predicting air pollution: a systematic review. Aerosol Air Qual Res 24:230151. https://doi.org/10.4209/aaqr.230151
94. Petrić V, Hussain H, Časni K, Vuckovic M, Schopper A, Ujević Andrijić Ž, Kecorius S, Madueno L, Kern R, Lovrić M (2024) Ensemble machine learning, deep learning, and time series forecasting: improving prediction accuracy for hourly concentrations of ambient air pollutants. aerosol air Qual. Res 24:230317. https://doi.org/10.4209/aaqr.230317

95. Hawkins DM (2004) The problem of overfitting. J Chem Inf Comput Sci 44:1–12. https://doi.org/10.1021/ci0342472
96. Westerhuis JA, Hoefsloot HCJ, Smit S, Vis DJ, Smilde AK, van Velzen EJJ, van Duijnhoven JPM, van Dorsten FA (2008) Assessment of PLSDA cross validation. Metabolomics 4:81–89. https://doi.org/10.1007/s11306-007-0099-6
97. Stone M (1974) Cross-validatory choice and assessment of statistical predictions. J R Stat Soc Ser B Stat Methodol 36:111–133. https://doi.org/10.1111/j.2517-6161.1974.tb00994.x
98. Varma S, Simon R (2006) Bias in error estimation when using cross-validation for model selection. BMC Bioinformatics 7:91. https://doi.org/10.1186/1471-2105-7-91
99. Charlton AJ, Wrobel MS, Stanimirova I, Daszykowski M, Grundy HH, Walczak B (2010) Multivariate discrimination of wines with respect to their grape varieties and vintages. Eur Food Res Technol 231:733–743. https://doi.org/10.1007/s00217-010-1299-2
100. Fawcett T (2006) An introduction to ROC analysis. Pattern Recogn Lett 27:861–874. https://doi.org/10.1016/j.patrec.2005.10.010
101. Sokolova M, Lapalme G (2009) A systematic analysis of performance measures for classification tasks. Inf Process Manag 45:427–437. https://doi.org/10.1016/j.ipm.2009.03.002
102. Chou CL, Haya K, Paon LA, Moffatt JD (2004) A regression model using sediment chemistry for the evaluation of marine environmental impacts associated with salmon aquaculture cage wastes. Mar Pollut Bull 49:465–472. https://doi.org/10.1016/j.marpolbul.2004.02.039
103. Brix KV, Tear L, Santore RC, Croteau K, DeForest DK (2021) Comparative performance of multiple linear regression and biotic ligand models for estimating the bioavailability of copper in freshwater. Environ Toxicol Chem 40:1649–1661. https://doi.org/10.1002/etc.5012
104. Price GAV, Stauber JL, Jolley DF, Koppel DJ, Van Genderen EJ, Ryan AC, Holland A (2023) Development and validation of multiple linear regression models for predicting chronic zinc toxicity to freshwater microalgae. Environ Toxicol Chem 42:2630–2641. https://doi.org/10.1002/etc.5749
105. Olivieri AC (2018) Introduction to multivariate calibration. Springer International Publishing, Cham. https://doi.org/10.1007/978-3-319-97097-4
106. Wold S, Ruhe A, Wold H, Dunn III WJ (1984) The collinearity problem in linear regression. The partial least squares (PLS) approach to generalized inverses. SIAM J Sci Stat Comput 5:735–743. https://doi.org/10.1137/0905052
107. Massy WF (1965) Principal components regression in exploratory statistical research. J Am Stat Assoc 60:234–256. https://doi.org/10.1080/01621459.1965.10480787
108. Vigneau E, Devaux MF, Qannari EM, Robert P (1997) Principal component regression, ridge regression and ridge principal component regression in spectroscopy calibration. J Chemom 11:239–249. https://doi.org/10.1002/(SICI)1099-128X(199705)11:3<239::AID-CEM470>3.0.CO;2-A
109. Sun J (1995) A correlation principal component regression analysis of NIR data. J Chemom 9:21–29. https://doi.org/10.1002/cem.1180090104
110. Gowen AA, Downey G, Esquerre C, O'Donnell CP (2011) Preventing over-fitting in PLS calibration models of near-infrared (NIR) spectroscopy data using regression coefficients. J Chemom 25:375–381. https://doi.org/10.1002/cem.1349
111. Takahama S, Dillner AM (2015) Model selection for partial least squares calibration and implications for analysis of atmospheric organic aerosol samples with mid-infrared spectroscopy. J Chemom 29:659–668. https://doi.org/10.1002/cem.2761
112. Wold S, Sjöström M, Eriksson L (2001) PLS-regression: a basic tool of chemometrics. Chemom Intell Lab Syst 58:109–130. https://doi.org/10.1016/S0169-7439(01)00155-1
113. Wold S, Martens H, Wold H (1983) The multivariate calibration problem in chemistry solved by the PLS method, pp. 286–293. https://doi.org/10.1007/BFb0062108
114. Hau N-X, Tuan N-T, Trung L-Q, Chi T-T (2024) Estimation of soil organic carbon content using visible and near-infrared spectroscopy in the Red River Delta, Vietnam. Chemom Intell Lab Syst 255:105253. https://doi.org/10.1016/j.chemolab.2024.105253
115. Stenberg B, Viscarra Rossel RA, Mouazen AM, Wetterlind J (2010) Visible and near infrared spectroscopy in soil science: pp. 163–215. https://doi.org/10.1016/S0065-2113(10)07005-7

116. Zhang L, Yang XM, Drury CF, Chantigny M, Gregorich E, Miller J, Bittman S, Reynolds W, Yang J (2017) Infrared spectroscopy prediction of organic carbon and total nitrogen in soil and particulate organic matter from diverse Canadian agricultural regions. Can J Soil Sci:CJSS-2017-0070. https://doi.org/10.1139/CJSS-2017-0070
117. Janik LJ, Forrester ST, Rawson A (2009) The prediction of soil chemical and physical properties from mid-infrared spectroscopy and combined partial least-squares regression and neural networks (PLS-NN) analysis. Chemom Intell Lab Syst 97:179–188. https://doi.org/10.1016/j.chemolab.2009.04.005
118. Forrester ST, Janik LJ, McLaughlin MJ, Soriano-Disla JM, Stewart R, Dearman B (2013) Total petroleum hydrocarbon concentration prediction in soils using diffuse reflectance infrared spectroscopy. Soil Sci Soc Am J 77:450–460. https://doi.org/10.2136/sssaj2012.0201
119. Heibati M, Stedmon CA, Stenroth K, Rauch S, Toljander J, Säve-Söderbergh M, Murphy KR (2017) Assessment of drinking water quality at the tap using fluorescence spectroscopy. Water Res 125:1–10. https://doi.org/10.1016/j.watres.2017.08.020
120. Guo Y, Liu C, Ye R, Duan Q (2020) Advances on water quality detection by UV-Vis spectroscopy. Appl Sci 10:6874. https://doi.org/10.3390/app10196874
121. Zhang H, Zhang L, Wang S, Zhang L (2022) Online water quality monitoring based on UV–Vis spectrometry and artificial neural networks in a river confluence near Sherfield-on-Loddon. Environ Monit Assess 194:630. https://doi.org/10.1007/s10661-022-10118-4
122. LeCun Y, Bengio Y, Hinton G (2015) Deep learning. Nature 521:436–444. https://doi.org/10.1038/nature14539
123. Galbo Engstrøm O, Albano-Gaglio M, Schou Dreier E, Bouzembrak Y, Font-i-Furnols M, Mishra P, Steenstrup Pedersen K (2025) Transforming hyperspectral images into chemical maps: a novel end-to-end deep learning approach. J Chemom 39:1–17. https://doi.org/10.1002/cem.70041
124. Guerri MF, Distante C, Spagnolo P, Bougourzi F, Taleb-Ahmed A (2024) Deep learning techniques for hyperspectral image analysis in agriculture: a review. ISPRS Open J Photogramm Remote Sens 12:100062. https://doi.org/10.1016/j.ophoto.2024.100062
125. Jia S, Jiang S, Lin Z, Li N, Xu M, Yu S (2021) A survey: deep learning for hyperspectral image classification with few labeled samples. Neurocomputing 448:179–204. https://doi.org/10.1016/j.neucom.2021.03.035
126. Ball JE, Anderson DT, Chan CS (2017) Comprehensive survey of deep learning in remote sensing: theories, tools, and challenges for the community. J Appl Remote Sens 11:1. https://doi.org/10.1117/1.JRS.11.042609
127. Zhai Y, Zhou L, Qi H, Gao P, Zhang C (2023) Application of visible/near-infrared spectroscopy and hyperspectral imaging with machine learning for high-throughput plant heavy metal stress phenotyping: a review. Plant Phenomics 5:0124. https://doi.org/10.34133/plantphenomics.0124
128. Lowe A, Harrison N, French AP (2017) Hyperspectral image analysis techniques for the detection and classification of the early onset of plant disease and stress. Plant Methods 13:80. https://doi.org/10.1186/s13007-017-0233-z
129. Li X, Peng L, Yao X, Cui S, Hu Y, You C, Chi T (2017) Long short-term memory neural network for air pollutant concentration predictions: method development and evaluation. Environ Pollut 231:997–1004. https://doi.org/10.1016/j.envpol.2017.08.114
130. Lyman SN, Tran HNQ, Mansfield ML, Bowers R, Smith A (2020) Strong temporal variability in methane fluxes from natural gas well pad soils. Atmos. Pollut Res 11:1386–1395. https://doi.org/10.1016/j.apr.2020.05.011
131. Hochreiter S, Schmidhuber J (1997) Long short-term memory. Neural Comput 9:1735–1780. https://doi.org/10.1162/neco.1997.9.8.1735
132. Nicholaus IT, Park JR, Jung K, Lee JS, Kang D-K (2021) Anomaly detection of water level using deep autoencoder. Sensors 21:6679. https://doi.org/10.3390/s21196679
133. Hinton GE, Salakhutdinov RR (2006) Reducing the dimensionality of data with neural networks. Science (80-.) 313:504–507. https://doi.org/10.1126/science.1127647

134. Esmaeili F, Cassie E, Nguyen HPT, Plank NOV, Unsworth CP, Wang A (2023) Anomaly detection for sensor signals utilizing deep learning autoencoder-based neural networks. Bioengineering 10:405. https://doi.org/10.3390/bioengineering10040405
135. Wang R, Ma L, He G, Johnson B, Yan Z, Chang M, Liang Y (2024) Transformers for remote sensing: a systematic review and analysis. Sensors 24:3495. https://doi.org/10.3390/s24113495
136. Rußwurm M, Körner M (2020) Self-attention for raw optical satellite time series classification. ISPRS J Photogramm Remote Sens 169:421–435. https://doi.org/10.1016/j.isprsjprs.2020.06.006
137. Zhang Z, Zhang S, Zhao X, Chen L, Yao J (2022) Temporal difference-based graph transformer networks for air quality PM2.5 prediction: a case study in China. Front Environ Sci 10. https://doi.org/10.3389/fenvs.2022.924986
138. Samek W, Müller K-R (2019) Towards explainable artificial intelligence, pp. 5–22. https://doi.org/10.1007/978-3-030-28954-6_1
139. Cheng G, Xie X, Han J, Guo L, Xia G-S (2020) Remote sensing image scene classification meets deep learning: challenges, methods, benchmarks, and opportunities. IEEE J Sel Top Appl Earth Obs Remote Sens 13:3735–3756. https://doi.org/10.1109/JSTARS.2020.3005403
140. Reichstein M, Camps-Valls G, Stevens B, Jung M, Denzler J, Carvalhais N, Prabhat (2019) deep learning and process understanding for data-driven earth system science. Nature 566:195–204. https://doi.org/10.1038/s41586-019-0912-1
141. Bejani MM, Ghatee M (2021) A systematic review on overfitting control in shallow and deep neural networks. Artif Intell Rev 54:6391–6438. https://doi.org/10.1007/s10462-021-09975-1
142. Wilkinson MD, Dumontier M, Aalbersberg IJ, Appleton G, Axton M, Baak A, Blomberg N, Boiten J-W, da Silva Santos LB, Bourne PE, Bouwman J, Brookes AJ, Clark T, Crosas M, Dillo I, Dumon O, Edmunds S, Evelo CT, Finkers R, Gonzalez-Beltran A, Gray AJG, Groth P, Goble C, Grethe JS, Heringa J, 't Hoen PA, Hooft R, Kuhn T, Kok R, Kok J, Lusher SJ, Martone ME, Mons A, Packer AL, Persson B, Rocca-Serra P, Roos M, van Schaik R, Sansone S-A, Schultes E, Sengstag T, Slater T, Strawn G, Swertz MA, Thompson M, van der Lei J, van Mulligen E, Velterop J, Waagmeester A, Wittenburg P, Wolstencroft K, Zhao J, Mons B (2016) The FAIR guiding principles for scientific data management and stewardship. Sci Data 3:160018. https://doi.org/10.1038/sdata.2016.18
143. Lahat D, Adali T, Jutten C (2015) Multimodal data fusion: an overview of methods, challenges, and prospects. Proc IEEE 103:1449–1477. https://doi.org/10.1109/JPROC.2015.2460697
144. Gibert K, Horsburgh JS, Athanasiadis IN, Holmes G (2018) Environmental Data Science. Environ Model Softw 106:4–12. https://doi.org/10.1016/j.envsoft.2018.04.005
145. Roscher R, Bohn B, Duarte MF, Garcke J (2020) Explainable machine learning for scientific insights and discoveries. IEEE Access 8:42200–42216. https://doi.org/10.1109/ACCESS.2020.2976199
146. Barredo Arrieta A, Díaz-Rodríguez N, Del Ser J, Bennetot A, Tabik S, Barbado A, Garcia S, Gil-Lopez S, Molina D, Benjamins R, Chatila R, Herrera F (2020) Explainable artificial intelligence (XAI): concepts, taxonomies, opportunities and challenges toward responsible AI. Inf Fusion 58:82–115. https://doi.org/10.1016/j.inffus.2019.12.012

Decoding Source Apportionment: From Traditional Linear Methods to Nonlinear AI Approaches

Pierluigi Barbieri, Stefano Fornasaro, and Sabina Licen

Abstract

This chapter reviews the development of source apportionment (SA) methods for environmental pollution, highlighting the transition from classical receptor models to modern machine learning (ML) approaches, particularly Self-Organizing Maps (SOM) and their integration with Positive Matrix Factorization (PMF). It begins with the mass-balance formulation that relates measured concentrations at receptors to contributions from multiple sources and summarizes the fundamentals of chemical mass balance (CMB) and PMF, detailing their assumptions, constraints, strengths, and evaluation criteria. The role of recent guidance documents is addressed, emphasizing their role in standardizing data requirements, quality assurance, and reporting for regulatory applications. The chapter introduces ML methods, specifically focusing on SOM as an unsupervised neural network. It explores the application of SOM in various environmental contexts, also considering clustering algorithm application for detecting characteristic regimes—or states—of the complex systems under study. The combination of SOM and PMF is reviewed, which facilitates the identification of pollution patterns and supports data integration from multiple sources. Finally, the chapter introduces challenges and future perspectives for incorporating interpretable ML into operational SA practices, stressing the need for consistent guidelines, quality criteria, and robust workflows to enhance environmental management and policy decisions.

P. Barbieri (✉) · S. Fornasaro · S. Licen
Department of Chemical and Pharmaceutical Sciences, University of Trieste, Trieste, Italy
e-mail: barbierp@units.it

M. Minella (ed.), *Complexity in Environmental Chemistry*,
https://doi.org/10.1007/978-3-032-31774-2_8

Keywords

Environmental monitoring · Source apportionment · Machine learning · CMB · PMF · SOM

Chapter Learning Goals

- The conceptual foundations of source apportionment, including the mass-balance framework that links observed concentrations at receptor sites to contributions from multiple emission sources.
- The main characteristics, assumptions, and limitations of classical receptor models such as chemical mass balance and Positive Matrix Factorization.
- The basic principles of machine learning in environmental pollution studies, including the distinction between supervised and unsupervised methods.
- Self-Organizing Maps peculiar features and how SOM can be used to identify spatial and temporal pollution patterns, also combined with other clustering algorithms.
- Examples of combined SOM–PMF workflows applied to soils, sediments, and air.
- Emerging perspectives and open challenges in integrating interpretable ML approaches into operational source apportionment.

1 Introduction: Complexity in Pollution Monitoring and Source Apportionment

The study and management of chemical pollution in the environment, understood as an undesired alteration in the composition of real environmental media such as air, water, soil, sediments, and biota, frequently requires the identification of emission sources and the quantification of their contributions to observed contamination levels at specific sites. This analytical and modelling practice is generally referred to as source apportionment (SA) and represents a crucial step for multiple purposes, including the attribution of responsibilities, the prioritization of reduction measures, and the design of effective regulatory strategies. In contemporary air quality management, SA is closely linked to legal frameworks and policy instruments that demand evidence-based assessments of how different source categories contribute to ambient concentrations of key pollutants, in particular particulate matter (PM) and its chemical constituents [1].

From a modelling perspective, SA can be pursued through distinct approaches depending on the availability of information on emissions, atmospheric transport, and transformation processes [1, 2]. When the relevant emission sources are known and characterized in advance and emission inventories provide reasonably accurate estimates of their temporal and spatial patterns, source-oriented chemistry–transport models (SO-CTM) can be used to simulate the dispersion, chemical transformation, and deposition of pollutants in the atmosphere. These models propagate emission inventories through meteorological fields and atmospheric chemistry schemes to estimate concentrations and deposition fluxes at receptor locations, and they can be used in forward or adjoint configurations to evaluate source contributions within scenario analyses or regulatory impact assessments. However, the construction and maintenance of comprehensive emission inventories are resource-intensive, and in many situations the number and profiles of relevant sources, including emerging or irregular contributors, are not sufficiently known a priori.

In such cases, receptor-oriented models or receptor models (RM) provide an alternative and complementary framework. Rather than starting from emissions and transport, receptor models operate directly on chemical composition data measured at one or more receptor sites and attempt to infer the underlying sources and their contributions through a combination of mass balance analysis and multivariate statistical techniques. The core idea is that, for a given sampling location, the measured concentration of each chemical species in each sample can be expressed as a linear combination of the contributions from a finite number of independent sources, each characterized by a specific profile, plus an error term accounting for measurement uncertainties and model imperfections. This conceptual model can be written as a mass balance equation linking an ambient concentration matrix to the product of a source contribution matrix and a source profile matrix, and different computational methods have been developed to solve this equation under appropriate physical and statistical constraints [3].

Receptor-model-based SA has been widely applied across environmental media, reflecting both the versatility of the mass balance framework and the availability of compositional monitoring data. In air quality studies, RM have been used predominantly for particulate matter and, to a lesser extent, for gaseous pollutants, where high temporal variability and complex atmospheric dynamics pose stringent demands on model performance. At the same time, RM have been successfully used for more conservative matrices such as soils and sediments, where temporal changes are slower, but spatial heterogeneity and legacy contamination may be substantial. The spectrum of available tools spans from relatively simple techniques based on elementary mathematical calculations and basic physical assumptions, such as enrichment factor analysis, to more complex models involving multivariate data analysis and, more recently, machine learning approaches.

1.1 Historical Evolution of Receptor Models

The historical development of receptor modelling for atmospheric PM illustrates how methodological sophistication has grown in response to both scientific and regulatory needs [3]. Early applications relied on straightforward correlations between co-emitted species, as in the first quantitative apportionment of automobile emissions based on joint measurements of CO and Pb, and on exploratory factor analysis to extract latent source-related factors from particle composition data collected in multiple cities. These initial studies demonstrated that compositional information at the receptor could indeed be used to infer source types and contributions, but they also highlighted limitations in terms of quantitative interpretation and uncertainty assessment. Subsequent work introduced explicit atmospheric mass balance formulations in which specific elements were associated with candidate source types, and ordinary least squares (OLS) fitting was used to estimate the mass contributions of each source to the observed particle concentrations. As the limitations of OLS in the presence of unequal and non-negligible measurement errors became apparent, weighted least squares and effective variance least squares (EVLS) were developed to incorporate both ambient and source profile uncertainties into the objective function, leading to the widely used chemical mass balance (CMB) model [4]. In parallel, factor analytic methods were refined and adapted to the receptor context, giving rise to approaches such as target transformation factor analysis (TTFA), which enabled more quantitative interpretation of factors in terms of source contributions by comparing them with hypothesised source profiles [5]. Later, positive matrix factorization (PMF) emerged as a powerful multivariate receptor approach in which both source profiles and contributions are derived from the ambient data alone, under non-negativity constraints and with uncertainties explicitly entering the weighted least-squares objective function [6]. PMF and related models such as UNMIX have become standard tools in SA, and their use is reflected in international guidelines and technical documents that provide recommendations on model setup, performance evaluation, and uncertainty assessment for receptor-based SA of particulate matter. At the European level, dedicated guidance documents and technical specifications have been published to harmonise RM applications and define performance tests and quality assurance procedures, particularly in the context of the ambient air quality directives [7, 8]. These developments have consolidated RM as a key component of the SA toolkit alongside SO-CTM, especially when detailed emission inventories are lacking or uncertain.

1.2 Linear Models and Their Limitations

Despite their success and institutional recognition, linear receptor models face intrinsic challenges when confronted with the full complexity of real-world environmental systems [2]. SA is both a complicated and a complex task: it must account for numerous anthropogenic and natural sources whose emissions may vary in time and space, and it must consider that the resulting alterations in environmental composition are not simply additive but may involve non-linear, synergistic, or antagonistic interactions among pollutants [9–11]. In the context of atmospheric PM, for instance, interactions among pollutants from different sources can modify the oxidative potential (OP) or other toxicological properties in a non-linear manner, such that the overall health-relevant effect cannot be captured by a simple linear superposition of source contributions.

In practice, significant deviations from linear model assumptions have been observed in air pollution applications, where goodness-of-fit indicators for linear models may remain modest despite careful model setup. Several factors contribute to these deviations, including non-stationarity of emission patterns, complex atmospheric chemistry, secondary formation processes, and interactions between sources that alter the resulting concentration–response relationships. Linear receptor models such as CMB and PMF can reflect the average relationships between covariates and pollution sources under the mass balance hypothesis, but they cannot fully represent the complex non-linear relationships between different variables and cannot fully explain the influence of environmental impact factors on source contributions [2].

Moreover, traditional SA methods entail specific sources of uncertainty beyond pure measurement error. For CMB, uncertainties are linked to the completeness and representativeness of the source profile database, the stability of source compositions over time, and the selection of species included in the model. For PMF, key uncertainty components arise from the choice of the number of factors, the handling of below-detection-limit and missing data, the specification of uncertainty estimates, and the subjective interpretation of factor profiles in terms of physical sources [12]. In particular, the association between PMF-derived factors and actual emission sources often relies on expert judgement and the identification of marker species, which can introduce subjectivity and potential biases when source profiles are similar or overlapping [13].

Recent perspectives in environmental data science have emphasized that, in complex environmental settings, linear models cannot handle dynamic, heterogeneous, and spatially expansive data in which variables are strongly interacting and relationships are non-linear and context dependent. This recognition has stimulated growing interest in non-linear approaches, including ML methods, to complement traditional receptor models and to better capture the structure of high-dimensional environmental datasets [2].

1.3 Environmental Systems as Complex Systems

The conceptualization of environmental systems as complex systems provides an additional lens through which to understand the limitations of purely linear approaches in SA. Complex systems are composed of many interacting components whose collective behaviour is intrinsically difficult to model and predict, and they often exhibit emergent properties that cannot be straightforwardly inferred from the properties of individual components. Examples include the Earth's climate, ecosystems, the human brain, engineered infrastructures such as power grids and transport networks, and, for many authors, the biosphere or even the entire universe. Environmental pollution processes—spanning emission, transport, transformation, and accumulation—unfold within and across such complex systems, where feedbacks, thresholds, and multi-scale interactions are the rule rather than the exception [9].

Within this perspective, air pollution can be seen as the outcome of intertwined processes involving primary emissions from diverse sources, chemical reactions forming secondary pollutants, meteorological variability, land use patterns, and socio-economic activities. The resulting temporal and spatial patterns of pollutant concentrations reflect non-linear dynamics and path dependence, making it difficult to capture all relevant behaviours through linear relationships. Similar considerations apply to aquatic systems, where hydrodynamics, biogeochemical cycles, and anthropogenic inputs interact, and to soils and sediments, where historical contaminant deposition, sorption processes, and remobilization events combine to shape current contamination profiles [14, 15].

ML has been proposed as a promising framework to address some of these challenges by providing flexible, data-driven models that can adapt to complex patterns without requiring explicit specification of all governing equations. ML has been defined as a classical form of artificial intelligence that teaches a program to recognize patterns in data, enabling it to discover latent relationships, realize the predictability of trends, and uncover new knowledge hidden in large and heterogeneous datasets. In environmental pollution science, ML methods have already been applied to tasks such as forecasting pollution levels, evaluating disease risks, predicting adsorption of organic contaminants, and implementing intelligent control strategies in wastewater treatment.

However, the integration of ML into SA itself is still at an early stage, and the use of interpretable ML tools in this context is notably scarce. Recent studies have demonstrated that combining PMF with non-linear models such as multilayer perceptron (MLP) can help quantify source contributions to health-relevant metrics like OP, capturing synergistic and antagonistic effects among PM sources that elude strictly linear formulations [10, 11]. Other investigations have explored the potential of clustering algorithms such as k-means and spectral clustering as receptor models for PM source apportionment, with results comparable to PMF in some settings and added value for regional source differentiation [16, 17]. These developments suggest substantial potential for future advancements in the application of ML techniques to SA, while also highlighting the importance of interpretability, reproducibility, and integration with existing regulatory frameworks.

Within this evolving landscape, self-organizing maps (SOMs), or Kohonen neural networks, stand out as an unsupervised ML method particularly well suited to exploratory analysis, pattern recognition, and variable relationship assessment in complex environmental monitoring datasets [18–20]. SOMs provide non-linear dimensionality reduction and clustering on a regular grid, enabling intuitive visualization of high-dimensional structures and facilitating the identification of source-related patterns, temporal regimes, and spatial gradients. Their previous applications in environmental chemistry, including water quality assessment, marine biogeochemistry, air particulate characterization, and odour monitoring, provide a strong foundation for considering SOM as a complementary tool for SA and knowledge extraction [14, 15, 20–23]. The following sections build on this foundation to present the fundamentals of receptor modelling, the role of ML in environmental pollution analysis, and the specific contribution that SOM can make to SA in complex environmental systems.

2 Foundations of Receptor Modelling

Receptor models are grounded in a mass balance formulation that links measured pollutant concentrations at a receptor site to the contributions of multiple emission sources, each characterized by a specific chemical profile. This framework assumes that the material collected at the receptor is a linear combination of contributions from a finite number of independent sources plus an error term accounting for measurement uncertainties and model imperfections. The goal of receptor modelling is to estimate both the source contributions and, where not known a priori, the source profiles, subject to physical and statistical constraints such as non-negativity and consistency with the observed data [1, 3].

In matrix form, the basic receptor model can be written as a mass balance equation involving three key components: the ambient data matrix, the source contribution matrix, and the source profile matrix. The ambient data matrix contains the measured concentrations of m chemical species across n samples, while the source contribution matrix contains the contributions of p sources to each sample, and the source profile matrix contains the composition of each source in terms of the m species. Different receptor modelling methods correspond to different ways of solving this mass balance equation depending on the available a priori information about sources and their chemical characteristics (Fig. 1).

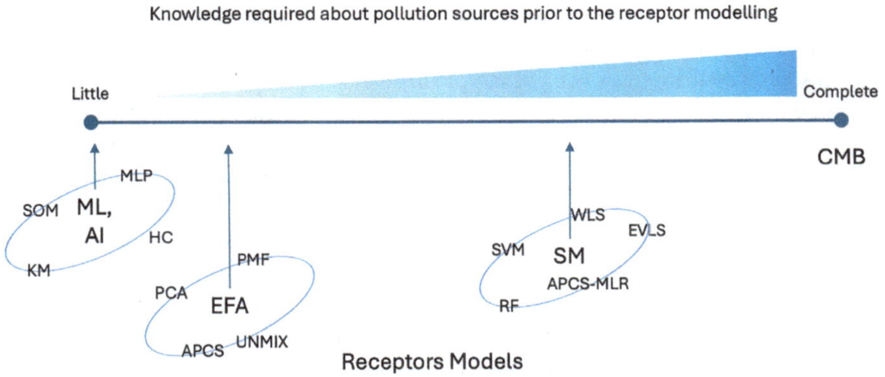

Fig. 1 Different approaches for estimating pollution source contributions using receptor models. *AI* artificial intelligence, *APCS* absolute principal component scores, *CFA* confirmatory factor analysis, *CMB* chemical mass balance, *EFA* exploratory factor analysis, *EVLS* effective variance least squares, *HC* hierarchical clustering, *KM* k-means clustering, *ML* machine learning, *MLP* multilayer perceptron; *MLR* multiple linear regression, *PCA* principal components analysis, *PMF*, positive matrix factorization, *RF* random forest, *SOM* self-organizing map, *SM*, supervised (machine learning) models, *SVM* supported vector machines, *WLS* weighted least squares. See text for more details

2.1 Mass Balance Formulation and Constraints

Let x_{ij} denote the concentration of the j-th chemical species measured in the i-th sample at a receptor site. In the basic receptor model, $x_{i,j}$ is expressed as the sum over p independent sources of the product of the contribution of source k to sample i, denoted $g_{i,k}$, and the concentration of species j in material from source k, denoted $f_{k,j}$, plus a residual term. The general mass balance equation can be written as

$$x_{i,j} = \sum_{k=1}^{p} g_{i,k} f_{k,j} + e_{i,j}$$

where e_{ij} represents the unmodelled portion of the variation for species j in sample i. In matrix notation, this relation can be expressed as the general bilinear model

$$\mathbf{X} = \mathbf{GF} + \mathbf{E}$$

where $\mathbf{X}$ is the $n \times m$ data matrix of ambient concentrations, $\mathbf{G}$ is the $n \times p$ matrix of source contributions, $\mathbf{F}$ is the $p \times m$ matrix of source profiles, and $\mathbf{E}$ is the $n \times m$ residual matrix (Fig. 2).

For the receptor model to be physically meaningful, several fundamental constraints must be satisfied [24]. First, the model must reproduce the original data within the limits of measurement and model error; that is, the fitted values $\grave{x}_{i,j} = \sum_k \grave{g}_{i,k} \grave{f}_{k,j}$ should provide an adequate representation of the observed concentrations. Second, the predicted source compositions must be non-negative, because a source cannot have a negative percentage of a chemical species. Third, the predicted source contributions to each sample must also be non-negative, since a source cannot emit a negative mass. Fourth, the sum of the predicted mass contributions of all sources to each species should be less than or equal to the total measured mass for that species, reflecting the principle that the whole is greater than or equal to the sum of its parts.

Depending on the availability of a priori information, different strategies are used to solve the mass balance equation. If the number of sources and their profiles are known from independent measurements, the problem is formulated as a chemical mass balance (CMB) model in which $\mathbf{F}$ is treated as known and $\mathbf{G}$ is estimated for each sample [4]. If the number and nature of sources are unknown, multivariate factorization methods such as positive matrix factorization (PMF) or UNMIX are applied to estimate both $\mathbf{G}$ and $\mathbf{F}$ from the ambient data under non-negativity and

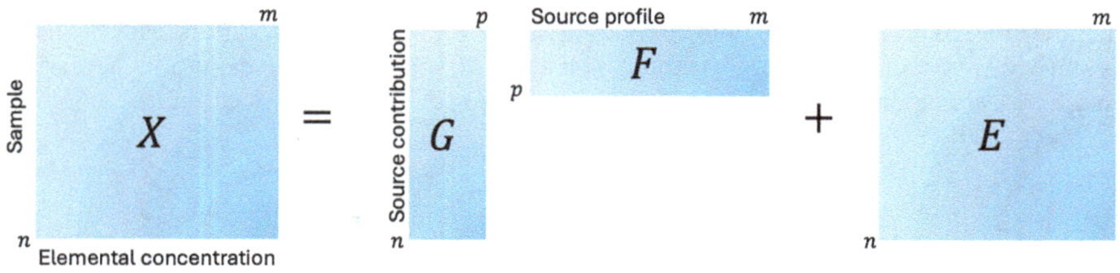

Fig. 2 Graphical representation of the bilinear factor decomposition model

other constraints [6, 12]. In all cases, the estimation procedure must account for the uncertainties in the measurements and, where relevant, in the source profiles.

2.2 Chemical Mass Balance (CMB) Models

In the CMB framework, it is assumed that the relevant sources have been identified and that their chemical composition profiles have been measured at or near the sources, with associated uncertainty estimates.

$$x_j = \sum_{k=1}^{p} g_k f_{kj} + e_{i,j}$$

where x_j is the concentration of species j in the sample of interest i, f_{kj} is the concentration of chemical species j in material from source k, g_k is the mass contribution of source k to the sample of interest, and $e_{i,j}$ is the unmodelled portion.

Early implementations of CMB used ordinary least squares (OLS) to estimate **g** by minimizing the sum of squared residuals between observed and modelled concentrations. However, because measurement errors vary among species and between ambient and source profile data, OLS does not provide an optimal solution and can give biased estimates when errors are heteroscedastic [3, 4]. To address this issue, effective variance least squares (EVLS) was introduced, in which each term in the objective function is weighted by the combined variance of the ambient and source profile measurements. EVLS requires an uncertainty estimate for each ambient concentration and each source profile concentration and propagates these errors through the iterative solution to obtain both the contribution estimates and their associated uncertainties.

Operationally, CMB models require several inputs and assumptions: a set of chemical species that are sufficiently specific to distinguish among sources; a database of source profiles measured under conditions representative of the study domain; estimates of measurement uncertainties for both ambient and source data; and an assessment of whether the selected sources adequately represent all major contributors. When these conditions are met, CMB can provide detailed and interpretable source contribution estimates with quantified uncertainties, and it has therefore been widely used in regulatory and research applications. Nonetheless, CMB is constrained by the availability and representativeness of source profiles, and it cannot directly reveal unknown sources or account for substantial variability in source compositions over time and space.

2.3 Multivariate Receptor Models and Positive Matrix Factorization

When the number and nature of the sources are not known a priori, multivariate factorization methods are employed to extract source-related patterns directly from the ambient data. In this context, the basic mass balance model is retained, but both the source contribution matrix **G** and the source profile matrix **F** are treated as unknowns to be estimated from **X**. Among the various multivariate approaches developed, positive matrix factorization (PMF) has become one of the most widely used in environmental SA.

PMF solves the receptor modelling problem by minimizing a weighted objective function that accounts for measurement uncertainties and imposes non-negativity on both **G** and **F**. The objective function typically has the form

$$Q=\sum_{i=1}^{n}\sum_{j=1}^{m}\left(\frac{x_{ij}-\sum_{k=1}^{p}g_{ik}f_{kj}}{s_{ij}}\right)^2$$

where s_{ij} is an estimate of the uncertainty for species j in sample i. This uncertainty combines measurement error and natural variability in the source profile value, recognizing that source profiles are not fundamental constants but may vary due to process conditions and other factors. By weighting residuals by their uncertainties, PMF downweighs less reliable data and focuses the fit on more informative observations, leading to more robust factor solutions.

A key advantage of PMF over unconstrained factor analysis is the enforcement of non-negativity, which reflects the physical reality that source contributions and species concentrations cannot be negative. PMF also allows the handling of missing and below-detection-limit data through appropriate uncertainty assignments, and it can incorporate a priori information in the form of constraints on specific elements of **G** or **F** when warranted. The output of PMF consists of a set of factor profiles and corresponding time series of factor contributions, which must then be interpreted as potential source types based on their chemical signatures, temporal patterns, and auxiliary information such as meteorology or activity data.

Despite these strengths, PMF also presents challenges and sources of subjectivity. The choice of the number of factors is not uniquely determined and typically requires exploratory analysis, diagnostics, and expert judgement, with different solutions potentially leading to different source interpretations. Too few factors will combine sources of different nature together. Too many factors will make a real factor dissociate into two or more non-existing sources. The association between factors and actual emission sources is based on qualitative and semi-quantitative criteria, including the presence of tracer species and the resemblance to known source profiles, and can be ambiguous when sources share similar chemical characteristics [13]. Furthermore, PMF is fundamentally a linear model, and therefore it cannot directly capture non-linear interactions among sources or between sources and environmental conditions, although its factors may implicitly reflect such interactions. Moreover, PMF is infeasible for very large datasets [25].

PMF has been extensively applied to particulate matter SA, and it is explicitly recommended in European guidance documents on receptor modelling for air pollution, where performance indicators, uncertainty evaluation procedures, and intercomparison exercises have been described [1, 8]. It has also been applied to other media such as soils and sediments, often in combination with complementary techniques such as enrichment factor analysis or clustering methods [14, 21].

Among receptor models, absolute principal component scores–multiple linear regression (APCS-MLR) has been proposed by Thurston and Spengler in 1985 [26] and is also widely applied to quantify contributions of natural and anthropogenic sources, especially for soils, groundwater, and surface waters [27, 28]. By transforming principal component scores into "absolute" scores using a zero-concentration sample and then regressing pollutant concentrations on these APCS, the method provides quantitative source contributions with relatively low computational cost and straightforward implementation, which explains its popularity in many environmental source apportionment case studies [29]. However, when moving from general research applications to regulatory practice for ambient particulate matter, PMF has gained a much stronger and more explicit recognition than APCS-MLR. The European Guide on Air Pollution Source Apportionment with Receptor Models, developed by the Joint Research Centre and FAIRMODE, identifies PMF—together with CMB and, to a lesser extent, UNMIX—as the primary state-of-the-art receptor models for PM_{10} and $PM_{2.5}$ source apportionment in support of the EU Air Quality Directive, and many worked examples and QA/QC recommendations are formulated specifically around PMF applications (https://source-apportionment.jrc.ec.europa.eu). Similarly, the US Environmental Protection Agency explicitly lists PMF among the official receptor models developed "for use in air quality management", alongside CMB and UNMIX, and provides dedicated user guides and protocols for their application and validation (https://www.epa.gov/scram/air-pollutant-receptor-modeling). In China, national and regional assessments of $PM_{2.5}$ routinely adopt PMF as the preferred receptor model for identifying and quantifying sources in support of air-quality planning and control, and recent reviews describe CMB and PMF as "the most commonly used methods of source apportionment" for airborne particles [30].

2.4 Methods Integrating Meteorological Information

Beyond purely chemical receptor models, several methods have been developed to integrate meteorological information (e.g. wind data and air mass back trajectories), geophysical parameters, road network locations, and satellite data into SA. Approaches that use local wind data, such as conditional probability functions or bivariate polar plots, relate source contributions or pollutant concentrations to wind speed and direction, thereby providing directional information that helps identify likely source regions. Other methods incorporate back trajectory analyses to link receptor measurements with air masses arriving from upwind locations, enabling the separation of local, regional, and remote contributions and supporting the interpretation of receptor model outputs [31].

These meteorologically informed methods are not receptor models in the strict chemical mass balance sense, but they enhance SA by providing spatial and dynamical context to chemically derived source contributions. In practical applications, tools like CMB and PMF are often used in conjunction with wind and trajectory analyses to yield a more comprehensive understanding of source impacts, especially in complex environments with multiple potential source regions. The integration of such information is also reflected in technical specifications that describe performance assessment and quality assurance for receptor-oriented SA, where meteorological consistency and plausibility of source attributions are important evaluation criteria [8]. Together, the mass balance formulation, CMB, PMF, and meteorology-enhanced methods constitute the core of classical receptor modelling and provide the conceptual and operational foundation upon which newer, ML-based approaches can build.

3 Machine Learning in Environmental Pollution and Source Apportionment

ML has rapidly gained prominence in environmental sciences as a flexible framework for extracting patterns and building predictive models from complex, high-dimensional datasets without requiring explicit specification of all governing equations. In the context of environmental pollution, ML has been used to forecast pollutant concentrations, estimate exposure–response relationships, predict physicochemical properties such as adsorption coefficients, and implement intelligent control strategies in engineered systems such as wastewater treatment plants [32, 33]. Although these applications are not, strictly speaking, SA in the receptor-model sense, they demonstrate the capacity of ML to capture non-linear relationships, handle collinearity among predictors, and exploit large heterogeneous datasets, all of which are relevant to SA problems.

Recent reviews have explicitly examined the gains and problems associated with ML in environmental pollution, with particular attention to SA [2, 34]. On the positive side, ML techniques offer several advantages over traditional linear SA methods: they can address collinearity among pollution sources, mitigate the subjectivity inherent in factor interpretation, and better capture non-linear relationships between sources, pollutants, and environmental impact factors. ML can also handle large, heterogeneous datasets, including high-frequency sensor data, remote sensing products, and ancillary variables such as meteorology and land-use indicators, facilitating the integration of information across scales and platforms.

However, ML methods also come with limitations and challenges. Many powerful ML algorithms, especially deep learning models, behave as "black boxes" in which the internal representation of relationships is not easily interpretable, which can hinder their acceptance in regulatory SA applications where transparency and traceability are required. Overfitting is a significant risk when models with high capacity are trained on limited data, and careful validation is necessary to ensure robust performance. In addition, the lack of standardized workflows, benchmarking datasets, and reporting practices for ML-based SA makes it difficult to compare results across studies and to establish best practices. For these reasons, recent literature emphasizes the need for interpretable ML in SA, the systematic evaluation of model uncertainty, and the development of guidelines that parallel those available for linear receptor models [34].

From a methodological standpoint, ML can be broadly divided into three main categories: supervised learning, unsupervised learning, and reinforcement learning. Supervised learning methods use labelled data—in which input features are paired with known outputs—to learn mappings that can be used for regression or classification tasks; unsupervised learning methods operate on unlabelled data to discover latent structure, clusters, or low-dimensional representations; and reinforcement learning focuses on sequential decision-making based on a reward–punishment paradigm, not relevant at present in the SA context. Within each category, both linear and non-linear algorithms are available, ranging from simple linear and polynomial regression to deep neural networks, with varying levels of complexity, interpretability, and data requirements. A recent review has pointed out benefits from application of ML methods to not-ordinary chemical features proposed in SA, as data from real-time aerosol mass spectrometry [35].

3.1 Supervised Learning

Supervised machine learning has been used in source-apportionment mainly to (i) support or extend receptor models such as APCS-MLR and PMF,and (ii) predict source contributions or labels from auxiliary data using regression and classification [36–38]. Supervised learning requires labelled targets, which in SA are typically source contributions, source categories, or factor labels derived from a receptor model. Models such as decision trees (DT), random forests (RF), support vector machines (SVM), and ensemble methods are then trained to predict these targets from environmental predictors (e.g. concentrations, soil properties, meteorology, land use) or to classify unlabelled factors into source types [39]. In many recent applications, supervised ML does not replace the receptor model but wraps around it: receptor models provide quantitative source information, and supervised ML learns mappings from environmental covariates to those source metrics, enabling spatial extrapolation, temporal prediction, or automated factor labelling. This hybrid strategy leverages the interpretability of receptor models and the predictive power of supervised ML [40].

Supervised machine learning has also been coupled with receptor models for prediction and SA of heavy metals in coastal reclaimed soils and related environments [37, 41–43]. In these works, receptor models provided baseline quantitative source contributions, while supervised regression algorithms (e.g. RF, SVM) were trained to predict those contributions across space from natural/environmental factors such as soil texture, hydrology, or land-use indicators. This framework allowed the authors to map source contributions and assess driving factors without running the receptor model at every location, effectively turning supervised ML into a spatial predictor of receptor-based SA outputs.

For organic aerosol, ensemble supervised models have been trained directly on PMF factor contributions to enable efficient prediction of aerosol sources. Using 6 years of aerosol chemical speciation monitor (ACSM) data from the Paris region, PMF was first applied to derive organic aerosol factors; then an ensemble machine-learning model was trained to predict the time series of these PMF source contributions from a simplified set of observations. Compared with individual algorithms, the ensemble reduced the error and increased correlation in predicting aerosol sources, thereby offering a near-real-time supervised SA tool grounded in receptor-model factors [44].

Supervised classification has also been used to label and interpret source factors derived from receptor models. In one study, factor analysis (FA) or PMF was first employed as a receptor model to obtain quantitative source contributions, and then a machine-learning classification model was trained to assign these factors to known source categories based on their chemical fingerprints and contextual information [36]. This approach helps standardize and automate the factor-labelling step in SA, reducing subjectivity and improving comparability across sites and studies.

Hybrid frameworks that couple PMF with supervised ML have also been developed for water quality and riverine pollution sources. In these systems, PMF is used to decompose water-quality indicators into source-related factors, while random forest and other supervised models are trained on hydrological, land-use, and seasonal variables to predict factor contributions and quantify seasonal variations in source impacts. Such PMF–ML combinations illustrate how supervised learning can help track and forecast source-specific pollution dynamics beyond the original receptor-model domain [45, 46].

Supervised neural-network like multilayer perceptron (MLP) have been used to probe non-linear relationships between source contributions and health-relevant metrics, including the oxidative potential (OP) of particulate matter [10, 11]. In this framework, PMF was first applied to chemical composition data to obtain time-resolved source contributions, which were then used as input features for MLP models that predict OP measured by different acellular assays. Borlaza and co-authors analysed a unique multi-site dataset from Grenoble, where $PM_{2.5}$ source contributions derived from PMF were combined with multiple OP assays and modelled using both classical multiple linear regression (MLR) and MLP neural networks [47]. While MLR captured a large fraction of the variance in OP and allowed estimation of intrinsic OP for each PM source, the MLP models revealed departures from linear additivity, highlighting potential synergistic or antagonistic interactions between sources in driving OP, especially for assays more sensitive to combined effects of organics and metals. In practical terms, the MLP analysis enabled the identification of episodes and regimes where the observed OP could not be explained by a simple linear sum of source contributions, suggesting that atmospheric processing or co-emission of specific species modifies the toxicological properties of the aerosol beyond what is implied by mass alone [48].

3.2 Unsupervised Learning, Clustering, and Dimensionality Reduction

Cluster analysis, or "clustering," groups unlabeled observations, data points, or feature vectors into cluster with analogous features, reducing intra-cluster dissimilarity and maximizing inter-cluster dissimilarity. Initially proposed in 1930, it gained popularity in the 1960s, with applications to environmental data emerging in the 1980s, particularly with *k*-means (KM) and hierarchical clustering (HC) agglomerative approaches [19, 49, 50]. More recently, spectral clustering has been introduced as a potential receptor model technique for SA, by directly applying clustering on mass concentration data [17]. Clustering algorithms can be useful to identify source profiles, but their application to environmental data shows some drawbacks, such as (i) handling of large datasets can be very time-consuming; (ii) samples with extremely low (noise) or high values can affect the final model and, therefore, the result interpretation; (iii) KM is usually randomly initialized thus potentially leading to rather different outcomes when applied to this type of data; (iv) the outcomes are difficult to visualize: KM result is simply a list of cluster assignment, and HC outcomes can be arranged in a dendrogram that can became very crowded and, thus, poorly interpretable when dealing with more than 100–200 of samples.

Other unsupervised methods include Self-Organizing Maps (SOM) [18–20], which combine vector quantization, topological mapping, and non-linear dimensionality reduction on a regular grid, and other neural or manifold learning techniques such as autoencoders [51, 52] and t-distributed stochastic neighbour embedding (t-SNE) [53, 54]. SOM, in particular, have been widely applied in environmental chemistry for tasks ranging from water quality pattern recognition to the analysis of particle size distributions and odour monitoring, usually coupled with KM or HC algorithms fed with SOM outputs. Their specific role in SA and data fusion will be discussed in detail in later sections [14, 15, 20–22, 55–57].

3.2.1 Self-Organizing Map Features

Self-Organizing Maps (SOMs), also known as Kohonen maps, are unsupervised artificial neural networks designed for automatic data analysis, vector quantization, clustering, and non-linear dimensionality reduction on a usually two-dimensional grid [18]. SOM specific features include the ability of easily handling big data, tolerance to outlier values, and data noise reduction. The latter is a property that comes from the way the algorithm works.

In SOM, the samples, which are represented as n-dimensional vectors, are grouped by a similarity measure (e.g. Euclidean distance), and in the output they are represented by a so-called neuron, which is an n-dimensional vector composed by modelled values of the n variables. The model has to be initialized selecting how many neurons will represent the samples. The neurons are arranged in a two-dimensional grid. The neuron number and grid dimensions are usually evaluated using Vesanto's heuristic rules [23]. The neuron vector values are either randomly initialized or, more frequently, initialized by using the first two principal components derived by PCA applied to the dataset. Then the algorithm selects a sample and finds the more similar neuron, which is called best matching unit (BMU). Next the BMU values and the values of the neighbour neurons are adjusted to be "more similar" to the sample. This process is iterated several times (epochs) for all the samples. In this way the model "learns" from the experimental dataset finally representing its variability with fewer vectors. The outcome of the model is a list of vectors representing the neurons and thus the values for the modelled variables. This list is called codebook.

The key property of SOM is that the neurons are automatically organized on a grid in such a way that more similar neurons become associated together, whereas less similar ones are placed farther apart. The neurons are presented either as squares or hexagons.

Unlike linear methods such as PCA, SOMs perform non-linear dimensionality reduction, in the sense that the mapping from input space to the two-dimensional grid can represent curved manifolds and complex structures that are not well captured by linear projections. Distances between nodes in the grid are not predetermined or homogeneous but emerge from the training process, depending on similarities in the input data; this allows SOM to approximate non-linear manifolds where similar samples are grouped and mapped together in a bidimensional representation. As a result, SOM can reveal clusters and transitions in high-dimensional environmental data that might remain hidden or distorted in purely linear projections (Fig. 3).

A powerful diagnostic tool derived from SOM is the set of **component planes**, which display, for each input variable (e.g. each chemical species), the values of that variable across all neurons on the map. By examining component planes side by side, one can assess how different variables co-vary across the map and identify groups of variables that share similar spatial patterns, which may correspond to source-related signatures or environmental regimes.

Another widely used SOM output is the **U-matrix** (unified distance matrix), which represents the distances between neighbour neurons. In the U-matrix, high values indicate large dissimilarities between neighbouring neurons and thus potential cluster borders, while low values indicate regions of relatively homogeneous neurons corresponding to clusters [19] (Fig. 4).

Each sample can be assigned to the neuron corresponding to its BMU; this mapping enables direct visualisation of the distribution of samples (hits) across the two-dimensional grid. Outlier detection can be achieved exploring the distances of the samples from their BMU: the more the distance, the more the possibility that the specific sample can be an outlier also with respect to the other samples with the same BMU [55, 58] (Fig. 5).

In many applications, cluster analysis, such as KM or HC, is applied to the codebook neuron derived from SOM to obtain a discrete partition of the map into clusters, each representing a group of similar neurons and, by extension, the samples mapped by those BMUs (hits). One of the main advantages of applying clustering algorithms to the SOM output is that the algorithms are fed with data already cleaned both from noise and outliers, thus leading to a clearer interpretation of cluster profiles.!! This process yields to combined SOM–clustering models that enhance interpretability and allow clear association of map regions with distinct patterns represented by clusters built (Fig. 6).

3.2.2 Applications of Self-Organizing Maps in Environmental Pollution and Source Apportionment

For environmental monitoring and SA, SOM offer several advantages that motivate their use alongside traditional receptor models and other ML methods. They can

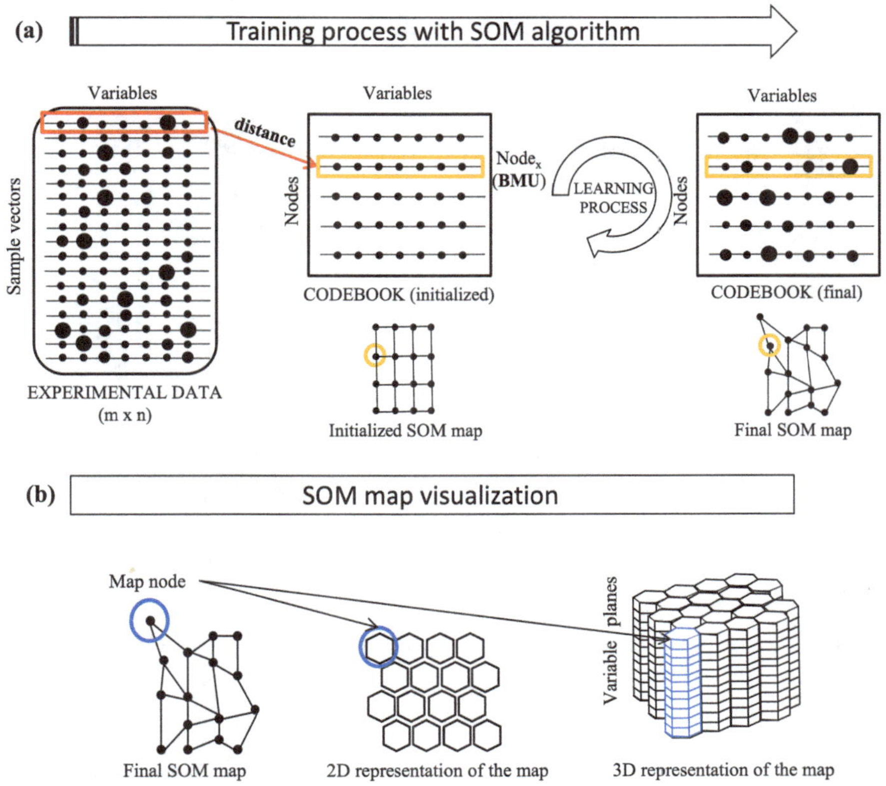

Fig. 3 SOM learning process a), and output visualization b). Node, neuron; BMU, Best Matching Unit. (Reproduced from Licen et al. [20])

handle high-dimensional datasets, including numerous chemical species, size-resolved particle counts, and auxiliary variables, compressing them into a manageable number of prototypes and a two-dimensional representation. They enable intuitive visualisation of patterns, such as temporal regimes or spatial gradients, and can be calibrated with known classes (e.g. source types) to act as semi-supervised classifiers for new samples.

Moreover, when combined with clustering and external information such as meteorology, SOM can support SA by grouping samples according to their compositional similarity and relating map regions to specific source influences or environmental scenarios [23]. In subsequent sections, examples from water quality, marine systems, air particulate characterisation, and odour monitoring will illustrate how SOM-based workflows have been used to identify spatial and temporal pollution patterns, compare outcomes with PMF and other multivariate methods, and perform data fusion across high-frequency and low-frequency measurements in complex monitoring networks.

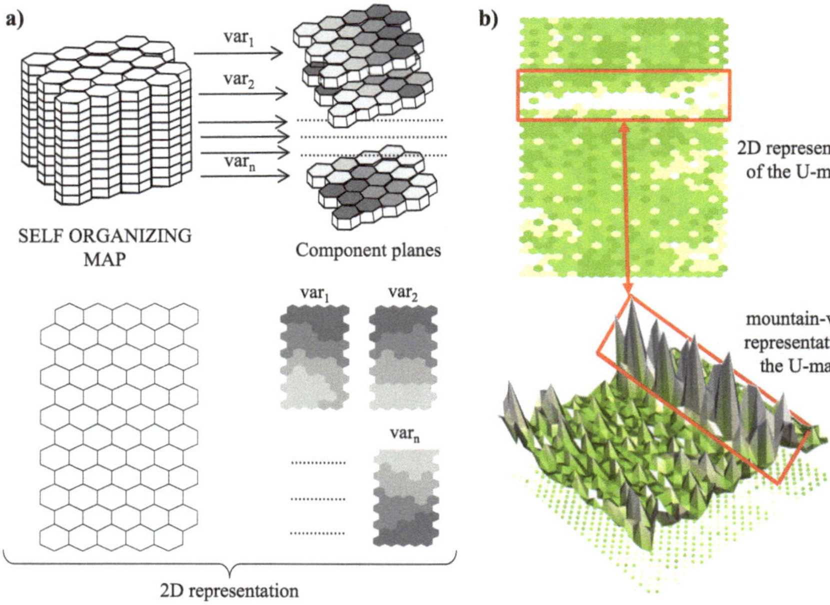

Fig. 4 SOM component planes and U-matrix representation. (Reproduced from Licen et al. [20])

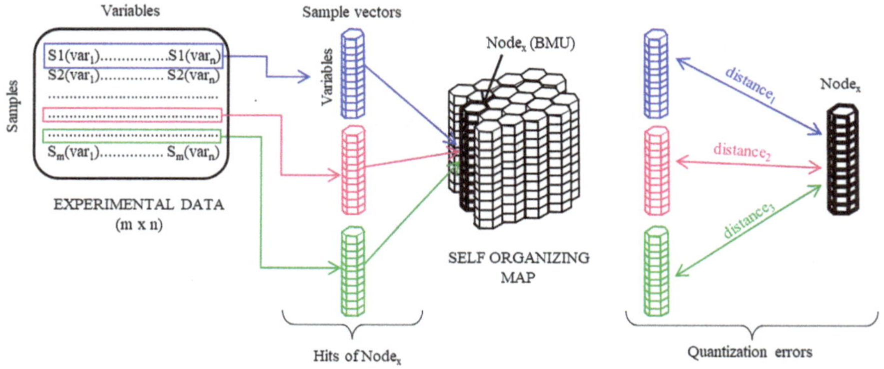

Fig. 5 SOM output features: hits and quantization errors. Node, neuron; BMU, best matching unit. (Reproduced from Licen et al. [20])

SOM in Waters and Aquatic Systems

Barbieri and co-workers have used SOM to analyse high-dimensional monitoring data from coastal and transitional waters, grouping sampling stations and periods according to multivariate signatures of nutrients, metals, and other contaminants [15]. In these studies, SOM maps made it possible to distinguish regimes associated with riverine inputs, offshore waters, and locally impacted coastal zones, thereby supporting the identification of pollution hot spots and their likely drivers in complex marine and lagoon environments.

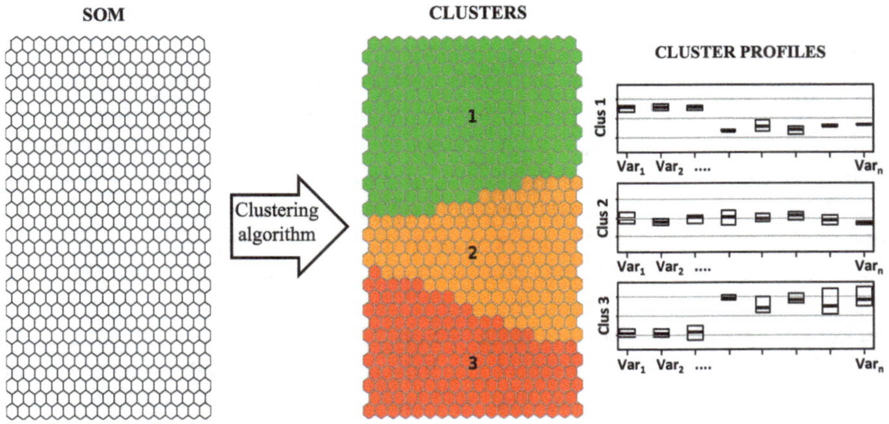

Fig. 6 Clustering algorithm applied to SOM output

In river and freshwater systems, SOM has been applied to time series of physicochemical parameters and contaminant concentrations to disentangle seasonal patterns, flow-related changes, and pollution episodes along river networks. Jiang et al. assessed the water quality of two different urban rivers in China identifying different sources using SOM implemented by KM [59]. The resulting maps typically reveal gradients from pristine or weakly impacted upstream segments to more contaminated downstream reaches, offering an intuitive visual framework for linking spatial classes to land use, wastewater discharges, and diffuse agricultural inputs.

Gelao et al. elaborated untargeted wastewater analysis results of both urban and industrial treatment plants, and, using SOM and HC, they were able to identify industrial source intrusion in urban plants even if considering common compounds to both plants only, thus without previous knowledge on industrial fingerprints [60].

SOM in Sediment and Soil Systems

SOM combined with further clustering has been successfully used by Vignati et al. to achieve pollutant source identification and distribution in river sediments being able to discriminate natural and anthropogenic sources. Hossain Bhuyian et al. were able to distinguish different pollution sources in soil, namely, sewage irrigation, atmospheric deposition, and industrial specific fingerprints, such as leather tanning [61].

SOM for Air Quality Fingerprinting

Licen and colleagues have employed SOM to analyse long-term datasets of particulate matter size distributions and composition, as well as related air quality variables, using the maps to cluster measurement days or sampling sites into recurrent "air types" characterised by distinct multi-pollutant profiles [23]. In these applications, SOM has helped to separate regimes dominated by secondary aerosols from those dominated by primary combustion sources, to detect site-specific signatures, and to identify atypical events or transitions that might be missed by traditional univariate indicators.

Beyond descriptive clustering, SOM-based analyses of atmospheric particulate matter have been used to support or interpret source apportionment by highlighting temporal regimes consistent with specific source patterns, such as traffic-dominated winter days or secondary-aerosol summer episodes [62]. This regime-based organisation of data facilitates the comparison of SOM clusters with factor contributions obtained from receptor models and can guide the selection of periods or subsets of data for separate PMF runs when non-stationary behaviour is suspected [63].

SOM in Odour Monitoring and Data Fusion

In environmental odour studies, SOM has been used as a core component of "odour control maps" constructed from electronic-nose signals collected near industrial or waste management facilities [22]. In these cases, raw e-nose sensor responses are projected onto a SOM to identify recurrent multivariate patterns, which are then grouped into a limited number of "air types" corresponding to different odour situations at the receptor (e.g. background, moderate odour, or intense odour linked to specific sources).

Barbieri, Licen and co-authors have extended these approaches by integrating heterogeneous data streams—such as e-nose outputs, gaseous pollutant concentrations, particulate indices, and meteorological variables—within a single SOM-based framework for data fusion. This joint organisation enables the identification of regimes where combinations of odour intensity, chemical markers, and meteorological conditions co-occur, supporting both diagnostic assessments of odour episodes and the design of targeted mitigation actions [22].

Combined SOM–PMF Approaches

In soil pollution studies, PMF remains the reference tool for quantitative source apportionment of potentially toxic elements, providing estimates of source profiles and contributions that can be directly related to risk indicators and regulatory thresholds. Recent hybrid approaches have combined SOM and PMF with additional machine learning components to capture spatial patterns and non-linear relationships, for example, in urban or industrial soils contaminated by heavy metals where SOM is used to characterise spatial regimes, while PMF quantifies the contributions of industrial, traffic, and natural sources [64].

In these settings, SOM-derived clusters of sampling locations or soil types often show strong correspondence with the spatial distribution of PMF factors: areas classified by SOM as "industrial influence" or "traffic-dominated" tend to coincide with zones where PMF assigns high contributions to industrial or vehicular sources, respectively. Such agreement strengthens the environmental plausibility of both methods, while discrepancies can highlight zones where additional processes (e.g. legacy contamination, sediment transport, or unaccounted sources) require further investigation or refined modelling [65].

In water body and groundwater studies, results obtained using SOM followed by cluster analysis were compared to PMF results to achieve independent and validated assessment of source profiles of perfluorinated compounds [66] and inorganic compounds [67]. The outcomes were used for health risk assessment and therefore pollution prevention and control.

For air quality applications, combined SOM–PMF workflows have been proposed to disentangle multi-annual, multi-site datasets in which temporal and spatial heterogeneity complicates direct receptor modelling. One strategy is to use SOM to organise days or observation periods into a low-dimensional map of multipollutant regimes and then to apply PMF either to the full dataset or to subsets associated with specific SOM clusters, improving interpretability and stability of factor solutions [63].

Studies that jointly use SOM and PMF in air quality typically report that SOM-identified regimes dominated by certain combinations of pollutants (e.g. traffic-related species, secondary aerosol markers, or biomass-burning tracers) align well with PMF source factors representing traffic, secondary inorganic aerosol, or biomass burning. When SOM classes and PMF factors converge in this way—across different compartments such as air, soils, and sediments—the resulting cross-validation supports the robustness of the inferred source contributions and enhances confidence in using these outcomes for regulatory assessment and management decisions [68, 69].

4 Perspectives and Open Challenges

Machine learning methods, and SOM in particular, are increasingly recognised as valuable complements to receptor models in environmental source apportionment, yet their integration into standard practice is still at an early stage. Future developments will need to reconcile the flexibility of ML with the transparency, comparability, and quality assurance requirements that already underpin receptor-model guidelines in air quality management.

4.1 Interpretable ML and Hybrid Workflows

A central challenge is to ensure interpretability of ML models used alongside or on top of receptor approaches, so that inferred patterns and regimes can be robustly linked to physically meaningful sources and processes. SOM offers an advantage in this respect, because its low-dimensional maps can be directly inspected and related to environmental variables, but systematic strategies are still needed to connect SOM units and clusters to quantitative source contributions, health-relevant metrics, and policy indicators.

Hybrid SOM–PMF workflows illustrate a promising direction, using SOM to organise complex datasets and PMF (or other receptor models) to provide quantitative source apportionment within or across SOM-defined regimes. However, best practices for designing, training, and validating such combined systems—especially in multi-site, multi-pollutant, or multi-compartment applications—have not yet been codified, and questions remain about how to propagate uncertainties coherently between the ML and receptor-model components.

4.2 Harmonised Guidelines and QA/QC

For receptor models such as PMF, European guidance documents and technical specifications already define minimum requirements for input data quality, model configuration, uncertainty analysis, and reporting [30, 70]. By contrast, ML-based tools—including SOM—are usually applied in an exploratory or case-study-specific manner, with heterogeneous choices of preprocessing, map size and topology, training parameters, and post-processing, which hampers reproducibility and comparability across studies.

Recent reviews of SOM applications in environmental compartments have started to formulate practical recommendations on documenting algorithm settings, validation strategies, and interpretation procedures, but these suggestions are not yet embedded in regulatory or consensus frameworks. An important open challenge is therefore to extend the existing receptor-model QA/QC and reporting culture to hybrid ML–receptor workflows, including clear criteria for model selection, sensitivity analyses, and performance evaluation that can be routinely applied in policy-relevant source-apportionment studies.

4.3 Data Fusion, Emerging Data Streams, and Non-stationarity

Another key perspective concerns the exploitation of new data streams—such as high-frequency sensor networks, satellite products, online mass-spectrometry measurements, and citizen-science observations—and their integration with classical receptor-model inputs. SOM-based data fusion frameworks are naturally suited to organise heterogeneous variables and sampling frequencies into coherent regimes, but it remains challenging to translate these regimes into quantitative source contributions compatible with existing regulatory indicators and concentration-based targets.

Non-stationarity in emissions, atmospheric processes, and environmental regulations introduces additional complexity: source contributions inferred for one period or regime may not apply to others, and models must be able to track and explain such changes. Hybrid SOM–PMF or more general ML–receptor approaches provide a framework for handling regime-dependent behaviour, yet practical methods for detecting regime shifts, updating models, and communicating time-varying source apportionment to stakeholders remain active areas of research.

4.4 Towards Operational Use in Policy and Management

To move from exploratory analyses to operational use in environmental policy and management, SOM-based and other ML-enhanced approaches must demonstrate added value over existing receptor-model workflows in terms of robustness, diagnostic power, and decision support. This includes showing that hybrid methods can improve attribution of sources under complex pollution scenarios, better quantify the impact of emission-control strategies, and support exposure and health impact assessments while remaining transparent enough to be scrutinised by regulators and stakeholders.

Developing open, well-documented software tools, benchmark datasets, and intercomparison exercises for SOM-based and ML-enhanced source apportionment would help consolidate the field and foster convergence towards shared standards. In this context, SOM can be seen as a bridge between classical multivariate statistics and more complex ML architectures, offering an interpretable platform on which to build integrated, quality-assured approaches to source apportionment and environmental decision- making in the coming years.

Questions

- Which are the conceptual foundations of source apportionment? How the information from receptor sites is linked to emission sources?
- Which are the main characteristics and limitations of receptor models such as CMB and PMF?
- How is machine learning used in environmental pollution studies? Which is the difference between supervised and unsupervised methods?
- How SOM algorithm works and which are the useful peculiarities of its outcomes that are useful for identifying spatial and temporal pollution patterns?
- Why is it useful to combine SOM and PMF workflows for assessing environmental pollution sources?
- Which are the emerging perspectives and open challenges in integrating interpretable ML approaches into operational source apportionment?

Glossary

- **Best matching unit (BMU)**
  Neuron on the SOM grid whose prototype vector is most similar to a given input sample, used to drive learning and mapping.
- **Chemical mass balance (CMB)**
  Deterministic receptor model that fits observed concentrations as a linear combination of known source profiles using weighted least squares, assuming profiles and number of sources are known.

(continued)

- **Clustering**
  Unsupervised technique that groups observations into classes so that objects in the same group are more similar to each other than to those in other groups.
- **Data fusion**
  Integration of heterogeneous datasets (e.g. chemical, particulate, odour, meteorological data) into a unified analytical framework to extract joint patterns or regimes.
- **Machine learning (ML)**
  Family of algorithms that learn patterns and relationships from data, used here for clustering, regression, classification, and feature extraction in environmental datasets.
- **Multilayer perceptron (MLP)**
  Feedforward neural network with one or more hidden layers and non-linear activation functions, used to model complex, non-linear relationships between inputs and outputs.
- **Neuron vector (codebook vector)**
  Weight vector associated with a SOM neuron that represents a local average pattern in the input space.
- **Non-stationarity**
  Condition where statistical properties or source contributions change over time, requiring models that can handle regime shifts and evolving patterns.
- **Oxidative potential (OP)**
  Effect-based metric of particulate matter that quantifies its capacity to generate reactive oxygen species in acellular or cellular assays.
- **Positive matrix factorization (PMF)**
  Factor-analytic receptor model that decomposes a concentration matrix into non-negative source profiles and contributions, optimising fit using uncertainties of data points.
- **Quality assurance/quality control (QA/QC)**
  Procedures and criteria ensuring that data, models, and results meet agreed standards of accuracy, consistency and transparency.
- **Receptor model**
  Model that infers source contributions from pollutant concentrations measured at receptors, without explicitly simulating transport and chemistry from emission sources.
- **Regime (air quality regime)**
  Recurrent multipollutant pattern or state of the system, identified by clustering or SOM, characterized by a specific combination of concentrations and conditions.
- **Self-Organizing Map (SOM)**
  Unsupervised neural network that projects high-dimensional data onto a usually two-dimensional grid, preserving topology and supporting visualisation and clustering.

- **Source apportionment (SA)**
  Statistical and modelling approaches used to quantify the contributions of different emission sources to measured pollutant levels at a receptor site.
- **Source contribution**
  Amount (mass or concentration) by which a given source type contributes to pollutant levels in each sample or time step.
- **Source profile (fingerprint)**
  Characteristic composition of emissions from a single source type, expressed as relative or absolute abundances of chemical species.
- **Supervised learning**
  ML approach where models are trained using input–output pairs (features and labelled targets) to make predictions for new data.
- **Uncertainty (in receptor modelling)**
  Quantitative estimate of measurement and modelling error used to weight observations in PMF and to evaluate robustness of factor solutions.
- **Unsupervised learning**
  ML approach that discovers structure (clusters, manifolds, regimes) in data without predefined labels, e.g. clustering and SOM.

Bibliography

1. Belis C, Favez O, Mircea M, Diapouli E, Manousakas M-I, Vratolis S, Gilardoni S, Paglione M, Decesari S, Mocnik G et al (2019) European guide on air pollution source apportionment with receptor models, EUR 29816 EN; OP KJ-NA-29816-EN-N. Publications Office of the European Union, Luxembourg. https://doi.org/10.2760/439106
2. Liu X, Lu D, Zhang A, Liu Q, Jiang G (2022) Data-driven machine learning in environmental pollution: gains and problems. Environ Sci Technol 56(4):2124–2133. https://doi.org/10.1021/acs.est.1c06157
3. Hopke PK (2016) Review of receptor modeling methods for source apportionment. J Air Waste Manage Assoc 66(3):237–259. https://doi.org/10.1080/10962247.2016.1140693
4. Watson JG (1984) Overview of receptor model principles. J Air Pollut Control Assoc 34(6):619–623. https://doi.org/10.1080/00022470.1984.10465780
5. Hopke PK (1989) Target transformation factor analysis. Chemom Intell Lab Syst 6(1):7–19. https://doi.org/10.1016/0169-7439(89)80061-9
6. Paatero P, Tapper U (1994) Positive matrix factorization: a non-negative factor model with optimal utilization of error estimates of data values. Environmetrics 5(2):111–126. https://doi.org/10.1002/env.3170050203
7. Belis CA, Pernigotti D, Pirovano G, Favez O, Jaffrezo JL, Kuenen J, Denier Van Der Gon H, Reizer M, Riffault V, Alleman LY et al (2020) Evaluation of Receptor and Chemical Transport Models for PM10 Source Apportionment. Atmospheric Environment: X 5:100053. https://doi.org/10.1016/j.aeaoa.2019.100053
8. CEN/TS 17458:2020 Ambient Air—Methodology to assess the performance of receptor oriented source apportionment modelling applications for particulate matter, 2021
9. Ladyman J, Lambert J, Wiesner K (2013) What is a complex system? Eur J Philos Sci 3(1):33–67. https://doi.org/10.1007/s13194-012-0056-8

10. Borlaza LJS, Weber S, Jaffrezo J-L, Houdier S, Slama R, Rieux C, Albinet A, Micallef S, Trébluchon C, Uzu G (2021) Disparities in particulate matter (PM_{10}) origins and oxidative potential at a city scale (Grenoble, France)—part 2: sources of PM_{10} oxidative potential using multiple linear regression analysis and the predictive applicability of multilayer perceptron neural network analysis. Atmos Chem Phys 21(12):9719–9739. https://doi.org/10.5194/acp-21-9719-2021
11. Borlaza LJS, Weber S, Uzu G, Jacob V, Cañete T, Micallef S, Trébuchon C, Slama R, Favez O, Jaffrezo J-L (2021) Disparities in particulate matter (PM_{10}) origins and oxidative potential at a city scale (Grenoble, France)—part 1: source apportionment at three neighbouring sites. Atmos Chem Phys 21(7):5415–5437. https://doi.org/10.5194/acp-21-5415-2021
12. Paatero P (1997) Least squares formulation of robust non-negative factor analysis. Chemom Intell Lab Syst 37(1):23–35. https://doi.org/10.1016/S0169-7439(96)00044-5
13. Hemann JG, Brinkman GL, Dutton SJ, Hannigan MP, Milford JB, Miller SL (2009) Assessing positive matrix factorization model fit: a new method to estimate uncertainty and bias in factor contributions at the measurement time scale. Atmos Chem Phys 9(2):497–513. https://doi.org/10.5194/acp-9-497-2009
14. Astel A, Tsakovski S, Barbieri P, Simeonov V (2007) Comparison of self-organizing maps classification approach with cluster and principal components analysis for large environmental data sets. Water Res 41(19):4566–4578. https://doi.org/10.1016/j.watres.2007.06.030
15. Solidoro C, Bandelj V, Barbieri P, Cossarini G, Fonda Umani S (2007) Understanding dynamic of biogeochemical properties in the Northern Adriatic Sea by using self-organizing maps and K-means clustering. J Geophys Res 112(C7):2006JC003553. https://doi.org/10.1029/2006JC003553
16. Hsu C-Y, Soo J-C, Lin S-L, Wu C-D, Chi KH, Hsu W-C, Tseng C-C, Chen Y-C (2023) Using cluster algorithms with a machine learning technique and PMF models to quantify local-specific origins of PM2.5 and associated metals in Taiwan. Environ Pollut 316:120652. https://doi.org/10.1016/j.envpol.2022.120652
17. Kumar V, Sahu M, Biswas P (2022) Source apportionment of particulate matter by application of machine learning clustering algorithms. Aerosol Air Qual Res 22(3):210240. https://doi.org/10.4209/aaqr.210240
18. Kohonen T (2013) Essentials of the self-organizing map. Neural Netw 37:52–65. https://doi.org/10.1016/j.neunet.2012.09.018
19. Saxena A, Prasad M, Gupta A, Bharill N, Patel OP, Tiwari A, Er MJ, Ding W, Lin C-T (2017) A review of clustering techniques and developments. Neurocomputing 267:664–681. https://doi.org/10.1016/j.neucom.2017.06.053
20. Licen S, Astel A, Tsakovski S (2023) Self-organizing map algorithm for assessing spatial and temporal patterns of pollutants in environmental compartments: a review. Sci Total Environ 878:163084. https://doi.org/10.1016/j.scitotenv.2023.163084
21. Astel A, Tsakovski S, Simeonov V, Reisenhofer E, Piselli S, Barbieri P (2008) Multivariate classification and modeling in surface water pollution estimation. Anal Bioanal Chem 390(5):1283–1292. https://doi.org/10.1007/s00216-007-1700-6
22. Licen S, Barbieri G, Fabbris A, Briguglio SC, Pillon A, Stel F, Barbieri P (2018) Odor control map: self organizing map built from electronic nose signals and integrated by different instrumental and sensorial data to obtain an assessment tool for real environmental scenarios. Sensors Actuators B Chem 263:476–485. https://doi.org/10.1016/j.snb.2018.02.144
23. Licen S, Cozzutto S, Barbieri P (2020) Assessment and comparison of multi-annual size profiles of particulate matter monitored at an urban-industrial site by an optical particle counter with a Chemometric approach. Aerosol Air Qual Res 20(4):800–809. https://doi.org/10.4209/aaqr.2019.08.0414
24. Henry RC, Lewis CW, Hopke PK, Williamson HJ (**1984**) Review of receptor model fundamentals. Atmos Environ (1967) 18(8):1507–1515. https://doi.org/10.1016/0004-6981(84)90375-5
25. Hopke PK, Chen Y, Rich DQ, Mooibroek D, Sofowote UM (2023) The application of positive matrix factorization with diagnostics to BIG DATA. Chemom Intell Lab Syst 240:104885. https://doi.org/10.1016/j.chemolab.2023.104885

26. Thurston GD, Spengler JD (1985) A quantitative assessment of source contributions to inhalable particulate matter pollution in metropolitan Boston. Atmos Environ (1967) 19(1):9–25. https://doi.org/10.1016/0004-6981(85)90132-5
27. Mu D, Meng J, Wang S, Xiao S, Wang H, Sun X, Wu P (2024) Source apportionment, source-specific health risks, and control factors of heavy metals in water bodies of a typical Karst Basin in Southwestern China. PLoS One 19(8):e0309142. https://doi.org/10.1371/journal.pone.0309142
28. Haji Gholizadeh M, Melesse AM, Reddi L (2016) Water quality assessment and apportionment of pollution sources using APCS-MLR and PMF receptor modeling techniques in three major rivers of South Florida. Sci Total Environ 566–567:1552–1567. https://doi.org/10.1016/j.scitotenv.2016.06.046
29. Gong C, Xia X, Lan M, Shi Y, Lu H, Wang S, Chen Y (2024) Source identification and driving factor apportionment for soil potentially toxic elements via combining APCS-MLR, UNMIX, PMF and GDM. Sci Rep 14(1):10918. https://doi.org/10.1038/s41598-024-58673-9
30. HJ 1351-2024 Ambient air—source apportionment on particulate matter—technical specification for method with receptor models based on manual monitoring (English Version), 2024
31. Kumar S, Mishra S, Singh SK (2020) A machine learning-based model to estimate PM2.5 concentration levels in Delhi's atmosphere. Heliyon 6(11):e05618. https://doi.org/10.1016/j.heliyon.2020.e05618
32. Suzaimi ND, Abuhabib A, Darwish M, AlHadi AC, Mohammad AW, Hamzah S (2026) Machine learning in industrial wastewater treatment: critical insights into modelling approaches and digital futures. J Ind Eng Chem:S1226086X25008275. https://doi.org/10.1016/j.jiec.2025.12.035
33. Xiong Z, Cheng L, Yunus AI, Jian D, Zhang X, Igou T, Tong S, Chen Y (2025) Machine learning-assisted expert control in wastewater aeration treatment process application. ACS EST Eng 5(11):3140–3152. https://doi.org/10.1021/acsestengg.5c00493
34. Nkinahamira F, Feng A, Zhang L, Rong H, Ndagijimana P, Guo D, Cui B, Zhang H (2024) Machine learning approaches for monitoring environmental metal pollutants: recent advances in source apportionment, detection, quantification, and risk assessment. TrAC Trends Anal Chem 180:117980. https://doi.org/10.1016/j.trac.2024.117980
35. Wang Y, Xu W, Zhong H, Chen B, Yang L, Wang L, Ren J, Duan J, Lin C, Huang R-J (2025) Advancing aerosol chemistry with machine learning: a short review. ACS EST Air 2(11):2323–2341. https://doi.org/10.1021/acsestair.5c00001
36. Kumar V, Malyan V, Sahu M, Biswal B (2023) Machine learning classification model to label sources derived from factor analysis receptor models for source apportionment. Aerosol Air Qual Res 23(7):220386. https://doi.org/10.4209/aaqr.220386
37. Zhang H, Yin A, Yang X, Fan M, Shao S, Wu J, Wu P, Zhang M, Gao C (2021) Use of machine-learning and receptor models for prediction and source apportionment of heavy metals in coastal reclaimed soils. Ecol Indic 122:107233. https://doi.org/10.1016/j.ecolind.2020.107233
38. Zhang Y, Fang J, Meng Q, Ge X, Chebaicheb H, Favez O, Petit J-E (2025) An ensemble machine learning approach for predicting sources of organic aerosols measured by aerosol mass spectrometry. ACS EST Air 2(3):378–385. https://doi.org/10.1021/acsestair.4c00262
39. Makhdoomi A, Sarkhosh M, Ziaei S (2025) PM2.5 concentration prediction using machine learning algorithms: an approach to virtual monitoring stations. Sci Rep 15(1):8076. https://doi.org/10.1038/s41598-025-92019-3
40. Sapper BC, Youn S, Henze DK, Canagaratna M, Stark H, Jimenez JL (2025) Positive matrix factorization of large real-time atmospheric mass spectrometry datasets using error-weighted randomized hierarchical alternating least squares. Geosci Model Dev 18(10):2891–2919. https://doi.org/10.5194/gmd-18-2891-2025
41. Chao J, Gu H, Liao Q, Zuo W, Qi C, Liu J, Tian C, Lin Z (2025) Natural factor-based spatial prediction and source apportionment of typical heavy metals in Chinese surface soil: application of machine learning models. Environ Pollut 366:125373. https://doi.org/10.1016/j.envpol.2024.125373

42. Zheng J, Wang P, Shi H, Zhuang C, Deng Y, Yang X, Huang F, Xiao R (2023) Quantitative source apportionment and driver identification of soil heavy metals using advanced machine learning techniques. Sci Total Environ 873:162371. https://doi.org/10.1016/j.scitotenv.2023.162371
43. Liu Z, Zeng Y, Wang W, Zhang B, Liu J, Zhao X (2026) Coupling positive matrix factorization with machine learning to quantify contamination sources and map health risks of heavy metals in karst soils. Soil Environ Health 4(1):100193. https://doi.org/10.1016/j.seh.2026.100193
44. Zhang Y, Favez O, Petit J-E, Canonaco F, Truong F, Bonnaire N, Crenn V, Amodeo T, Prévôt ASH, Sciare J et al (2019) Six-year source apportionment of submicron organic aerosols from near-continuous highly time-resolved measurements at SIRTA (Paris area, France). Atmos Chem Phys 19(23):14755–14776. https://doi.org/10.5194/acp-19-14755-2019
45. Xu Y, Li P, Zhang M, Xiao L, Wang B, Zhang X, Wang Y, Shi P (2024) Quantifying seasonal variations in pollution sources with machine learning-enhanced positive matrix factorization. Ecol Indic 166:112543. https://doi.org/10.1016/j.ecolind.2024.112543
46. Zhang H, Ren X, Chen S, Xie G, Hu Y, Gao D, Tian X, Xiao J, Wang H (2024) Deep optimization of water quality index and positive matrix factorization models for water quality evaluation and pollution source apportionment using a random Forest model. Environ Pollut 347:123771. https://doi.org/10.1016/j.envpol.2024.123771
47. Borlaza LJS, Uzu G, Ouidir M, Lyon-Caen S, Marsal A, Weber S, Siroux V, Lepeule J, Boudier A, Jaffrezo J-L et al (2023) Personal exposure to PM2.5 oxidative potential and its association to birth outcomes. J Expo Sci Environ Epidemiol 33(3):416–426. https://doi.org/10.1038/s41370-022-00487-w
48. Weber S, Uzu G, Favez O, Borlaza LJS, Calas A, Salameh D, Chevrier F, Allard J, Besombes J-L, Albinet A et al (2021) Source apportionment of atmospheric PM_{10} oxidative potential: synthesis of 15 year-round urban datasets in France. Atmos Chem Phys 21(14):11353–11378. https://doi.org/10.5194/acp-21-11353-2021
49. Govender P, Sivakumar V (2020) Application of K-means and hierarchical clustering techniques for analysis of air pollution: a review (1980–2019). Atmos Pollut Res 11(1):40–56. https://doi.org/10.1016/j.apr.2019.09.009
50. Jain AK, Murty MN, Flynn PJ (1999) Data clustering: a review. ACM Comput Surv 31(3):264–323. https://doi.org/10.1145/331499.331504
51. Jouanny A, Upadhyay A, Jiang J, Vasilakos P, Via M, Cheng Y, Flueckiger B, Uzu G, Jaffrezo J-L, Voiron C et al (2025) Machine-learning-driven reconstruction of organic aerosol sources across dense monitoring networks in Europe. Environ Sci Technol Lett 12(11):1523–1531. https://doi.org/10.1021/acs.estlett.5c00771
52. Zhang S, Ding W (2025) Research on CatBoost model based on AutoEncoder dimensionality reduction in pollution source apportionment. Environ Geochem Health 47(12):543. https://doi.org/10.1007/s10653-025-02814-0
53. Liu P, Pan D, Jiao X-Y, Liu J-N, Du P-H, Li P-C, Xue M-Z, Jin Y-C, Wang C-S, Wang X-R et al (2025) Integrating non-target analysis and machine learning: a framework for contaminant source identification. Npj Clean Water 8(1):78. https://doi.org/10.1038/s41545-025-00504-z
54. Cejas MJ, Barrick RC (2021) Chemometric mapping of polychlorinated Dibenzo- *p* -dioxin (PCDD) and dibenzofuran (PCDF) congeners from the Passaic River, NJ: integrated application of RSIMCA, PVA, and t-SNE. Environ Forensic 22(1–2):155–171. https://doi.org/10.1080/15275922.2020.1834022
55. Licen S, Cozzutto S, Barbieri G, Crosera M, Adami G, Barbieri P (2019) Characterization of variability of air particulate matter size profiles recorded by optical particle counters near a complex emissive source by use of self-organizing map algorithm. Chemom Intell Lab Syst 190:48–54. https://doi.org/10.1016/j.chemolab.2019.05.008
56. Vraćko M, Szymoszek A, Barbieri P (2004) Structure-mutagenicity study of 12 trimethylimidazopyridine isomers using orbital energies and "Spectrum-like representation" as descriptors. J Chem Inf Comput Sci 44(2):352–358. https://doi.org/10.1021/ci030420i

57. Vračko M, Bandelj V, Barbieri P, Benfenati E, Chaudhry Q, Cronin M, Devillers J, Gallegos A, Gini G, Gramatica P et al (2006) Validation of counter propagation neural network models for predictive toxicology according to the OECD principles: a case study. SAR QSAR Environ Res 17(3):265–284. https://doi.org/10.1080/10659360600787650
58. Muñoz A, Muruzábal J (1998) Self-organizing maps for outlier detection. Neurocomputing 18(1–3):33–60. https://doi.org/10.1016/S0925-2312(97)00068-4
59. Jiang Y, Gui H, Li C, Chen J, Chen C, Wang C, Zhao H, Guo Y, Xu J, Li J et al (2021) Evaluation of the difference in water quality between urban and suburban Rivers based on self-organizing map. Acta Geophys 69(5):1855–1864. https://doi.org/10.1007/s11600-021-00631-4
60. Gelao V, Fornasaro S, Briguglio SC, Mattiussi M, De Martin S, Astel AM, Barbieri P, Licen S (2024) Self-organizing maps: an AI tool for identifying unexpected source signatures in non-target screening analysis of urban wastewater by HPLC-HRMS. Toxics 12(2):113. https://doi.org/10.3390/toxics12020113
61. Hossain Bhuiyan MA, Chandra Karmaker S, Bodrud-Doza M, Rakib MA, Saha BB (2021) Enrichment, sources and ecological risk mapping of heavy metals in agricultural soils of Dhaka District employing SOM, PMF and GIS methods. Chemosphere 263:128339. https://doi.org/10.1016/j.chemosphere.2020.128339
62. De Oliveira RH, Carneiro CDC, De Almeida FGV, De Oliveira BM, Nunes EHM, Dos Santos AS (2019) Multivariate air pollution classification in urban areas using Mobile sensors and self-organizing maps. Int J Environ Sci Technol 16(10):5475–5488. https://doi.org/10.1007/s13762-018-2060-9
63. Hassan MS, Bhuiyan MAH, Rahman MT (2023) Sources, pattern, and possible health impacts of PM2.5 in the central region of Bangladesh using PMF, SOM, and machine learning techniques. Case Stud Chem Environ Eng 8:100366. https://doi.org/10.1016/j.cscee.2023.100366
64. He Y, Han X, Ge J, Wang L (2022) Multivariate statistical analysis of potentially toxic elements in soils under different land uses: spatial relationship, ecological risk assessment, and source identification. Environ Geochem Health 44(3):847–860. https://doi.org/10.1007/s10653-021-00992-1
65. Li J, Wang G, Liu F, Cui L, Jiao Y (2021) Source apportionment and ecological-health risks assessment of heavy metals in topsoil near a factory, Central China. Expo Health 13(1):79–92. https://doi.org/10.1007/s12403-020-00363-8
66. Zeng J, Liu K, Liu X, Tang Z, Wang X, Fu R, Lin X, Liu N, Qiu J (2024) Driving factor, source identification, and health risk of PFAS contamination in groundwater based on the self-organizing map. Water Res 267:122458. https://doi.org/10.1016/j.watres.2024.122458
67. Hu B, Li P, Chen R, Zheng Y, Li J, Li Y (2024) The new region demarcation framework for implementing the joint prevention and control of groundwater pollution: a case study in Western of Bohai Bay, China. J Hazard Mater 478:135565. https://doi.org/10.1016/j.jhazmat.2024.135565
68. Zhang Y, Zhang Q, Chen W, Shi W, Cui Y, Chen L, Shao J (2023) Source apportionment and migration characteristics of heavy metal(loid)s in soil and groundwater of contaminated site. Environ Pollut 338:122584. https://doi.org/10.1016/j.envpol.2023.122584
69. Fornasaro S, Astel A, Barbieri P, Licen S (2025) Disentangling multiannual air quality profiles aided by self-organizing map and positive matrix factorization. Toxics 13(2):137. https://doi.org/10.3390/toxics13020137
70. European Commission. Joint Research Centre. Institute for Environment and Sustainability (2008) The Krakow receptor modelling inter-comparison exercise. Publications Office: LU. https://doi.org/10.2788/33914

Environmental Chemistry Complexity in Practice

Quantum Chemistry Applied to Environmental Research

Federica Lauria and Andrea Maranzana

Abstract

Theoretical and computational chemistry is a powerful tool for supporting experimental results. Quantum calculations can be used to describe the molecular structures in detail, estimate the relative energies and activation barriers, and obtain information about the chemical and physical properties of the materials under examination. *Density functional theory* (DFT) is the most widely used tool because of its good compromise between accuracy and computational cost.

Among the many possible examples, some applications of computational chemistry in environmental chemistry are presented. In particular, the reactivity of butenedial in the troposphere is discussed by comparing experimental data with data obtained from quantum mechanical calculations.

Keywords

Ab initio · DFT · Reaction mechanism · Theoretical calculations · Photochemistry · Environmental chemistry

F. Lauria · A. Maranzana (✉)
Department of Chemistry, University of Turin, Torino, Italy
e-mail: andrea.maranzana@unito.it

M. Minella (ed.), *Complexity in Environmental Chemistry*,
https://doi.org/10.1007/978-3-032-31774-2_9

Chapter Learning Goals

- Fundamental principles of computational chemistry and their role as complementary tools to experimental chemistry.
- Quantum mechanical methods used to describe molecular structure, energetics, and reactivity.
- The potential energy surface (PES) for understanding reaction mechanisms and activation energies.
- Differences among the main computational approaches (molecular mechanics, semi-empirical methods, DFT, and post-Hartree–Fock methods).
- Choice of basis set and DFT functional.
- Application of computational chemistry to photochemical and oxidative processes in the troposphere. As a case study, we selected 2-butenedial.
- Integrating theoretical results with experimental data to clarify reaction mechanisms, identify intermediates, and explain the formation of reaction products.

1 Overview of What Computational Chemistry Is

Environmental chemistry is intrinsically complex because it studies systems that are constantly evolving. It is a branch of chemistry that can involve different phases (gas, liquid, solid, interfaces, aerosols, mineral surfaces, and ice) and molecules of various types and reactivities. Numerous substances can interact with one another. Environmental processes occur at different temporal and spatial scales and are strongly influenced by human activity, temperature, pH, and sunlight.

Computational chemistry provides valuable support, for example, by allowing the identification of highly reactive intermediates (which are difficult to isolate experimentally) and reaction products, determining reaction mechanisms, and estimating the energetics and kinetics of each reaction involved.

In this chapter, after a brief introduction to the basic concepts of the various theoretical tools available to the computational chemist, some examples are presented to show how a computational study can help in understanding an environmental process and complement experimental data.

1.1 Why Perform Calculations?

In the last 10 years, the use of computational methods in environmental chemistry has increased significantly. Until a few decades ago, this field of research was confined to basic computational methods or very small molecular systems. However, with the advent of modern computers, it became feasible to perform ab initio

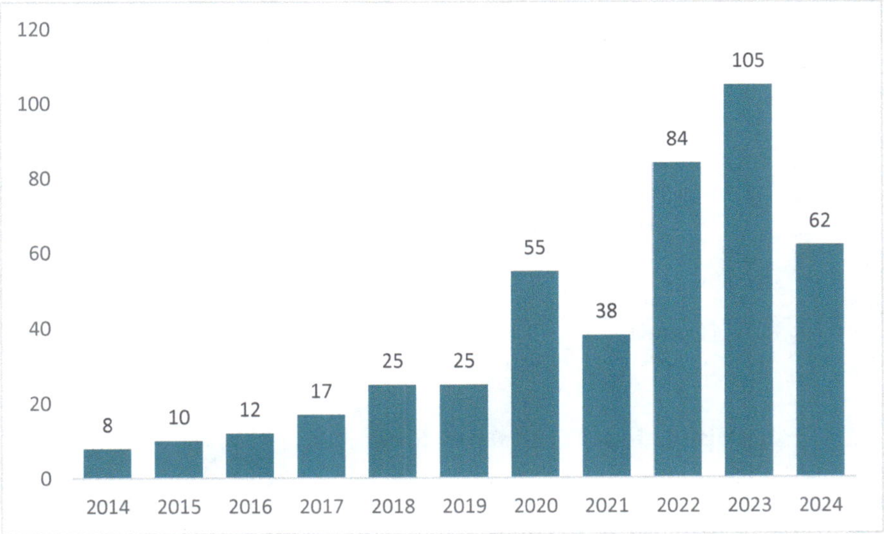

Fig. 1 Approximate number of articles containing the DFT keyword in *Environmental Science and Pollution Research*

calculations on large molecules and subsequently achieve highly accurate electronic structure treatments [1]. This trend reflects advances in both software and hardware, particularly High-Performance Computing (HPC), which today allows us to make reliable predictions on the behavior of matter in relatively short times. Annual counts show (Fig. 1) an increase in the number of articles that include density functional theory (DFT) as an analytical or interpretative tool [2]; however, this measure should be interpreted with caution because it depends heavily on the indexing policy of journals and the definition of "use of DFT" (whether it is a central application or an ancillary calculation).

Computational chemistry is a discipline that applies mathematical methods to describe molecular properties. This approach provides complementary information that is often not directly accessible in laboratories. In practice, computational chemistry can be used to investigate different types of energies (such as potential, enthalpy, free, and zero-point energies) for both molecules and transition states. It also allows the determination of molecular geometries, provides the equilibrium positions of atoms, and explores chemical reactivity by predicting the electron density of molecules in the gas phase and in solution. Furthermore, computational methods enable the study of light-induced processes, making it possible to investigate photochemistry, including the energies and geometries of excited states. Finally, computational chemistry can be employed to simulate various spectra, such as IR, Raman, NMR/EPR, UV-Vis absorption and emission, and circular dichroism, and in some cases even mass spectrometry, offering a direct link between theoretical prediction and experimental observables.

Table 1 Synoptic table of the computational methodologies. The choice of method depends on the size of the system and the physical property of interest (i.e., structure or reactivity)

Method	What is considered?	Instruments	Computational cost	Accuracy
Molecular mechanics	Atoms as spheres	Force fields	Low (seconds)	Geometry only (no reactivity)
Semiempirical	Valence electrons	Parametrization	Low (minutes)	Good for general trends
Post HF	Wavefunction (Ψ)	Wavefunction method + basis set	Very high	Accuracy and precision (benchmark)
DFT	Electron density (ρ)	Functional + basis set	Medium/high	Best compromise often

1.2 Computational Chemistry Tools

Several techniques are available to acquire all possible information about our system. The main methods can be classified into four categories, as summarized in Table 1.

1. Molecular mechanics (MM): is governed by classical mechanics. It does not describe electrons, but only the positions of atoms. It is used for very large systems such as proteins.
2. Semi-empirical methods are quantum-mechanical approaches in which some integrals are replaced with empirically fitted parameters. They start from fundamental equations (such as Hartree–Fock) but introduce empirical parameters derived from experimental data. They are used for rapid geometry screening.
3. HF and post-HF: Based on quantum mechanics, the Schrödinger equation is solved, and depending on the method, highly accurate results can be obtained. These methods include Hartree–Fock (HF) methods and post-HF methods (MP2, MP4, coupled-cluster).
4. Density functional theory (DFT). This method computes the electronic structure using the electron density instead of the many-electron wavefunction.

DFT is currently the most widely used approach capable of describing the electronic structure (and therefore bond breaking and formation) with satisfactory chemical accuracy while maintaining reasonable calculation times.

Having defined the method of choice, the chapter now proceeds following a "general to specific" logic. First, the concept of the potential energy surface (PES) is introduced. It is essential to emphasize that PES is a universal theoretical construct, transversal to any computational approach (from molecular mechanics to ab initio methods), as it defines the fundamental energy topology on which chemical systems operate. It should be highlighted that the numerical shape of the PES depends on the level of theory used for its computation.

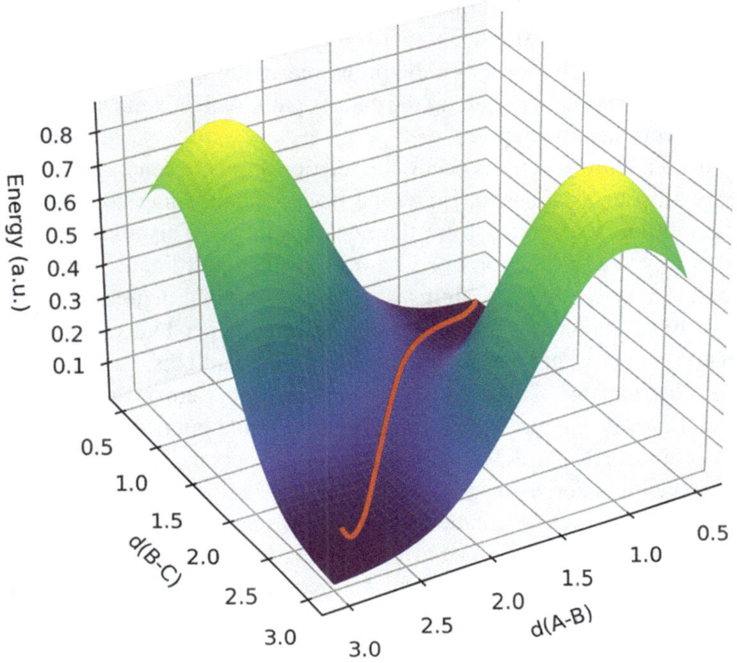

Fig. 2 Simplified potential energy surface as a function of two generic reaction coordinates (*x*- and *y*- axes), here chosen as simplified interatomic distances d(A-B) and d(B-C). The potential energy is represented by the vertical axes (*z*); the valleys (minima) correspond to stable molecular species, while the "pass" (first-order saddle point) between the two peaks (second-order saddle points) represents the transition states along the reaction pathway

Once this common frame of reference was established, the specific tools needed to calculate this energy were discussed, and the nature of the functionals and criteria for selecting the basis set were analyzed in detail.

2 Potential Energy Surface

While computational chemistry yields precise numerical data, the sheer volume of output (matrices of coordinates and energy values) can be abstract and difficult to interpret directly. To make sense of these abstract results, a quantity known as the potential energy surface (PES) was employed. The reduced dimensionality PES (Fig. 2 and Fig. 4) allows the mathematical relationship between molecular geometry and energy to be visualized in a graphical and intuitive form.

When a reaction mechanism is theoretically investigated, a simplified PES is constructed (Fig. 2). Molecules do not cross the highest peaks (high-energy maxima); rather, they "relax" in the valleys. These valleys correspond to energy minima and stable structures as reagents, products, or intermediates. To move from one valley to another (as in a chemical reaction), a molecule must cross a barrier, following the easiest route available (minimum energy path (MEP), red line in Fig. 2) between the reactant and product valleys. From a computational perspective, the peak of the pass corresponds to a first-order saddle point, representing a transition state (TS)—the point of maximum energy along the easiest path, which is the minimum energy required to overcome the barrier.

To understand how reactions occur, the relationship between molecular structure and energy must be established. Under the Born–Oppenheimer approximation, the electronic motion is separated from the nuclear motion. This allows to define the potential energy E as a function of a vector of nuclear coordinates **R**.

A molecular system with N atoms has $3\,N - 6$ internal degrees of freedom (or $3\,N - 5$ for linear molecules), which correspond to independent vibrational coordinates, that is, the only geometric deformations (bond lengths, valence angles, and dihedral angles) capable of modifying the energy of the system on the PES. Therefore, the potential energy E is not a function of each variable but depends on a vector of nuclear coordinates **R**. The $E(\mathbf{R})$ function defines a hypersurface in a $3\,N$-dimensional space, the potential energy surface.

It is important for a computational chemist to mathematically identify the *stationary point* within the PES. The critical points of interest are minima and first-order saddle points. A stationary point is a geometric configuration in which the energy gradient with respect to the nuclear coordinates, R, is zero.

$$\nabla E = 0$$

To distinguish between these points (minima or transition states), it is necessary to analyze the curvature of the surface, which is described by the *Hessian matrix* (the matrix of second derivatives of energy). Mathematically, the *Hessian matrix* is a square matrix of second-order partial derivatives of a function with multiple variables. In other words, it provides information on the mutual coupling of various geometric coordinates that define the position of the nuclei. This mathematical definition may be more familiar if it is directly translated into the concept of stiffness, where the elements of the Hessian are essentially related to the force constants (k) of a harmonic oscillator (Hooke's law). Therefore, this matrix provides a complete picture of internal forces. The diagonal elements are the force constants along specific coordinates, that is, how resistant a bond is to stretching or bending. The non-diagonal elements of the Hessian matrix measure the mutual coupling between the motions, for example, how a bending vibration affects a nearby bending angle or torsional rotation.

Based on this topological analysis (Fig. 3), when the energy gradient is zero, we classify the stationary points as minima (P_1) and saddle points (P_2). However, the stationary condition $\left(\frac{dE}{dx} = 0\right)$ alone only identifies the critical points where the force on the atoms is zero, but it does not allow us to distinguish between a minimum (P_1) and n-order saddle point P_2. Therefore, to resolve this ambiguity, the second derivative matrix must be examined: a positive second derivative (more precisely, all positive eigenvalues of the Hessian matrix) indicates an upward concavity

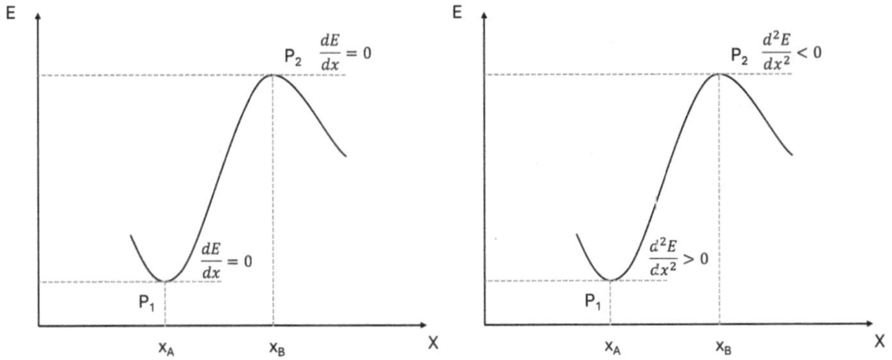

Fig. 3 Differential analysis of critical points P_1 and P_2. The zero first derivative (left graph) identifies stationary points, and the sign of the second derivative (right graph) discriminates their topological nature (local minimum vs. local maximum)

typical of a stable species (minimum), whereas a negative derivative (single negative eigenvalue of the Hessian matrix) indicates a downward concavity typical of the transition state. In chemical terms, we can attribute the minima to the chemical compounds that correspond to reagents, products, or intermediates.

Once the topology of the potential energy surface has been mapped, two fundamental quantities that quantify the behavior of the system must be extracted: the activation energy $\Delta E^{\ddagger}$ and reaction energy ΔE. These correspond to the differences in height between the identified stationary points (Fig. 4).

The activation energy defines the reaction kinetics. It corresponds to the difference between the transition state (TS) energy and the reagent energy:

$$\Delta E^{\ddagger} = E_{\mathrm{TS}} - E_{\mathrm{Reagent}}$$

In physical terms, it refers to the energy cost required for the system to reach the geometry of the activated complex. According to the transition state theory [3], once the reagents reach the transition state configuration, they continue to form products. The higher this barrier, the slower the reaction will be (or the higher the temperature required for it to occur).

Once the kinetic behavior of the system has been defined, it is also important to assess its thermodynamic properties through the reaction energy, defined as

$$\Delta E = E_{\mathrm{Product}} - E_{\mathrm{Reagent}}$$

This quantity is a state function; it depends only on the start and end points, completely ignoring the path or barrier overcome. If ΔE is greater than zero, the reaction is endoergic, whereas if it is less than zero, the reaction is exoergic.

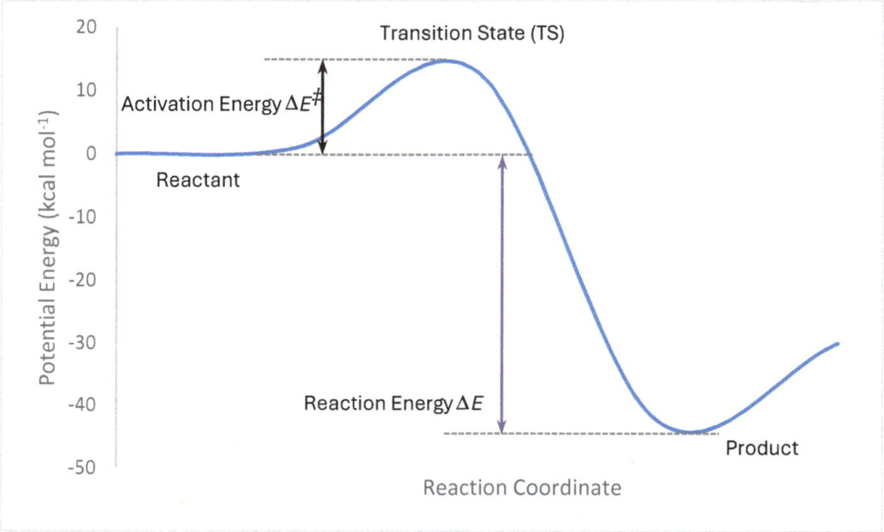

Fig. 4 Two-dimensional projection of the potential energy surface (PES). The minima of the curve correspond to stable species (reagents and products), while the saddle point corresponds to the transition state

3 Solving the Schrödinger Equation

To calculate the energy E of a molecule, the Schrödinger equation must be solved. In this section, some fundamental aspects underlying the quantum treatment of molecules are described, with the aim of recalling the concepts of quantum mechanics that are useful for the molecular electronic description.

3.1 The Hartree–Fock Approximation

According to quantum theory, solving the Schrödinger equation allows the determination of many properties, such as energy. The time-independent Schrödinger equation is defined as

$$\hat{H}\Psi = E\Psi$$

where Ψ is the wavefunction that describes the state of a system, E is the energy, and $\hat{H}$ represents the Hamiltonian operator consisting of kinetic ($\hat{T}$) and potential ($\hat{V}$) energies:

$$\hat{H} = \hat{T} + \hat{V}$$

For a system with n electrons and N nuclei, the Hamiltonian consists of the kinetic energy of the electrons and nuclei and $\hat{V}$ of the potential energy due to the Coulombic interaction between all the particles (electrons and nuclei) in the system:

$$\hat{H} = \underbrace{-\sum_{A=1}^{M} \frac{\hbar^2}{2m_A} \nabla_A^2}_{\hat{T}_e} \underbrace{-\sum_{i=1}^{N} \frac{\hbar^2}{2m_e} \nabla_i^2}_{\hat{T}_e} \underbrace{-\sum_{i=1}^{N} \sum_{A=1}^{M} \frac{Z_A e^2}{r_{iA}}}_{\hat{V}_{ne}}$$

$$+ \underbrace{\sum_{i=1}^{N} \sum_{j>i}^{N} \frac{e^2}{r_{ij}}}_{\hat{V}_{ee}} + \underbrace{\sum_{A=1}^{M} \sum_{B>A}^{M} \frac{Z_A Z_B e^2}{r_{AB}}}_{\hat{V}_{nn}}$$

where m is the mass of the electron, e is the unit charge of the electron, Z_A is the atomic number of atom A, r_{iA} is the distance between electron i and nucleus A, and r_{ij} is the distance between electrons i and j. Finally, $\hbar$ is the reduced Planck constant, equal to the Planck constant h divided by 2π, and the dot product of ∇ with itself is the Laplacian ∇^2. In three dimensions, using Cartesian coordinates, the Laplace operator is

$$\nabla^2 = \frac{\partial^2}{\partial x^2} + \frac{\partial^2}{\partial y^2} + \frac{\partial^2}{\partial z^2}$$

The Hamiltonian operator comprises five distinct terms: $\hat{T}_n$ and $\hat{T}_e$ represent the kinetic energies of the nuclei and electrons, respectively. The term $\hat{V}_{ne}$ describes the electrostatic attraction between negatively charged electrons and positively charged nuclei. Finally, $\hat{V}_{ee}$ and $\hat{V}_{nn}$ represent the repulsive interactions between pairs of electrons and nuclei, respectively.

The system state, which is defined by the total wavefunction of a molecule, depends on the coordinates of both nuclei (**R**) and electrons (**r**):

$$\Psi(\boldsymbol{R}, \boldsymbol{r})$$

This dependence implies that the motions of the nuclei and electrons are coupled, and they dynamically influence each other. Solving this coupled equation is extremely difficult. However, a simplification based on mass can be provided. Because the masses of nuclei are thousands of times greater than those of electrons, the latter move much faster than the former. Therefore, according to the Born–Oppenheimer approximation, assuming that the nuclei are fixed (stationary) with respect to the electrons, the corresponding operator $\hat{T}_n$ is considered zero, and $\hat{V}_{nn}$ is treated as a constant because the nuclei are fixed at specific coordinates.

This defines the electronic Hamiltonian ($\hat{H}_{elec}$), which is a function of the electron coordinates only:

$$\hat{H}_{elec} = -\sum_{i=1}^{N} \frac{1}{2} \nabla_i^2 - \sum_{i=1}^{N} \sum_{A=1}^{M} \frac{Z_A}{r_{iA}} + \sum_{i=1}^{N} \sum_{j>i}^{N} \frac{1}{r_{ij}}$$

In the case of a hydrogen atom, the last term, the electron-electron repulsion, is zero and can be eliminated, allowing the wavefunction to be solved exactly for a single-electron system. However, because of this term, the variables cannot be separated in the case of larger systems.

Generally, the total wavefunction cannot be expressed simply as the product of two independent wavefunctions because solving the equation for electron 1 necessitates knowledge of the solution for electron 2, and vice versa. In quantum chemistry, this issue is referred to as the "many-body problem." To address this, one must consider the interaction between the particles as an average interaction, specifically using the Hartree–Fock (HF) approximation, which represents the simplest method for solving an ab initio calculation. The Hartree–Fock wavefunction, with a sufficiently large basis set, can account for 99% of the total energy. The missing 1% of the error might seem negligible; however, it is often larger than the energy scale of important chemical phenomena such as hydrogen bonds or reaction barriers. The difference between the HF energy and the exact energy is called the *electron correlation* (EC) energy.

The main methods developed to estimate the EC are configuration interaction (CI), Møller–Plesset (MP) perturbation theory, and the coupled cluster (CC) method. Density functional theory (DFT) considers EC in a different way (see Sect. 3.2.2 for details).

3.2 Other Methods

3.2.1 Post-Hartree–Fock Methods

To overcome the limitations of the mean-field approximation and recover the missing electron correlation, advanced computational methods have been developed previously. These techniques are collectively known as post-Hartree–Fock methods. This name reflects their strategy: they used the Hartree–Fock wavefunction as a starting point and mathematically added the missing electron–electron interactions.

In the configuration interaction, the total wavefunction is described as a linear combination of Hartree–Fock determinant (Ψ_{HF}) and various excited Slater determinants (ϕ). Then the expansion coefficients (d) to minimize the energy are calculated:

$$\Psi_{CI} = d_0 \Psi_{HF} + \sum_{i=1} d_i \phi_i$$

Full configuration interaction (full CI) uses all Slater determinants obtained by the excitation of all electrons in the available virtual orbitals (Fig. 5).

It provides an exact solution to the Schrödinger equation within the limitations of the chosen basis set. This means that the number of determinants required increases factorially as the number of electrons and orbitals increases; therefore, the number of determinants can be truncated to make the calculations feasible. A common approach is CISD (CI with single and double-excitations), which includes all determinants of S- and D-type. This technique is very accurate, but it is still computationally demanding and only applicable to atomic or molecular systems characterized by a dozen electrons.

Rather than solving for every interaction, the MP method considers electron correlation as a small perturbation to the HF system [4]. The energy is calculated at various orders of approximation: second-order (MP2), third-order (MP3), fourth-order (MP4), and so on.

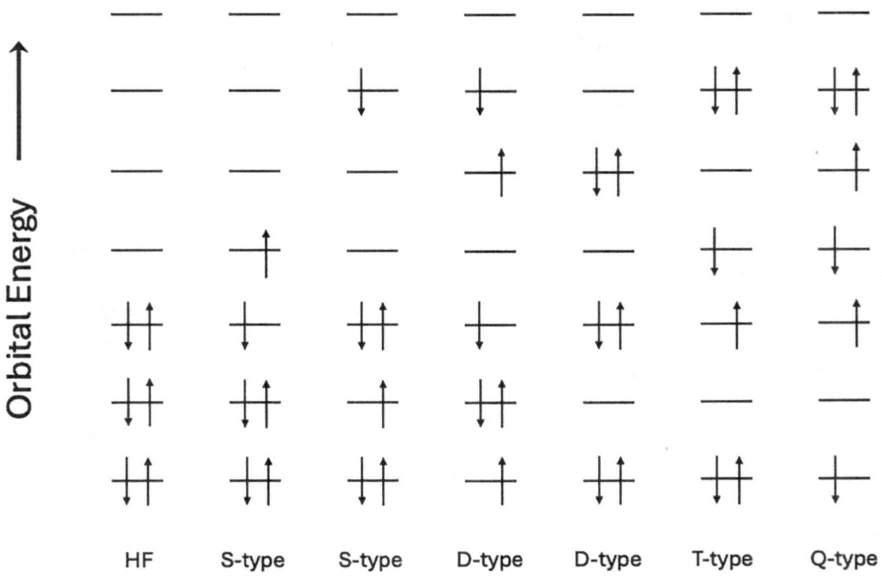

Fig. 5 Schematic representation of the Hartree–Fock reference determinant (HF) and examples of excited Slater determinants generated by promoting electrons from occupied to virtual orbitals. The labels denote the level of excitation: S-type (singles), D-type (doubles), T-type (triples), and Q-type (quadruples)

MP2 is the most common method, and it is significantly faster than CI; however, because it is not a variational method, the calculated energy could be lower than the true physical value.

The last correlated method is the coupled cluster [5], which is often considered the "gold standard" in modern quantum chemistry (specifically CCSD(T)). Mathematically, an exponential operator ($e^{\hat{T}}$) is introduced to construct the determinant's expansion. This method is robust and accurate in capturing the electron correlation without the factorial cost of Full CI. The standard variants are CCSD (which accounts for single and double excitation) and highly accurate CCSD(T), which includes a perturbative treatment of triple excitations.

3.2.2 DFT Methods

In all the methods discussed so far, the central variable is the wavefunction (Ψ). For a system with N electrons, Ψ depends on 3 N spatial coordinates. It is understandable that this is a large amount of data. However, Hohenberg and Kohn [6] proved, using their density functional theory, that the ground-state electronic energy is completely determined by the electron density (ρ). This can be visualized as a massive reduction in complexity. Mathematically, the electron density, ρ, is derived by squaring the wavefunction and integrating over N-1 electron coordinates. Consequently, the electron density is always a function of only three spatial variables (x, y, z). While the complexity of a wavefunction increases exponentially with the number of electrons, the electron density has the same number of variables, independent of system size.

Modern calculations rely on the Kohn-Sham formulation [7]. This approach maps the actual system of interacting electrons onto a fictitious system of non-interacting electrons (similar to the HF method) that moves in an effective potential. This allows the major part of the kinetic energy of the electrons to be calculated accurately using orbitals. The Kohn–Sham orbitals were expanded in a chosen basis set, and the Kohn–Sham equations were solved numerically. All complex quantum mechanical effects, including electron correlation, are swept into a single term called the *exchange-correlation* (E_{xc}), which is calculated using an exchange-correlation functional. In principle, if the exact form is known, DFT provides the exact solution to the Schrödinger equation (for the ground state). Excited states require additional formalisms (such as time-dependent DFT, TD-DFT). Unfortunately, the universal form of this functional is unknown; therefore, we must use approximate functionals to model the missing term. The accuracy of a DFT calculation depends on how well the chosen functional approximates the "unknown" physics. This has led to a hierarchy of approximations, known as "Jacob's Ladder of density functional approximations," introduced by Perdew and Schmidt to rank functionals according to the variables integrated into their design [8]. This ladder is composed of five rungs, from the bottom to the top:

- Local density approximation (LDA): This is the simplest approximation. It assumes that the electron density is locally uniform and behaves like a uniform electron gas. It is used for solid-state calculations but fails for molecules because it overestimates the bond strengths.
- Generalized Gradient Approximation, GGA: These functionals add the gradient of the local densities to the *exchange-correlation (*xc) functional. The concept is that real molecules are not uniform, and their electron density changes rapidly near the nuclei. Common GGA exchange functionals include the B88 functional [9] and the Perdew–Burke–Ernzerhof (PBE) exchange and correlation functional and is widely used in calculation on solids [10].
- Meta-GGA: An improvement of the latter which includes dependence on the Laplacian of the density ρ, in addition to the kinetic-energy density. Common mGGA functionals are TPSS, M06-L [11], and B97M-V [12].
- Hybrid (m)GGA Functional: These functionals mix GGA (or mGGA) and were first proposed by Becke. One of the leading representatives is B3LYP hybrid GGA [13]. M06-2X is another widely used functional [14].
- Double Hybrid: These sit at the top of the ladder (closest to chemical accuracy). It is mixed in a perturbative correction (similar to MP2) using virtual orbitals to improve the correlation energy. An example is the B2PLYP functional [15].

There is no single "universal" functional suitable for every type of calculation. A function that provides excellent results for organic compounds may perform poorly with inorganic transition metal complexes. Furthermore, the functionals differ in their predictive power. For example, a specific functional may provide accurate binding energies (thermodynamics) but fail to predict the correct activation energies (kinetics).

Currently, there are more than 300 DFT functionals, and none are accurate for all properties of interest. Therefore, the choice of functional must be carefully tailored to the specific chemical system and properties of interest [16].

A brief guide to navigating the choice of DFT functional, basis set, and the computational software has been written by Grimme et al. [17].

3.3 Basis Sets

The second essential tool for computational chemists is the basis set. While the method (such as DFT and HF) defines the physics behind the electron interaction, the basis set provides a mathematical description of *the behavior of* the electrons. The choice of method and basis set determines the accuracy of the results. Mathematically, a basis set is a set of functions combined linearly to model molecular orbitals. According to the *linear combination of atomic orbitals* (LCAO) approximation, we calculated the molecular orbitals by adding all the atomic orbitals of the individual atoms. The orbitals almost always have the following form:

$$\phi = Y_{\mathrm{lm}} \sum_i C_i \sum_j C_{ij} e^{-\xi_{ij} r^2}$$

where the Y_{lm} function provides the orbital symmetry (s, p, d, etc.). The exponential e^{-r^2} is a Gaussian primitive function, and C_{ij} and ξ_{ij} are obtained from a database of standard functions. This equation is referred to as the Gaussian-type orbital (GTO).

Ideally, electrons should be described by Slater-type orbitals (STO), which have the form $e^{-\xi r}$. This approach is closer to the exact solution of the Schrödinger equation for hydrogen atoms. However, the GTO basis set can be computed analytically, which is significantly faster than the numerical integrals over the STO functions. The STO basis set is sometimes used for high-accuracy calculations, but it is most common to use a GTO basis set even if it does not describe the shape of the electron cloud near the nucleus or far away as accurately as the STOs. To balance speed and accuracy, a linear combination of several Gaussian "primitives" is commonly used to mimic the shape of a single Slater orbital.

Using a GTO basis set means that the wavefunction is described by a finite number of functions. Theoretically, an infinite number of GTO functions are required to describe the wavefunction exactly. The differences in the results that arise purely due to the size or quality of the basis set are referred to as *basis set effects*. To avoid this problem, one can use a numeric basis set or a complete basis set (CBS), which performs a series of calculations with increasing basis set sizes to mathematically extrapolate the results to the "infinite basis set limit."

Basis sets are classified according to the number of functions used to describe the electrons:

- Minimal basis set: this is the smallest possible set. The most popular is STO-3G [18], which indicates that three Gaussian primitives are required to approximate the STO basis set. Minimal basis sets are used for very large molecules and for qualitative results.
- Split-Valence/Pople basis sets: These allow valence electrons to be more flexible than the core electrons. The notation 6–31G [19] indicates that the core orbitals are described by six primitive functions. The valence orbitals were divided into two contractions: one using three primitives and the other using only one.
- Polarization and Diffuse functions: Standard basis sets are too rigidly structured; therefore, it is common to add specific functions to improve the accuracy. Polarization (* or p,d) is useful when there are charges in the system whose atoms in the molecules are distorted by their neighbors. It adds functions with higher angular momentum to allow the orbital to change shape. In the notation, 6–31G* or 6–31G(d) adds d-functions to heavy atoms. However, 6–31G** adds p-functions to hydrogen. The diffuse functions (+) are primitives with small exponents that allow the orbitals to expand far from the nucleus. The notation 6–31 + G(d) indicates that a diffuse function was added to the heavy atoms. These basis sets are essential for anions (which have large electronic distributions) and for long-range interactions, such as van der Waals forces.
- Correlation-Consistent Sets: Created by Dunning [20], these basis sets are optimized specifically for calculation methods that include electron correlation (such as CI). An example is cc-pVDZ, which stands for *correlation-consistent polarized valence double zeta*. The aug-cc-pVDZ notation adds diffuse functions.
- Numerous other basis sets are available. For instance, the ANO (Atomic Natural Orbital) basis set is employed when high accuracy is required, particularly with post-Hartree–Fock methods [21]. Jensen's [22] and Karlsruhe's [23] basis sets are particularly well suited for DFT calculations. A comprehensive list of possible basis sets can be found online [24].

3.4 Properties

In conclusion, a wide range of information can be obtained using quantum-mechanical calculations, such as energetics, geometry, photochemistry, estimation of kinetic and equilibrium constants, as well as a function of temperature and pressure and electronic information (charge and spin distribution, nucleophilicity, and electrophilicity). Quantum calculations enable the study of a variety of short-lived intermediates (carbocations, carbanions, radicals, diradicals, and carbenes) that are difficult to isolate experimentally.

4 Reactivity of Dicarbonyl Compounds in the Troposphere

Some examples of the application of computational studies in environmental chemistry are presented in this section. Of course, there is a vast body of literature on computational studies; here, only the case study of some dialdehydes is presented in this section.

Dicarbonyl and unsaturated carbonyl compounds have been shown to negatively impact the environment and human health [25, 26]. Dicarbonyls can be released into the atmosphere from anthropogenic sources, such as the oxidation of aromatic compounds (benzene, toluene, and xylenes) or the combustion of fossil fuels, as well as from natural sources, such as the oxidation of furans resulting from biomass combustion [27, 28].

Once released into the atmosphere, these dicarbonyl compounds can be oxidized by the main oxidants present in the atmosphere or photochemically degraded.

In particular, unsaturated aldehydes, such as butenedial and hexadienedial, have been found to irritate the respiratory tract, eyes, and skin and may have genotoxic and potentially carcinogenic effects [29, 30]. Therefore, it is important to understand the formation reactions of these compounds and their atmospheric degradation pathways.

Butenedienal, hexadienedial and dialdehydes in general, are broad categories of dicarbonyl compounds generated during combustion or post-combustion oxidation. In the case of hexadienedial [31] and 2-butenedial [32], it has been shown that photochemical oxidation proceeds more rapidly than oxidation by the hydroxyl radical in the troposphere.

4.1 Photochemistry of Hexadienedial, Butenedial, and 4-Oxo-2-Pentenal. Experimental Results

In this chapter, we focus on three unsaturated aldehydes, and 2-butenedial is analyzed in detail. The aim is not to present an exhaustive literature review on the reactivity of dialdehydes in the troposphere but rather to show how experiments and computational calculations can complement each other.

Photolysis experiments on hexadienedial [31], butenedial, and 4-oxo-2-pentenal [32] were conducted in the EUPHORE smog chamber (European photo-reactor) in Valencia, Spain, which is one of the major outdoor simulation chambers, comprising two hemispherical domes of approximately 200 m^3. The concentrations of the species formed inside were monitored using FTIR spectrometry.

Hexadienedial Photolysis of the E,Z-2,4-hexadienedial isomer showed rapid conversion to the E,E-2,4-hexadienedial isomer, which is thermodynamically more stable [31]. The kinetic constants for the reaction with OH were k_{OH}(E,Z-2,4-hexadienedial) = $(7.4 \pm 1.9) \cdot 10^{-11}$ cm^3 molecule^{-1} s^{-1} and k_{OH}(E,E-2,4-hexadienedial) = $(7.6 \pm 0.8) \cdot 10^{-11}$ cm^3 molecule^{-1} s^{-1}. The kinetic constants for the reactions were determined by measuring the decrease in reactant concentrations over time. In a previous study [33], the rate constants were slightly higher, $(8.8 \pm 0.3) \cdot 10^{-11}$ cm^3 $molecule^{-1}$ s^{-1} for E,E-2,4-hexadienedial and $(10.9 \pm 1.1) \cdot 10^{-11}$ cm^3 $molecule^{-1}$ s^{-1} for E,Z-2,4-hexadienedial, and the difference was attributed to different spectral analysis software [31]. The kinetics of the photochemical reaction were an order of magnitude faster. In the troposphere, E,Z-2,4-hexadienedial undergoes photoisomerization to form the E,E isomer, which can then degrade to form as-yet-unidentified products.

2-Butenedial/4-Oxo-Pentenal For both 2-butenedial and 4-oxo-2-pentenal [32], photochemical removal was the main degradation process in the planetary boundary layer, with average lifetimes of 10–15 min. Degradation initiated by OH radicals has half-lives on the order of 3–7 h. Product formation occurs predominantly via primary photochemical activation rather than OH-initiated chemistry. The (E) isomers of 2-butenedial must first isomerize to reach the necessary (Z) configuration that allows for the abstraction of a hydrogen atom at the γ-position, which is considered the key step in this reaction mechanism. For both the E and Z isomers, the main identified products were 41% 2(3H)-furanone and 12–14% maleic anhydride from (E)-2-butenedial, 30–42% 5-methyl-2(3H)-furanone from 4-oxo-2-pentenal, and 10–20% CO from both compounds [32]. Secondary products from the degradation of 2-butenedial, maleic anhydride, 2(5H)-furanone, and an unidentified carbonyl compound were also detected.

The proposed mechanism involves the photoinduced formation of a ketene-enol intermediate (Scheme 1). This intermediate is formed following intramolecular H abstraction, which is only possible for (Z) isomers.

The ketene-enol that forms degrades rapidly, even in the absence of light, with average lifetimes of 465 s (2-butenedial) and 235 s (4-oxo-2-pentenal). It was hypothesized that the final products, furanone and maleic anhydride, were derived from this reaction. The latter is thought to result from the tautomerization of ketene-enol to a ketene-carbonyl, which can then undergo cyclization.

The subsequent degradation in the troposphere of the cyclic products (furanones and maleic anhydride), which occurs on a timescale of hours, could contribute to the formation of ozone and secondary organic aerosol [32]. The authors also tried closing and reopening the smog chamber, that is, turning the solar irradiation of the reagents in the chamber on and off. The trends of the various species detected are shown in Fig. 6. The concentration of butenedial remained approximately constant when the chamber was closed but decreased during irradiation. The disappearance of butenedial does not occur due to oxidants such as OH but is the result of photochemically initiated degradation.

The ketene-enol does not form in a closed chamber, but only during irradiation, and is then rapidly consumed when the chamber is closed. Its disappearance is hypothesized to be due to reactions leading to the formation of the main products (2(3H)-furanone and maleic anhydride) and secondary products (2(3H)-furanone and an unknown carbonyl compound). Maleic anhydride appears to form both in the presence of light and in the dark, similar to furanones.

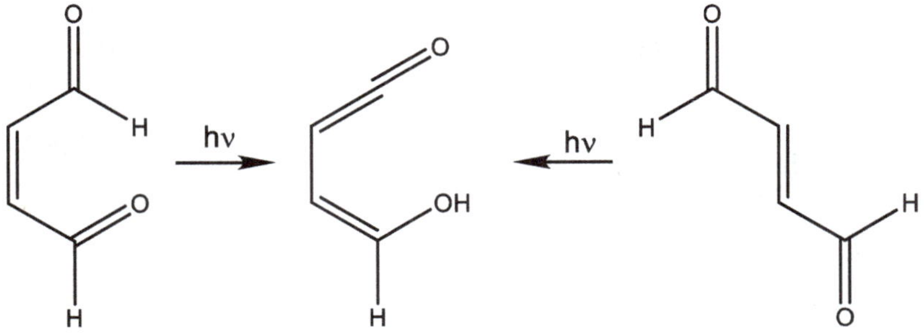

Scheme 1 Hypothesis for the formation of the ketene-enol species. (Adapted from reference [32])

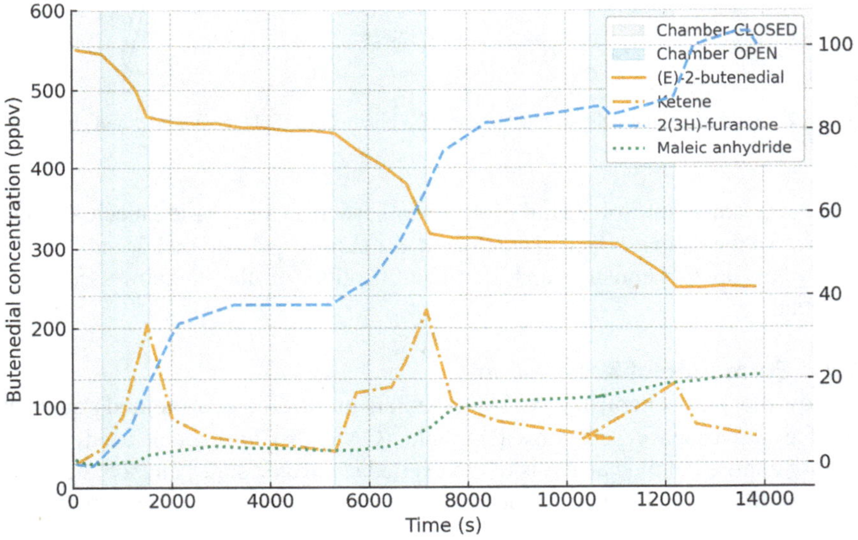

Fig. 6 Trends of butenedial, ketene-enol, furanone, and maleic anhydride over time in open and closed chambers. (Adapted from reference [32])

Newland et al. [32] proposed a mechanism (Scheme 2) involving the initial transfer of hydrogen at the gamma position, a Norrish type II-like intramolecular hydrogen transfer reaction, which results in the formation of ketene-enol.

The overall mechanism hypothesized by the authors is illustrated in Scheme 2; however, the mechanism of maleic anhydride formation remains unclear.

4.1.1 Mechanism of the Photochemistry of 2-Butenedial. Computational Study

The studies cited in the previous paragraph provide a wealth of important and detailed information on the oxidation of unsaturated aldehydes, including kinetic constants, yields, distribution of reaction products, and hypotheses regarding the reaction mechanisms.

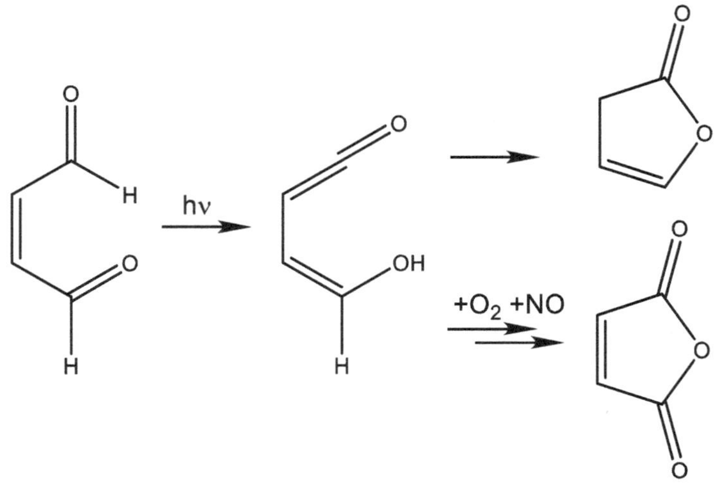

Scheme 2 General photochemical mechanism of 2-butenedial. (Scheme adapted and simplified from reference [32])

However, some unresolved questions remain, such as the detailed mechanism of formation of the main products (ketene-enol, furanones, CO, and maleic anhydride), the role of photochemistry, and the identification of the unknown carbonyl compound.

4.1.2 Formation of Ketene-Enol

To study the photochemistry of 2-butenedial, multireference methods such as complete active space self-consistent field (CASSCF) have been used, which allow for a more detailed analysis of the excited electronic states and a better definition of the conical intersections between the states than single-reference methods.

The active space used was (12,10) [35], which means that it included 12 electrons in 10 orbitals, as shown in Fig. 7. These orbitals comprise the entire π system (six orbitals, three filled and three empty), two lone pairs on the oxygen atoms, and $\sigma_{C\text{-}H}$ orbitals (one bonding and one anti-bonding).

The π^1–π^3, p_O–p_O, $p_O + p_O$, and $\sigma_{C\text{-}H}$ orbitals are occupied in the ground state. These electrons and orbitals, shown in Fig. 7, are the most directly involved in the electronic excitation processes promoted by the photon and H transfer (Norrish reaction).

CASSCF optimizations of the excited states, TS, and conical intersections were performed using the MOLPRO program with the 6-31G(d) and cc-pVTZ basis sets. Only the first two excited states (S_1 and S_2) were considered, as the third requires the absorption of light in the range of 200–220 nm. This wavelength is not available in the troposphere because O_2 and O_3 in the upper layers of the atmosphere absorb solar radiation at wavelengths below 290 nm [36].

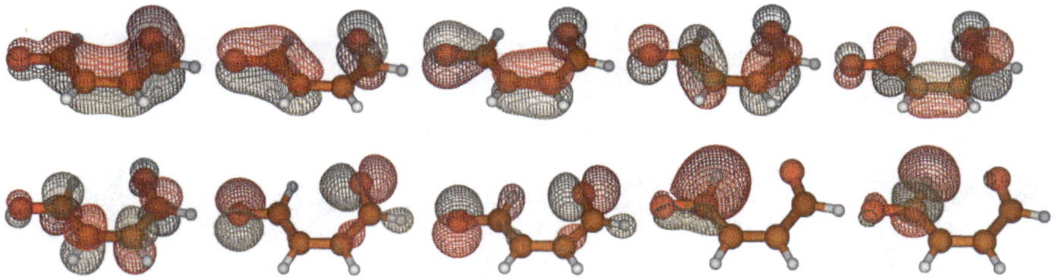

Fig. 7 Active space (12,10). Molecular orbitals were plotted using the MOLDEN program [34]. (Figure from reference [35])

On the S_2 surface (second excited state), hydrogen transfer can occur, leading to the formation of ketene-enol, with a barrier of approximately 29 kcal mol^{-1} (see Scheme 3). A cyclization with a barrier of approximately 15 kcal mol^{-1}, which would lead to a possible precursor of furanone, would be more favorable. However, a conical intersection between the S_2 and S_1 states was found at only 9 kcal mol^{-1}. Therefore, the internal conversion process, which passes through the conical intersection, is favored, causing S_2 to decay to the S_1 state via a nonradiative pathway.

On the S_1 surface, two reactions can occur: cyclization, characterized by a barrier of 29 kcal mol^{-1}, or intramolecular hydrogen transfer to form the ketene-enol, with a barrier of 6 kcal mol^{-1}. The latter does not exist as an intermediate in S_1 because a conical intersection between S_1/S_0 allows a nonradiative return to the ground state.

In conclusion, ketene-enol is formed rapidly on the S_1 surface via a Norrish type II reaction (intramolecular H-shift), while the cyclization reaction, which would lead to a precursor of furanones and maleic anhydride, is not a competitive process because of the high energy barriers; consequently, such products can only be formed in the ground state. These results indicate that furanones and maleic anhydride are formed on the ground-state surface after excited-state decay.

Ketene-enol has also been observed as an intermediate in other photochemical reactions [37], such as the photoisomerization of o-phthalaldehyde [38, 39]. In this case, the authors concluded that H transfer occurred on the S_1 surface, followed by a conical intersection S_1/S_0. Fröbel et al. [40], in an experimental study on o-phthalaldehyde, instead concluded that intersystem crossing S_1/T_1 is a fast process and that the role of the T_1 state cannot be excluded. With other aldehydes, the role of T_1, on the other hand, prevails [41, 42].

4.2 CO Formation

CO formation is accompanied by the formation of an unsaturated carbonyl species, such as acrolein or its precursor (e.g., the acroleyl radical), which is identified among the minor products of the reaction.

Decarbonylation cannot occur in the ground state at room temperature but can occur following the absorption of a quantum of light. In the literature, CO loss from

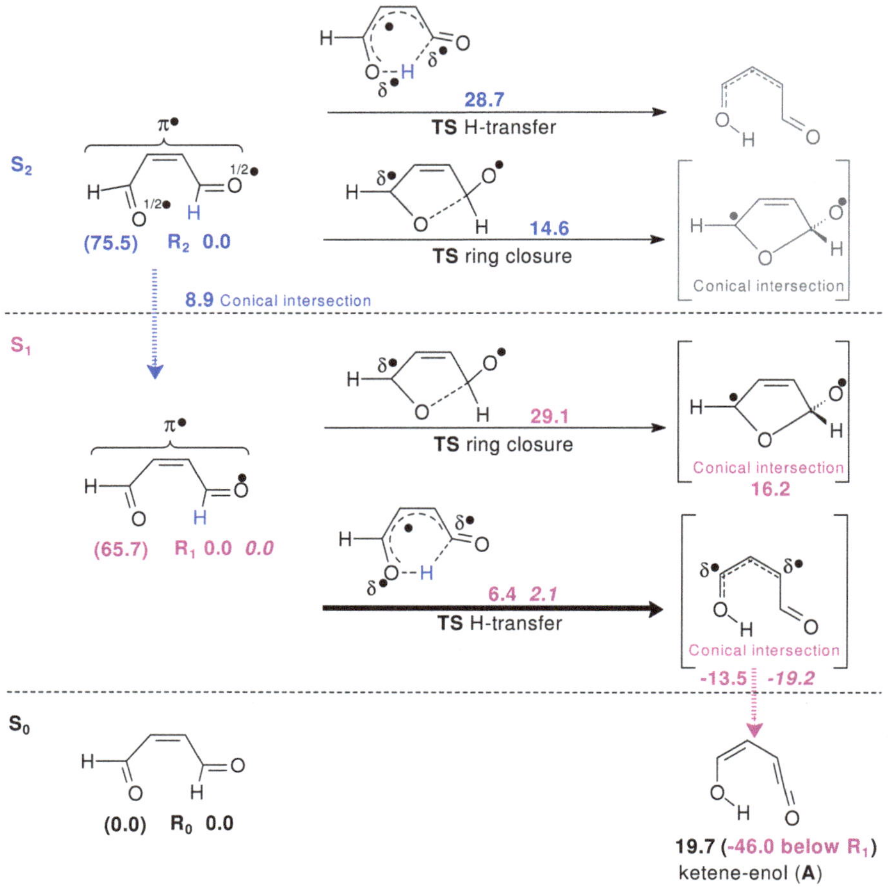

Scheme 3 Formation of ketene-enol. SA-CASSCF(12,10)/6–31G(d) level of theory. SA-CASSCF(12,10)/cc-pVTZ//CASSCF (12,10)/6–31G(d) energies in italics. In parentheses the abiabatic transition energies. The potential energies are expressed in kcal mol^{-1}. (Figure from reference [35])

the T_1 state has been previously reported [43–45]. First, we evaluated the rate of Intersystem Crossing S_1/T_n. The study [46] was conducted using TD-DFT with four different functionals (O3LYP, OLYP, PBE0, and B3LYP) and the cc-pVTZ basis set using the Orca program [47].

The kinetic constants of the S_1/T_1 transitions were found 2–4 orders of magnitude slower than those of the S_1/T_2 transitions (Table 2). The T_2 state results from the excitation of an electron from a p_O-p_O or $p_O + p_O$ orbital to the empty π^4 orbital. The rate of Intersystem Crossing is consistent with the experimental data for similar systems [48, 49]. The T_2/T_1 internal conversion is fast, characterized by a minimum energy crossing point (MECP) of approximately 1 kcal mol^{-1}.

The decarbonylation mechanism is illustrated in Scheme 4. Initial hydrogen migration forms a diradical intermediate, **L**, and upon cleavage of the C-C bond,

Table 2 ISC rates in s^{-1} and MECP barriers from T2 (kcal mol^{-1}). (from reference [46])

Functional[a]	S_1/T_1 ISC rate	S_1/T_2 ISC rate	T_2/T_1 MECP barrier
O3LYP	8.7×10^7	1.0×10^9	0.85
OLYP	1.6×10^6	1.1×10^{10}	1.01
PBE0	5.9×10^6	1.2×10^{10}	0.99
B3LYP	6.9×10^6	1.6×10^{10}	0.81

[a] cc-pVTZ basis set

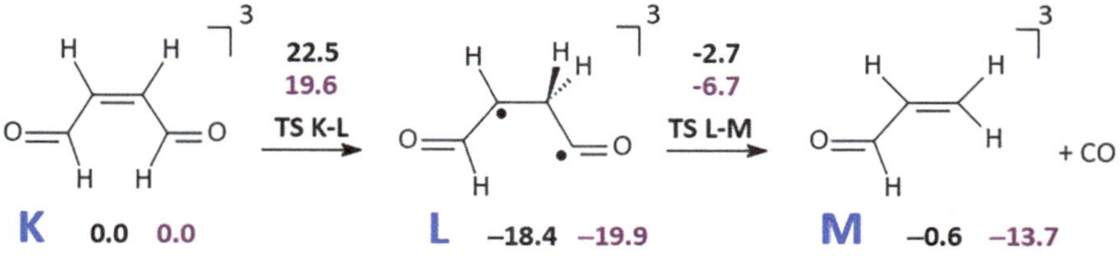

Scheme 4 Formation of carbon monoxide from 2-butenedial on the T_1 potential energy surface. Black, potential energies; purple, Gibbs Free Energies, in kcal mol^{-1} (Figure from reference [46])

CO and acrolein are formed. The latter undergoes rapid decay to the ground state [50]. The breaking of the C-C bond directly from **K** could lead to the release of a formyl radical (HCO) and an acroleyl radical; however, this step has a higher energy barrier than that of H migration.

These mechanistic details do not replace experimental evidence but provide a consistent framework for interpreting observations that cannot be directly accessed experimentally.

4.3 Degradation of Ketene-Enol

The other products observed experimentally, particularly furanones and maleic anhydride, were not formed directly by the excitation of the reagent but were derived from the ketene-enol, which formed after irradiation, consistent with the concentration trends shown in Fig. 6.

4.3.1 Furanones

Starting from 2-butenedial, 3H-furanone (41%) and 5H-furanone (4%) are formed, while from 4-oxo-2-pentenal, 5-methyl-2(3H)-furanone (30–42%) is formed [32].

The computational study of ground-state reactivity was carried out at the M06-2X/cc-pVTZ level, and the thermochemical data reported below were calculated at 298 K [14]. Calculations were performed using the Gaussian program [51].

The most intuitive hypothesis is an intramolecular nucleophilic attack of the hydroxy group from the ketene-enol on the carbonyl group. This reaction would form a zwitterionic intermediate which, through H migration, would give rise to a furanol (**B**, in Scheme 5), and from there, the formation of 2(3H)-furanone and

2(5H)-furanone would be possible. However, the energy barrier for this cyclization is extremely high (47 kcal mol^{-1}).

Any other attempt to cyclize the ketene-enol (**A**, Scheme 5) or to isomerize it failed because of the extremely high energy barriers (see Scheme 6).

A similar situation was encountered by Weingart et al. [52] while studying the photoexcitation of o-acetylbenzaldehyde. Following a Norrish type II reaction, a ketene-enol was obtained, and a lactone formation barrier of 40 kcal mol^{-1} was observed. The authors demonstrated that cyclization can occur when catalyzed by water molecules, which lowers the barrier to 2.5 kcal mol^{-1}.

This catalytic role of water is not new and has been shown to occur in other reactions as well. The number of water molecules involved in the process can also vary: sometimes a single molecule is not enough, and it is necessary to include two or more [53, 54].

This phenomenon is also known as the cooperative effect of water and facilitates the transfer of a hydrogen atom from one part of a molecule to another. In the case of **TS A-B** (Scheme 5 and Fig. 8), a new C–C bond is formed during cyclization, and the hydroxyl hydrogen is transferred to the first water molecule. This, in turn, donates a hydrogen atom to the adjacent water molecule, which then transfers another hydrogen atom to the carbonyl oxygen of the ketene-enol. The net result of the reaction is cyclization with the migration of H to form furanol (**B**), and the two water molecules remain the same.

The process is so effective that it lowers the reaction barrier for the **A-B** step from 46.6 kcal mol^{-1} (without water) to 14.8 kcal mol^{-1} (with just one water molecule) and 11.5 kcal mol^{-1} (with two water molecules).

Similarly, water catalyzes the transfer of the hydroxyl hydrogen of furanol (**B**) to the ring to form two possible furanones. Here again, in the absence of water molecules, the barriers would be insurmountable.

Note that the experiment by Newland et al. [32] was carried out under dry conditions, but even with a relative humidity <1%, as many as 10^{15} molecules cm^{-3} of water can be present.

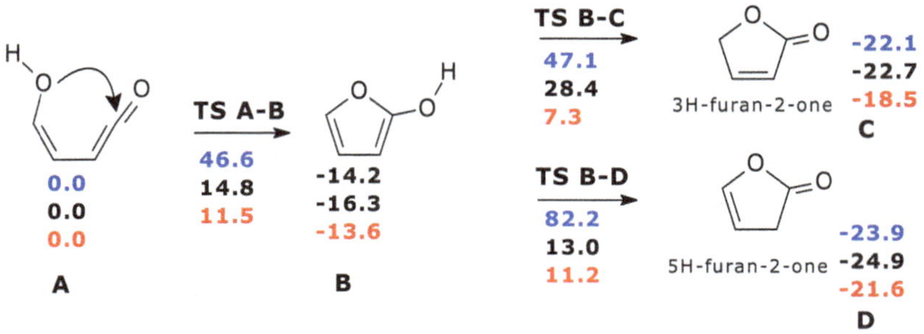

Scheme 5 Furanones formation mechanism from ketene-enols in the ground state. Blue, gas phase without explicit water molecules; black, gas phase with one explicit water molecule; red, gas phase with two explicit water molecules added to the system. For clarity, the two water molecules are not shown in this scheme. ΔG(298 K) in kcal mol^{-1}. (Scheme from reference [35])

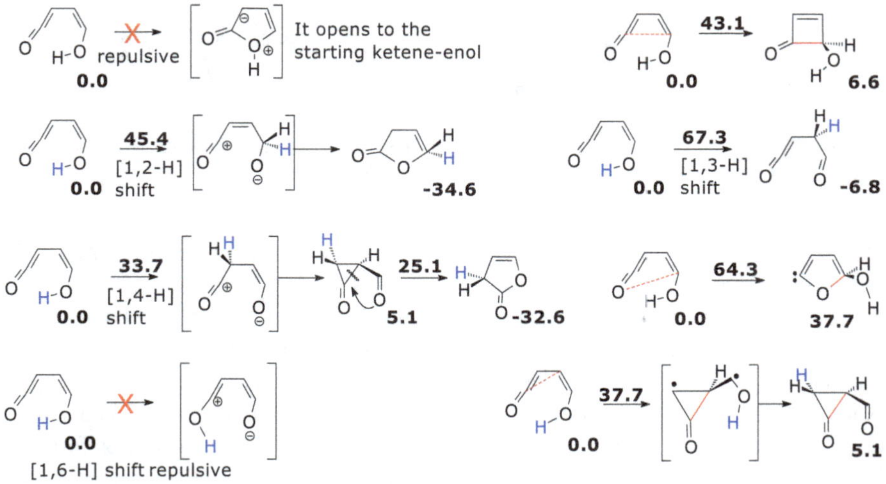

Scheme 6 Attempts to form furanones from ketene-enol in the ground state. The M06-2X/6–31G(d) potential energies are expressed in kcal mol^{-1}. (Scheme from reference [35]). In square brackets, the structures that do not exist as minima on the potential energy hypersurface

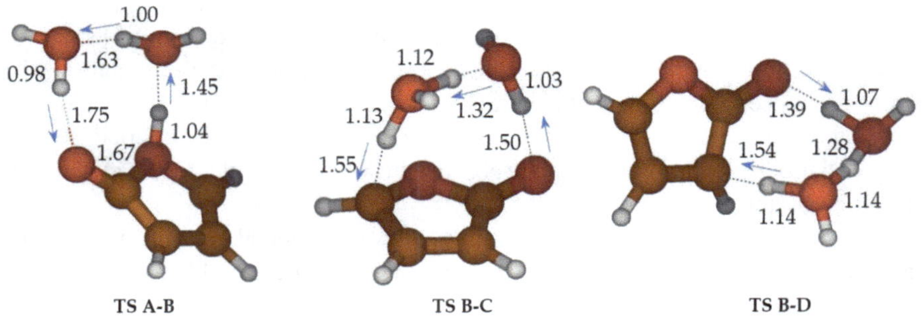

Fig. 8 Cyclization of ketene-enol TS A-B, hydrogen migrations TS B-C, and TS B-D catalyzed by two water molecules. Bond distances in Å. Scheme from reference [35]. Molecular structures created by the MOLDEN program [34]

Trace amounts of water, even under nominally dry tropospheric conditions, can qualitatively reshape reaction pathways.

4.3.2 Maleic Anhydride

Maleic anhydride is structurally similar to furanones but has one additional oxygen atom; therefore, it must be formed as a result of an oxidative process involving ketene-enol (**A**, Scheme 5) or furanol (**B**, Scheme 5). There is a well-known and commonly accepted mechanism of tropospheric oxidation, initiated by OH, as shown in Scheme 7.

Following the extraction of H by OH, a delocalized radical is formed, which adds O_2 to produce a peroxyl radical. In a polluted environment, NO extracts an oxygen

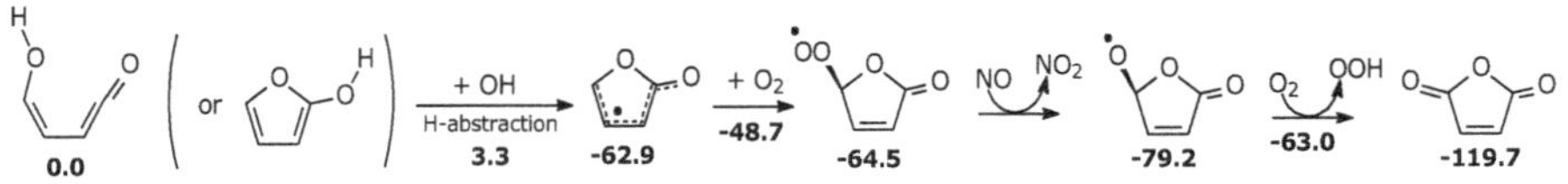

Scheme 7 Formation of maleic anhydride from ketene-enol **A** (or 2-furanol **B**) initiated by OH. ΔG(298 K) in kcal mol^{-1}. (Scheme from reference [35])

atom from peroxyl, forming an oxyl radical which, after losing an H (extracted by O_2), forms maleic anhydride.

The calculated barriers for each reaction step were relatively low and compatible with the reaction kinetics observed in the troposphere. However, this mechanism is inconsistent with experimental data. Newland et al. introduced an excess of isopropanol into the chamber as a radical scavenger. Isopropanol reacts with OH and lowers its concentration. The yield of maleic anhydride from the reaction was found to be approximately unchanged with (12 ± 2%) or without (14 ± 2%) scavenger (see Table 3).

Evidently, the reaction follows a mechanism that does not involve hydroxyl radicals. An alternative mechanism involves the participation of oxygen in the singlet $^1\Delta_g$ state, which is present in the troposphere at a concentration of 10^8 molecules cm^{-3} [55]. $^1\Delta_g$ oxygen can undergo Diels–Alder reactions and add to furanol **B** to form endoperoxide **E** (Scheme 8) with a barrier of less than 9 kcal mol^{-1} [35].

The O-O bond is labile, and its cleavage, concerted with the elimination of a water molecule, gives rise to maleic anhydride. In this case, the presence of two water molecules favors H transfer, lowering the barrier of the second step to 15.5 kcal mol^{-1} (instead of 25.7 in the absence of water molecules), thus making the reaction possible.

This mechanism highlights the importance of experimental data, even in computational studies, as it allows for the exploration of chemical problems beyond preconceived frameworks.

4.4 Unknown Carbonyl Compound

Among the products of the photo-oxidation of 2-butenedial and 4-oxo-2-pentenal carbonyl compounds characterized by a C=O stretching frequency of 1819 cm^{-1} (2-butenedial) or 1829 cm^{-1} (4-oxo-pentenal), with significant yields of 7–24%, respectively were also observed [32]. However, it was not possible to identify the specific species responsible for these absorptions based on the IR spectra. These carbonyls were formed with a delay, suggesting that they were derived from ketene-enol rather than from a direct photochemical reaction of the starting aldehydes.

Computationally, it is possible to make some hypotheses regarding the carbonyl with a C=O stretching frequency of 1819 cm^{-1}, the structure of which has not previously been identified [32]. First, it is necessary to identify a calculation method

Table 3 Yields (%) of products with and without OH scavenging. (Adapted from reference [32])

	2-butenedial without OH	2-butenedial with OH	(Z)-4-oxo-2-pentenal with OH	(Z)-4-oxo-2-pentenal without OH
2(3H)-furanone	41	41	–	–
5-methyl-2(3H)-furanone	–	–	30	42
Maleic anhydride	14	12	3	4
Primary CO	20	18	10	13
Unknown carbonyl (1819 cm^{-1})	9	7	–	–
Unknown carbonyl (1829 cm^{-1})	–	–	16	20

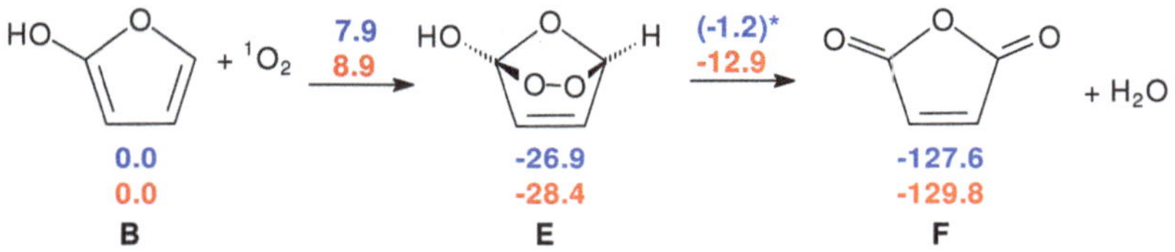

Scheme 8 Formation of maleic anhydride from 2-furanol B initiated by $^1O_2$. Blue, gas phase without explicit water molecules; red, gas phase with two explicit water molecules. For the sake of clarity, the two water molecules are not shown in this scheme. ΔG (298 K) in kcal mol^{-1}. (Scheme from reference [35])

suitable for reproducing vibrational frequencies, particularly the C=O bond stretching, as accurately as possible [46]. The DFT functional M06-2X, which was used to study the reaction mechanisms for the formation of furanones and maleic anhydride, was not sufficiently accurate in this case.

The aim was to compare the experimental C=O stretching frequencies with the calculated ones for a series of carbonyl compounds analogous to those that could form and for which experimental gas-phase IR spectra are known (Fig. 9).

The B2PLYP functional, together with the cc-pVTZ basis set, proved to be the best compromise between the computational cost and accuracy, with an average error of approximately 40 cm^{-1}.

Subsequently, we hypothesized the possible carbonyls that could form in the troposphere from the molecules present in the reaction environment: butenedial, ketene-enol, furanones, furanol, and maleic anhydride. The role of water molecules in catalyzing these reactions was also investigated.

Two candidates were identified, characterized by calculated stretching frequencies very close to the experimental value of 1819 cm^{-1}, as shown in Scheme 9.

Identifying two carbonyl compounds with a C=O stretching frequency similar to the experimental one is insufficient; these compounds must also be kinetically accessible. This implies that mechanisms must exist for the formation of such

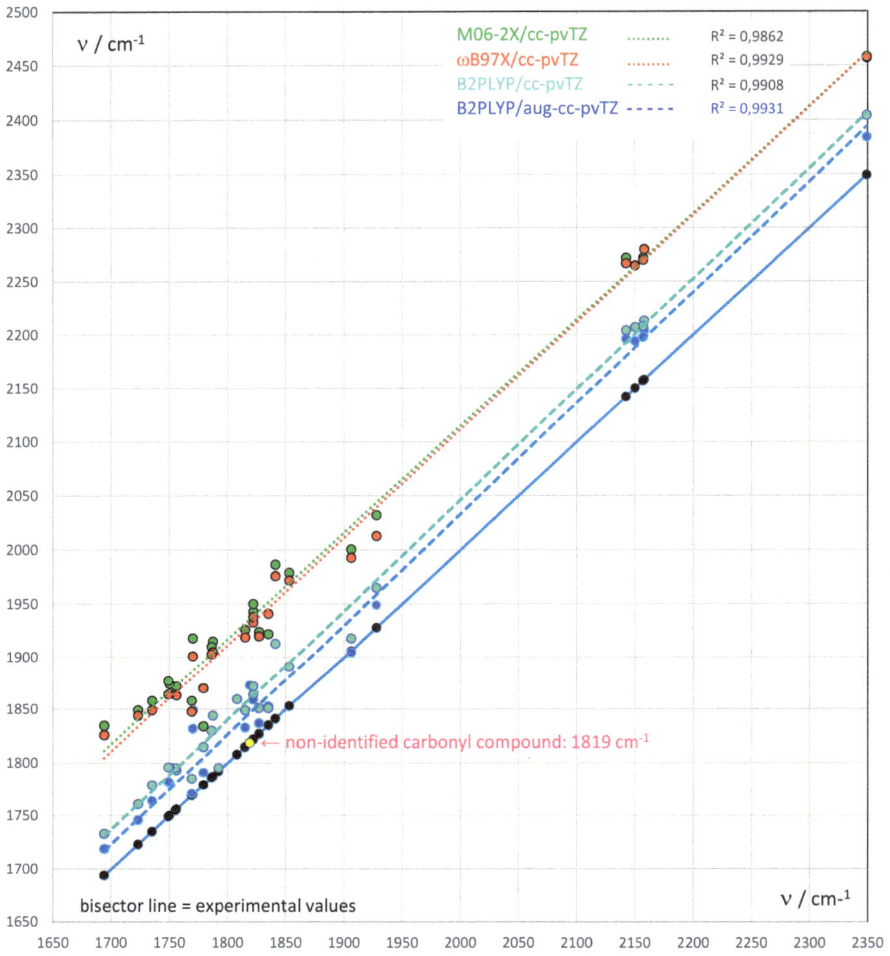

Fig. 9 C=O stretching frequencies $\nu_{C=O}$/cm^{-1}. The black bisector line collects the experimental data (black dots; yellow dot, unidentified carbonyl compound). Other colored dots correspond to the $\nu_{C=O}$ frequencies calculated using various DFT methods and basis sets. B2PLYP/aug-cc-pVTZ, blue dashed line; B2PLYP/cc-pVTZ, turquoise dashed line; B2PLYP/aug-cc-pVDZ light blue dashed line; comparison with the dotted lines: M06-2X/cc-pVTZ, green; ωB97XD/cc-pVTZ, pink; ωB97X/cc-pVTZ, red. (Figure from reference [46])

compounds, characterized by energy barriers and kinetics compatible with experimental observations.

Both lactone **F** and anhydride **J** can be formed from ketene-enol A through the addition of a water molecule and subsequent isomerization. In the case of lactone **F**, the slowest step of the reaction (rate-determining step) is characterized by a free energy barrier of approximately 20 kcal mol^{-1}. For anhydride **J**, the slow step of the reaction has a free energy barrier of 18 kcal mol^{-1}. Given the small energy

Scheme 9 Possible candidates for the unknown carbonyl compounds. (Data from reference [46])

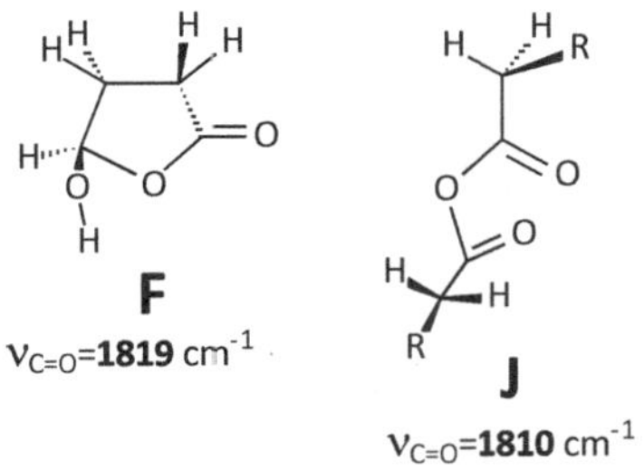

difference between the two processes, both compounds are potential candidates for the unidentified carbonyl product.

5 Conclusion

Ultimately, experimental data and data derived from quantum mechanical calculations can be integrated to provide a deeper understanding of the chemical reactions. Experimental data define the phenomenology of chemical processes, whereas quantum-mechanical calculations provide a molecular-level interpretation of the observations.

Experimental data are fundamental for those performing calculations, as they guide scientists in choosing the optimal calculation method and basis set in terms of accuracy and computational time, sometimes even prompting them to explore unusual reaction mechanisms (e.g., maleic anhydride formation). Therefore, it is extremely important to seek synergy between experimental and theoretical chemists when conducting environmental research.

Questions

- How does computational chemistry integrate with experimental data in studying environmental processes?
- Which quantum-chemical method is currently considered the best compromise between accuracy and computational cost, and why?
- What role do conical intersections play in ketene-enol formation?
- How does the presence of water molecules modify the energy barriers for ground-state furanone formation?
- Why is the term "cooperative effect" used to describe the role of water molecules in these reactions?
- Why was the mechanism of maleic anhydride formation mediated by the OH radical excluded, and which alternative oxidant was proposed based on the experimental results?

Glossary

- Density Functional Theory (DFT): a computational method based on electron density used to calculate the electronic structure and energies of molecules.
- Transition Structure (TS): the highest-energy structure along a reaction pathway connecting the reactants and products.
- Potential Energy Surface (PES): A surface that represents how a system's energy changes with molecular geometry.
- Intersystem Crossing (ISC): A nonradiative transition between electronic states with different spin multiplicities.
- Conical Intersection (CI): A point where two electronic potential energy surfaces of the same spin multiplicity intersect, allowing ultrafast nonradiative transitions between electronic states.

Further Reading

- Frank Jensen, Introduction to Computational Chemistry, Wiley, 2017.
- Errol G. Lewars, Computational Chemistry, Springer Nature, 2024.

Bibliography

1. Sandler I, Chen J, Taylor M, Sharma S, Ho J (2021) Accuracy of DLPNO-CCSD(T): effect of basis set and system size. J Phys Chem A 125(7):1553–1563. https://doi.org/10.1021/acs.jpca.0c11270
2. Springer Nature, L. https://link.springer.com/search
3. Eyring H (1935) The activated complex in chemical reactions. J Chem Phys 3(2):107–115. https://doi.org/10.1063/1.1749604
4. Møller C, Plesset MS (1934) Note on an approximation treatment for many-electron systems. Phys Rev 46(7):618–622. https://doi.org/10.1103/PhysRev.46.618
5. Čížek J (1966) On the correlation problem in atomic and molecular systems. Calculation of wavefunction components in Ursell-type expansion using quantum-field theoretical methods. J Chem Phys 45(11):4256–4266. https://doi.org/10.1063/1.1727484
6. Hohenberg P, Kohn W (1964) Inhomogeneous electron gas. Phys Rev 136(3B):B864–B871. https://doi.org/10.1103/PhysRev.136.B864
7. Kohn W, Sham LJ (1965) Self-consistent equations including exchange and correlation effects. Phys Rev 140(4A):A1133–A1138. https://doi.org/10.1103/PhysRev.140.A1133
8. Perdew J, Schmidt K (2001) Jacob's ladder of density functional approximations for the exchange-correlation energy. AIP Conf Proc 577:1–20. https://doi.org/10.1063/1.1390175
9. Becke AD (1988) Density-functional exchange-energy approximation with correct asymptotic behavior. Phys Rev A Gen Phys 38(6):3098–3100. https://doi.org/10.1103/physreva.38.3098
10. Perdew JP, Burke K, Ernzerhof M (1996) Generalized gradient approximation made simple. Phys Rev Lett 77(18):3865–3868. https://doi.org/10.1103/PhysRevLett.77.3865
11. Zhao Y, Truhlar DG (2006) A new local density functional for main-group thermochemistry, transition metal bonding, thermochemical kinetics, and noncovalent interactions. J Chem Phys 125(19):194101. https://doi.org/10.1063/1.2370993

12. Mardirossian N, Head-Gordon M (2015) Mapping the genome of meta-generalized gradient approximation density functionals: the search for B97M-V. J Chem Phys 142(7):074111. https://doi.org/10.1063/1.4907719
13. Becke AD (1993) Density-functional thermochemistry. III. The role of exact exchange. J Chem Phys 98(7):5648–5652. https://doi.org/10.1063/1.464913
14. Zhao Y, Truhlar DG (2007) The M06 suite of density functionals for main group thermochemistry, thermochemical kinetics, noncovalent interactions, excited states, and transition elements: two new functionals and systematic testing of four M06-class functionals and 12 other functionals. Theor Chem Accounts 120(1–3):215–241. https://doi.org/10.1007/s00214-007-0310-x
15. Grimme S (2006) Semiempirical hybrid density functional with perturbative second-order correlation. J Chem Phys 124(3):034108. https://doi.org/10.1063/1.2148954
16. Morgante P, Peverati R (2020) The devil in the details: a tutorial review on some undervalued aspects of density functional theory calculations. Int J Quantum Chem 120(18):e26332. https://doi.org/10.1002/qua.26332
17. Bursch M, Mewes JM, Hansen A, Grimme S (2022) Best-practice DFT protocols for basic molecular computational chemistry. Angew Chem Int Ed Engl 61(42):e202205735. https://doi.org/10.1002/anie.202205735
18. Collins JB, Schleyer P v R, Binkley JS, Pople JA (1976) Self-consistent molecular orbital methods. XVII. Geometries and binding energies of second-row molecules. A comparison of three basis sets. J Chem Phys 64(12):5142–5151. https://doi.org/10.1063/1.432189
19. Gordon MS, Binkley JS, Pople JA, Pietro WJ, Hehre WJ (2002) Self-consistent molecular-orbital methods. 22. Small split-valence basis sets for second-row elements. J Am Chem Soc 104(10):2797–2803. https://doi.org/10.1021/ja00374a017
20. Dunning TH (1989) Gaussian basis sets for use in correlated molecular calculations. I. The atoms boron through neon and hydrogen. J Chem Phys 90(2):1007–1023. https://doi.org/10.1063/1.456153
21. Almlöf J, Taylor PR (1987) General contraction of Gaussian basis sets. I. Atomic natural orbitals for first- and second-row atoms. J Chem Phys 86(7):4070–4077. https://doi.org/10.1063/1.451917
22. Jensen F (2001) Polarization consistent basis sets: principles. J Chem Phys 115(20):9113–9125. https://doi.org/10.1063/1.1413524
23. Weigend F, Ahlrichs R (2005) Balanced basis sets of split valence, triple zeta valence and quadruple zeta valence quality for H to Rn: design and assessment of accuracy. Phys Chem Chem Phys 7(18):3297–3305. https://doi.org/10.1039/b508541a
24. https://www.basissetexchange.org/
25. Vaskova J, Kovacova G, Pudelsky J, Palencar D, Mickova H (2025) Methylglyoxal formation-metabolic routes and consequences. Antioxidants (Basel) 14(2):212. https://doi.org/10.3390/antiox14020212
26. Du Z, Mo J, Zhang Y (2014) Risk assessment of population inhalation exposure to volatile organic compounds and carbonyls in urban China. Environ Int 73:33–45. https://doi.org/10.1016/j.envint.2014.06.014
27. Wang T, Huang RJ, Yang L, Dai W, Ni H, Gong Y, Guo J, Zhong H, Lin C, Xu W (2023) Direct emissions of particulate glyoxal and methylglyoxal from biomass burning and coal combustion. Sci Total Environ 862:160757. https://doi.org/10.1016/j.scitotenv.2022.160757
28. Pavuluri CM, Wang S, Fu PQ, Zhao W, Xu Z, Liu CQ (2020) Molecular distributions of diacids, oxoacids, and α-dicarbonyls in summer- and winter-time fine aerosols from Tianjin, North China: emissions from combustion sources and aqueous phase secondary formation. J Geophys Res Atmos 126(1):e2020JD032961. https://doi.org/10.1029/2020jd032961
29. Delikhoon M, Fazlzadeh M, Sorooshian A, Baghani AN, Golaki M, Ashournejad Q, Barkhordari A (2018) Characteristics and health effects of formaldehyde and acetaldehyde in an urban area in Iran. Environ Pollut 242(Pt A):938–951. https://doi.org/10.1016/j.envpol.2018.07.037

30. Marques MM, Beland FA, Lachenmeier DW, Phillips DH, Chung F-L, Dorman DC, Elmore SE, Hammond SK, Krstev S, Linhart I, Long AS, Mandrioli D, Ogawa K, Pappas JJ, Parra Morte JM, Talaska G, Tang M-s, Thakur N, van Tongeren M, Vineis P, Grosse Y, Benbrahim-Tallaa L, Suonio E, Turner MC, El Ghissassi F, Middleton D, Miranda-Filho A, Chung F, Liu Y, Vega S, Mattock H, Schubauer-Berigan MK, Guyton KZ (2021) Carcinogenicity of acrolein, crotonaldehyde, and arecoline. Lancet Oncol 22(1):19–20. https://doi.org/10.1016/s1470-2045(20)30727-0
31. Klotz B, Barnes I, Becker K-H (1999) Kinetic study of the gas-phase photolysis and OH radical reaction of E,Z- and E,E-2,4-Hexadienedial. Int J Chem Kinet 31(10):689–697. https://doi.org/10.1002/(sici)1097-4601(1999)31:10<689::Aid-jck2>3.0.Co;2-p
32. Newland MJ, Rea GJ, Thuner LP, Henderson AP, Golding BT, Rickard AR, Barnes I, Wenger J (2019) Photochemistry of 2-butenedial and 4-oxo-2-pentenal under atmospheric boundary layer conditions. Phys Chem Chem Phys 21(3):1160–1171. https://doi.org/10.1039/c8cp06437g
33. Klotz BG, Bierbach A, Barnes I, Becker KH (1995) Kinetic and mechanistic study of the atmospheric chemistry of muconaldehydes. Environ Sci Technol 29(9):2322–2332. https://doi.org/10.1021/es00009a026
34. Schaftenaar G, Noordik JH (2000) Molden: a pre- and post-processing program for molecular and electronic structures. J Comput Aided Mol Des 14(2):123–134. https://doi.org/10.1023/a:1008193805436
35. Maranzana A, Tonachini G (2023) Mechanism of the photochemical isomerization and oxidation of 2-Butenedial: a theoretical study. Molecules 28(13):4994. https://doi.org/10.3390/molecules28134994
36. Finlayson-Pitts BJ, Pitts JN (2000) Spectroscopy and photochemistry. In: Chemistry of the upper and lower atmosphere. Academic Press, Cambridge, MA
37. Rowell KN, Kable SH, Jordan MJT (2022) An assessment of the tropospherically accessible photo-initiated ground state chemistry of organic carbonyls. Atmos Chem Phys 22(2):929–949. https://doi.org/10.5194/acp-22-929-2022
38. Scaiano JC, Encinas MV, George MV (1980) Photochemistry of o-phthalaldehyde. J Chem Soc Perkin Trans 2 7(5):724–730. https://doi.org/10.1039/p29800000724
39. Li Q, Migani A, Blancafort L (2012) Irreversible phototautomerization of o-phthalaldehyde through electronic relocation. Phys Chem Chem Phys 14(18):6561–6568. https://doi.org/10.1039/c2cp40359e
40. Frobel S, Buschhaus L, Villnow T, Weingart O, Gilch P (2015) The photoformation of a phthalide: a ketene intermediate traced by FSRS. Phys Chem Chem Phys 17(1):376–386. https://doi.org/10.1039/c4cp03351e
41. He H-Y, Fang W-H, Phillips DL (2004) Photochemistry of butyrophenone: combined complete-active-space self-consistent field and density functional theory study of Norrish type I and II reactions. Chem A Eur J 108(25):5386–5392. https://doi.org/10.1021/jp0377351
42. Rowell K, Kable S, Jordan MJT (2020) Structural causes of singlet/triplet preferences of Norrish Type II reactions in carbonyls. ChemRxiv 0911. https://doi.org/10.26434/chemrxiv.12941702.v1
43. Tadic JM, Moortgat GK, Bera PP, Loewenstein M, Yates EL, Lee TJ (2012) Photochemistry and photophysics of n-butanal, 3-methylbutanal, and 3,3-dimethylbutanal: experimental and theoretical study. J Phys Chem A 116(24):5830–5839. https://doi.org/10.1021/jp208665v
44. Shemesh D, Nizkorodov SA, Gerber RB (2016) Photochemical reactions of cyclohexanone: mechanisms and dynamics. J Phys Chem A 120(36):7112–7120. https://doi.org/10.1021/acs.jpca.6b06184
45. Oelgemoller M, Hoffmann N (2016) Studies in organic and physical photochemistry—an interdisciplinary approach. Org Biomol Chem 14(31):7392–7442. https://doi.org/10.1039/c6ob00842a
46. Maranzana A, Tonachini G (2024) Tropospheric photochemistry of 2-butenedial: role of the triplet states, CO and Acrolein formation, and the experimentally unidentified carbonyl compound-theoretical study. Molecules 29(3):575. https://doi.org/10.3390/molecules29030575

47. Neese F (2022) Software update: the ORCA program system—version 5.0. WIREs Comput Mol Sci 12(5). https://doi.org/10.1002/wcms.1606
48. Voros D, Mai S (2023) Role of ultrafast internal conversion and intersystem crossing in the nonadiabatic relaxation dynamics of ortho-Nitrobenzaldehyde. J Phys Chem A 127(28):5872–5886. https://doi.org/10.1021/acs.jpca.3c02899
49. Cao J, Xie ZZ (2016) Internal conversion and intersystem crossing in alpha,beta-enones: a combination of electronic structure calculations and dynamics simulations. Phys Chem Chem Phys 18(9):6931–6945. https://doi.org/10.1039/c5cp06833a
50. Devaquet A, Salem L (1971) Potential energy- sheets for the $n\pi^*$ and $\pi\pi^*$ triplet states of Acrolein. Can J Chem 49(6):977–979. https://doi.org/10.1139/v71-160
51. Frisch MJ, Trucks GW, Schlegel HB, Scuseria GE, Robb MA, Cheeseman JR, Scalmani G, Barone V, Petersson GA, Nakatsuji H et al (2016) Gaussian 16 Rev. B.01. Gaussian Inc, Wallingford, CT
52. Weingart O, Reschke B, Marian CM (2018) Solvent mediated catalysis and proton-shuttling in the formation of 3-methylphthalide from a ketene intermediate. Chem Phys 515:750–756. https://doi.org/10.1016/j.chemphys.2018.05.016
53. Wei XG, Sun XM, Wu XP, Geng S, Ren Y, Wong NB, Li WK (2011) Cooperative effect of water molecules in the self-catalyzed neutral hydrolysis of isocyanic acid: a comprehensive theoretical study. J Mol Model 17(8):2069–2082. https://doi.org/10.1007/s00894-010-0917-z
54. Lewis M, Glaser R (1998) Synergism of catalysis and reaction center rehybridization. An ab initio study of the hydrolysis of the parent carbodiimide. J Am Chem Soc 120(33):8541–8542. https://doi.org/10.1021/ja980938g
55. Schurath U (1987) Metastable oxygen molecules in the troposphere. Free Radic Res Commun 3(1–5):173–184. https://doi.org/10.3109/10715768709069782

Environmental Speciation Study: The Case of the Soluble Fraction of Atmospheric Particulate Matter

Marafante Matteo, Bertinetti Stefano, and Berto Silvia

Abstract

Most environmental and biological systems are composed by aqueous solutions, and in many cases their complex chemistry can be modeled using the rules of equilibrium chemistry. On the basis of these rules, and given the total amount of solution components and the equilibrium constants of the species they can form, it is possible to define the exact composition of the solutions. Furthermore, recognized a model capable of describing the chemical system, it is also possible to predict what happens when the solution's composition varies.

The definition of an adequate chemical model must follow certain rules to make it representative of the real system. Since calculating species concentration requires to solve systems of nonlinear equations by iterative methods, the use of dedicated software is recommended. This allows for rapid solution of the mathematical problem but implies that the chemical equilibria must be represented with a formalism useful for the computational process. Moreover, the list of the species involved in the chemical system need to be exhaustive. Failure to consider a species when defining the model can lead to incorrect estimate of the concentration of other species if the species involve a common component. A further fundamental stage of any equilibrium modelling calculation is the selection of equilibrium constants that quantify the thermodynamic stability of the species. This chapter provides explanations and suggestions for each step of speciation studies, with the aim of guiding the reader through the chemical modelling process. Databases for finding equilibrium constants and useful software for plotting speciation diagrams are listed and described. Tips are provided for avoiding overlooking relevant species and for using the most appropriate equilibrium constants.

M. Matteo · B. Stefano · B. Silvia (✉)
Department of Chemistry, University of Turin, Torino, Italy
e-mail: silvia.berto@unito.it

M. Minella (ed.), *Complexity in Environmental Chemistry*,
https://doi.org/10.1007/978-3-032-31774-2_10

The application of the chemical modelling to the soluble fraction of atmospheric particulate matter is then described in order to highlight the rational of the scientific investigation and the analytical process that leads to define the species present in a real sample.

Keywords

Speciation · Solution equilibria · Stability constants · Species distribution diagram · Metal complexes · Soluble fraction

Chapter Learning Goals

IUPAC defines “speciation” as the distribution of an element among defined chemical species in a system [1]. Species can be distinguished based on the isotopic composition, the electronic or oxidation state, and the involvement of the element in inorganic compounds and complexes, organometallic compounds, and organic and macromolecular complexes. This chapter will be limited to deal the speciation which concerns the distribution of elements among different inorganic compounds and inorganic or organic complexes in solution. On the other hand, this speciation process, based on the chemical equilibria laws, can be useful for defining which species form in water solution when several chemical components, with known concentration, are present in the mixture and can interact with each other, and most environmental and biological systems belong to this type of chemical system. The complex chemistry of an environmental system is closely related to the composition of the water that is part of it. The composition of a natural water will be defined by the reactions occurred between its components, as well as the matter fluxes from and to atmosphere, suspended particles (i.e., particulate matter or suspended sediments), other liquids, biota, and soil. Knowing the composition of the water solution being part of a natural system is a fundamental step in describing the chemistry of the entire system, and chemical equilibria modelling is undoubtedly one of the most useful tools for defining the composition of an aqueous solution.

The chapter was written for a reader who knows the basic concepts of equilibrium chemistry, and that is able to manage its rules and formalism. Some simple mathematical equations used to describe the chemical equilibria, such as mass balance equation, global or stepwise stability constants of chemical species, and solubility product of low-soluble species, will be defined here. If the reader needs to recall some concepts, they can be easily found in any fundamental books of analytical chemistry.

The compounds that can be formed in water solution exhibit variable thermodynamic stabilities and lability. It is possible to observe species that can be considered inert—therefore, sufficiently stable in time—and others that form and dissociate rapidly. In the latter case, such species cannot be separated from each other without affecting their distribution. This means that the amount of species formed can only be estimated through the application of chemical-physical laws, such as the mass balance equations and the law of mass action. Moreover, the distribution of species in a system depends on factors such as total concentrations, stoichiometry, pH, ionic strength, and temperature [2]. Consequently, species distribution is typically presented as two- or three-dimensional plots showing the amount of the formed species around a set of conditions of interest. The most common species distribution diagrams represent the concentration of the species as a function of pX, where $pX = -\log[X]$ and [X] is the concentration of a selected species or component of the system. In other words, this diagram shows the effect that the variation of the concentration of X has on the concentration of the species considered in the model.

In order to plot a species distribution diagram, it is essential to know the total concentrations of components in the mixture, the identity (i.e., the stoichiometry) of the various species that may be formed from these components, and the corresponding stability constants that describe the thermodynamic stability of the species, hence the energy gain due to the formation of the species from the individual components.

The aims of this chapter are to:

- provide tips for obtaining reliable speciation models,
- suggest some tools for finding stability constants and drawing species distribution diagrams,
- show how this approach can be effectively applied to real systems.

1 Species Distribution Diagrams

The definition of the concentration of the species in a multicomponent solution can be obtained by the resolution of a complex set of linear equations based on the law of conservation of mass and the law of mass action. Thankfully, the computation of these concentrations at equilibrium and, eventually, the plotting of speciation diagrams can be performed by some devoted software. In fact, the calculation of the concentration of species can be described by algorithms. If the free concentration in solution of a given species, or component, is known (for instance, the concentration of H^+), that of other components and soluble species considered in the chemical model can be obtained by solving mass balance equations by iterative methods, provided the total concentrations of the components and the stability constants of the species. To perform calculations, chemical equilibria must be written using a formalism that is suitable for the algorithms. Moreover, in order to obtain a species distribution diagram able to accurately describe the real system under study, all the species involved in the chemical system must be considered. Failure to consider a species in the model definition may lead to overestimation of the concentration of another species if the missing one competes with the second one for a component of the system.

The following paragraphs will provide suggestions on how to describe the chemical system with a notation useful for calculations, and the rules for preparing a complete list of species formed in solution from a mixture of defined components. Moreover, since stability constant values are fundamental to quantify the species in solution, a list of databases that collect these values will be proposed.

1.1 Description of the Chemical System

To describe a chemical system, it is necessary to define the chemical species in solution.

In a chemical system where metal cations, M^{m+}; inorganic or organic anionic components L^{z-}; and the proton, H^+, are present, many chemical equilibria can occur, leading to the formation of numerous species. The anionic components can form various protonated species. Metal cations, on the other hand, can form hydrolytic species and can interact with anions, in their various protonated forms, to form complex species or ion pairs.

The Morel's table formalism [3] is commonly used to calculate species concentration. A chemical system in which a N_S number of soluble chemical species S_i, and eventually a N_P number of precipitable P_l species, are present in solution can be represented as a linear combination of N_C number of C_j components. The components are the chemical units, such as inorganic or organic cations, anions, and neutral molecules, that are stable in solution and that can interact each other to form new species. In aqueous solution, H^+ and OH^- is always considered a component. If the component shows acidic functions, for the Morel's table formalism, it must be considered its completely deprotonated form. The stoichiometric coefficients of every component are the coefficients of the linear combination for the definition of the species S_i. Therefore, a generic species S_i, with a net charge z_i, can be reported according to the Morel's description (Eq. 1):

$$S_i \equiv \left\{ \otimes_{j=1}^{N_c} \left(C_j \right)_{p_{ij}} \right\}^{z_i} = \sum_{j=1}^{N_c} p_{ij} \cdot C_j \tag{1}$$

$$z_i = \sum_{j=1}^{N_c} p_{ij} \cdot z_j \tag{2}$$

where the symbol $\otimes$ describes the operation of concatenation of strings referring to the different components C_j and $p_{i,j}$ ($0 \rightarrow \infty$, $\in$ N) are the stoichiometric coefficients that the component C_j has in the species S_i. The complete chemical system can be represented with a matrix where on the rows each S_i species is described by the stoichiometric coefficient of each C_j component that are reported on the columns. Table 1 shows an example of matrix for a generic metal-ligand-proton system. The rows corresponding to the components are usually neglected by the software.

Table 1 Morel's table for a metal-ligand-proton system. To define the stoichiometric coefficients

	Components (C_j)		
Species (S_i)	M^{m+}	L^{z-}	H^+
M^{m+}	1	0	0
L^{z-}	0	1	0
H^+	0	0	1
OH^-	0	0	−1
$L_qH_r^{r-qz}$	0	q	r
$[M_pH_r]^{pm+r}$	p	0	r
$[M_pL_qH_r]^{pm+r-qz}$	p	q	r

Morel's formalism gives a qualitative description of the composition of the chemical system. Additional quantitative information is needed to compute the concentration of every species S_i at equilibrium. The stability constants allow to relate the concentration of components in equilibrium condition $[C_j]$ to the concentration of each species $[S_i]$, and the total concentration of components $[C_j]_T$ can be related to the stability constants through the mass balance equation, as described by Eqs. 3–5. Therefore, it is possible to define the concentration of S_i if the total concentration of each component $[C_j]_T$ and the value of the stability constant for each soluble species S_i or precipitable species $[P_l]$ are known.

$$\text{law of mass action} \quad \left[S_i\right] = \beta_i \Pi_k^{N_C} \left[C_j\right]^{p_{i,j}} \tag{3}$$

$$\text{mass balance equation} \quad \left[C_j\right]_T = \Sigma_i^{N_S} p_{i,j} \left[S_i\right] + \Sigma_l^{N_P} b_{l,j} \left[P_l\right] \tag{4}$$

$$\left[C_j\right]_T = \Sigma_i^{N_S} p_{i,j} \beta_i \Pi_k^{N_C} \left[C_j\right]^{p_{i,j}} + \Sigma_l^{N_P} b_{l,j} \left[P_l\right] \tag{5}$$

In the mass balance equation, Eq. 4, $[C_j]_T$ corresponds to the sum of the free concentration of the soluble S_i and/or precipitate $[P_l]$ species containing that component, adjusted for its stoichiometric coefficient: $p_{i,j}$, for the ith species in solution, or $b_{l,j}$, for the lth in the precipitate. To define the concentration of S_i,a system of nonlinear equations needs to be solved. For this purpose, the Newton-Raphson method, an iterative root-finding algorithm, can be applied.

The stability constants used for the calculation, β_i, are the overall stability constants (Eq. 6) that refer to the generic equilibrium (Eq. 7) and, if precipitable species can be formed in the chemical system, also the solubility product values (Eq. 8) must be known. The stability constant associated to the species OH^- is the water dissociation constant (K_w).

$$\beta = \frac{\left[M_pL_qH_r\right]}{[M]^p [L]^q [H]^r} \tag{6}$$

$$pM^{m+} + qL^{z-} + rH^{+} \rightleftarrows \left[\left(M^{m+}\right)_p \left(L^{z-}\right)_q \left(H^{+}\right)_r \right]^{pm+r-qz} \tag{7}$$

$$K_s = [M]^p [L]^q [H]^r \tag{8}$$

Using this formalism, the stoichiometric coefficient r becomes negative in case the proton is a product of the reaction. This occurs if a non-acidic proton is released, as in case of hydrolytic species where the proton derives from a water molecule or in case of non-acidic protons released by the ligand when it interacts with the metal cation. Protons can be, for example, released by alcoholic non-acidic moieties of the ligand if these groups participate to the metal cation coordination.

For soluble species containing hydroxide ions, the formation constants are frequently defined as in Eq. 9, but to be coherent with the formalism used by the Morel's table, in which the hydroxide ion is represented as the loss of a proton from water, they must be written as in Eq. 10. For precipitable species the same reasoning applies with the solubility product of species containing hydroxide ions (Eqs. 11 and 12).

$$\beta = \frac{\left[M_p L_q (OH)_r\right]}{[M]^p [L]^q [OH]^r} \tag{9}$$

$$\beta^* = \frac{\left[M_p L_q (OH)_r\right]}{[M]^p [L]^q K_w [H]^{-r}} \tag{10}$$

$$K_s = [M]^p [OH]^r \tag{11}$$

$$K_s^* = [M]^p K_w [H]^{-r} \tag{12}$$

In literature, both the formation constants β and β^* are reported for many species, but the solubility products are usually reported as K_s. If it is necessary to make a conversion, this can be achieved quite easily exploiting the logarithms properties, as reported in Eqs. 13 and 14:

$$\log \beta^* = \log \beta + r \log K_w \tag{13}$$

$$\log K_s^* = \log K_s - r \log K_w \tag{14}$$

where r is the stoichiometric coefficient of the hydroxide ion in the species. The numerical difference between β and β^*, or K_s and K_s^*, is very high; therefore it is essential to correctly associate the value of stability constant with the right equation.

Once we know how to describe the species present in solution, it is fundamental to enter in the chemical model all the species that can be formed by the components. In order to avoid to forget some species, it would be useful to follow some rules to compile the Morel's table:

1. The water dissociation constant must be always considered. Since it corresponds to the constant of OH^- formation, the value to enter in the table is the K_w or $\log K_w$ ($I = 0$ M, $T = 298.2$ K, $\log K_w = -13.99$ [4]).
2. The interaction of the components with the proton must be taken into account. The constant values commonly used by the software are those corresponding to the protonation equilibria and expressed as overall protonation constants. Remember that the Morel's table describes the formation equilibrium of species, starting from the deprotonated components.
3. If metal cations are present in the chemical system, both the soluble and insoluble hydrolytic species must be involved in the computation, and the $\beta^*$and K_s^* need to be included in the calculation.
4. Both ion pair interactions with stability constants lower than 10^{-4} and complex species have to be considered in the chemical model, and, for each equilibrium involved, the overall formation constants need to be entered.
5. Mixed species, i.e., those complexes that can be formed by two different ligands (generic stoichiometry, $[M(L_1)_q(L_2)_r]$), or polynuclear species (generic stoichiometry, $[(M_1)_q(M_2)_rL]$) should be taken into account.

The relevance of some type of species in the chemical system can be estimated on the basis of some general considerations: (i) soluble hydrolytic species of metal cations must be always entered in the chemical model; they will be especially important in systems where the cation concentration is relatively low, and, in this condition, the pH range over which such species can be present will be relatively wide [5], but their relevance depends also on the metal/ligand(s) concentration ratio, since hydrolytic species mostly form when not enough ligand is present to engage all the metal ion; (ii) the formation of protonated, $M(L_1)_q(H)_r$, or hydroxylated complexes, $M(L_1)_q(OH)_r = M(L_1)_q(H)_{-r}$, depend on the pH of the solution; (iii) if an excess of metal is present in the solution, polynuclear complexes can be formed; (iv) in dilute solution, mononuclear complexes are preferentially formed; (v) for mononuclear complexes, the average number of ligands coordinated to the central atom most often depends on the concentration of the ligand and the values of the formation constants [2]; (vi) in the case of polydentate ligands (e.g., EDTA), only 1:1 species are often formed, also in the presence of a ligand excess.

On the basis of these considerations, it is clear that to make the model useful to describe a chemical system in which the total component concentration may be comprised in a wide range, it would be better to not neglect any possible species. The model will have a high predictive capacity, the more complete the list of species considered is, and the more the stability constants used are consistent with the ionic strength and temperature conditions of the real chemical system. Since the great advantage of optimizing a model of a complex natural system is to gain predictive power, it is often worth spending time formulating a model with general validity rather than settling for a model that is valid only under limited conditions.

1.2 IT Tools for Speciation Studies

Essential tools for speciation studies are software able to compute the species concentration at equilibrium in solution and to draw the speciation diagrams. The most used and/or recently developed software are PyES [6], ReactLab™ EQSIM [7], Hyperquad Simulation and Speciation (HySS) program [8], ES4ECI (ES4) [9], Medusa [10], PHREEQC [11], and MINEQL+ [12]. In order to estimate the concentration of each species in a mixture, the free concentration in solution of a given species, or component, needs to be fixed (e.g., H^+), and it is used as independent variable. The concentration of other components and soluble species considered can be obtained from the total concentrations of components by solving their mass balance equations for each value of the independent variable by an iterative method. Stability constants of each species and the total concentration of components must be given as input values. The estimated concentration of the species, or the percentage of the species with respect to the total concentration of one component, can be plotted as a function of the independent variable to obtain the species distribution diagrams.

All the software mentioned before are able to perform these calculations, but we would introduce PyES to the readers because it is an open-source software recently developed. Moreover, it is freely downloadable and it shows some novelties. In particular, it is able to take into account the variation of the ionic strength in the system during the calculation. The variation of the ionic strength may be due to the variation of the ionic species concentration in solution either as a function of the independent species concentration or during the titration process. The possibility to take into account the real ionic strength value of the chemical system turns out particularly useful for those systems characterized by highly charged species or for solutions with low ionic strength, for which modest concentration variations correspond to a marked change of the solution ionic strength.

This functionality can also be used in case the stability constants collected in the scientific literature and databases do not exactly refer to the ionic strength conditions of the chemical system under study. Commonly available stability constants are expressed as stoichiometric constants, i.e., they are expressed in terms of concentration instead of activities and refer to a specific ionic strength, or as thermodynamic equilibrium constants that refer to infinite dilution conditions. Since it is possible that the ionic strength of the real chemical system for which you wish to plot the species distribution diagram may differ from that used to estimate the stability constants, it is useful to adjust the stability constants for real conditions by applying equations able to calculate the activity coefficients. In this case, the stability constant value can be adjusted before the concentration estimation, but the ionic strength of the solution will be approximated to the ionic strength due to the total concentration of ionic components.

There is some software able to adjust the stability constants at different values of ionic strength. For example a freely available online software is the Aqueous Solutions package [13]. It is composed of programs for the quantitative treatment of a wide range of equilibria in solution. It can correct thermodynamic parameters on the basis of background ion type and concentration and of temperature and pressure. Different approaches can be used to explain the dependence of the stability constants on the ionic strength. The most well-known theories are those of Debye-Hückel [14], Davies [15], Bromley [16], Pitzer [17, 18], and the Specific ion Interaction Theory (SIT) [19, 20]. Each approach corresponds to a mathematical expression that explains the relation between the activity coefficients and the ionic strength of the solution. Each model is valid in a well-defined ionic strength range. For a strong electrolyte solution with stoichiometry 1:1 (e.g., NaCl, KCl, $NaClO_4$, etc.), the upper concentration limit varies from about 0.001 mol kg^{-1} for the Debye-Hückel equation to about ca. 0.1 mol kg^{-1} for the Davies equation, to 3.5 mol kg^{-1} for the SIT, and up to saturation for the Pitzer model. The equations refer to concentrations expressed in the molal scale, but these equations can be safely used also in the most common molar scale up to 0.1 mol dm^{-3}.

PyES is able to calculate the stability constant value at different ionic strength applying an Extended Debye-Hückel equation which allows for modeling the stability constant dependence for ionic strength $I \leq 1.0$ mol dm^{-3}. The calculation process is integrated into the concentration estimation process. Differently, Aqueous Solutions package adjusts separately the stability constants for changes in ionic strength using SIT (Specific Interaction Theory). Moreover, it can estimate electrolyte activity coefficients, osmotic coefficients, and the activity of water using Pitzer and the Lin-Tseng-Lee equations. The effect of the temperature variation on the stability constants and oxygen solubility can be predicted with the same tool on the basis of different models and equations.

It is now clear that a fundamental part of any equilibrium modelling calculation is the selection of constants that quantify the thermodynamic stability of the species. Finding suitable stability constants is not an easy task. In the past, compilations of equilibrium constant values have been published, but their maintenance is difficult, and many of them are not still updated. However, the data there collected can be consulted. Some compilations evaluated the stability constants' reliability on the basis of different parameters and classified them in different categories as a function of their consistency. The digital version of NIST Critically Selected Stability Constants of Metal Complexes Database can be freely downloaded [21] or can be consulted by the Stability Constant Explorer searching program [22]. A lot of solubility data are collected in the IUPAC-NIST Solubility Database, Version 1.1 [23]. Moreover, the Nuclear Energy Agency (NEA) of the Organization for Economic Co-operation and Development (OECD) developed a critically evaluated chemical thermodynamic database for a number of elements and ligands of interest to nuclear technology. Fifteen volumes of the Chemical Thermodynamics Series [24] have been produced, and all volumes in the series can be downloaded from the NEA website. Moreover, an interactive TDB electronic database is available [25]. Registration with the NEA is required in order to obtain access. Furthermore, stability constants of hydrolytic species can be found in some references [4, 5], and a website reports a compilation [26].

Having defined the stability constants of all the possible species that can be formed with the components of the mixture, they can be entered into the software

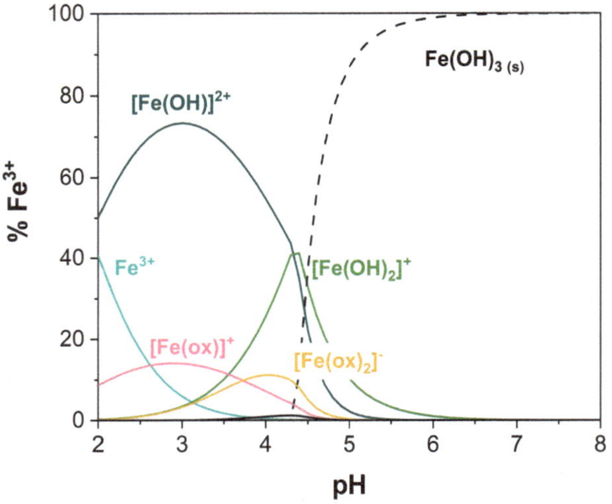

Fig. 1 Species distribution diagram of a solution of Fe^{3+} 7.9 × 10^{-7} mol dm^{-3}, oxalate 3.6 × 10^{-6} mol dm^{-3} (I = 0 mol dm^{-3}, T = 298.2 K)

with the total concentration of components, and the calculation can be finally performed. The result is the estimation of the concentration of each species involved in the chemical model as a function of pX (-log[X]), and it can be graphically represented with a species distribution diagram. An example is reported in Fig. 1.

2 Chemical Species in Aerosol

Aerosol components are the particles (solid or liquid) and associated substances suspended in the atmosphere. The atmospheric particulate matter (PM) is a type of aerosol, and its composition varies widely based on location and season. The composition mainly includes organic compounds, elemental carbon, inorganic anions, and metal cations. The chemical components are derived from various natural and anthropogenic sources, such as soil or road dust, vehicle non-exhaust emissions, biomass and fuel combustion, sea salt, and forest fires, and may have great spatio-temporal variations. Atmospheric particles strongly affect the environment: (i) they affect the surface energy budget [27], both scattering and absorbing the solar radiation producing the heating or cooling of the atmosphere according to the composition of the particles; (ii) they interact with the coexisting gaseous and liquid phases and can work as cloud condensation nuclei (CCN) [28]; (iii) they affect the geochemical cycle of elements [29]; and (iv) they impact on the health of living organisms [30]. Inhalation of PM is generally associated with increased risks of cardiopulmonary diseases and stroke and increased susceptibility to inflammation-associated pathologies such as respiratory diseases and lung cancer [31]. In particular, the soluble fraction of metal cations in atmospheric depositions is recognized as a key factor to describe both the metal geochemical cycle [32, 33] and the health risk [34–36]. The solubility of metal cations, and particularly iron, is in relation with low-molecular-weight inorganic and organic components of particulate matter

[14, 15]. These compounds can in fact interact with metal cations promoting the formation of coordination compounds and/or the activation of photo-reactivity processes that affect the oxidation state of the metal and therefore change their mobility in the environment.

The chemical processes occurring in the liquid phase in contact with the particles, such as raindrops, liquid aerosol, or a biological fluid, present features that strongly affect the chemistry of the system: (i) homogenous chemical reactions in the liquid phase are usually faster than those in the gas phase, and some reactions can only take place in the liquid phase (e.g., reactions that involve ionic species); (ii) photochemical processes in solution are usually more efficient than those in air; and (iii) interactions between solid and liquid phases affect both heterogeneous reactions and the reactivity of the liquid phase. In this contest, defining the chemistry of the PM soluble fraction is fundamental to increase the knowledge about the PM reactivity and the definition of the species that actually occur in solution is the first step to model the more complex heterogeneous system.

2.1 Rational of the Speciation Study of Aerosol

Aerosol can be removed from atmosphere by deposition processes that transfer atmospheric material from the air to surfaces. The deposition may be classified dry or wet, according to whether it is triggered by raindrops, snowflakes, fog droplets, or just gravity and diffusion processes, respectively, and can be collected by passive samplers exposed to the atmosphere for a defined period. If dry deposition is collected, ultrapure water can be added to collect and re-suspend any particles attached to the inner surfaces of the sampler. Samples are then shaken and filtered through 0.45 μm membrane filters [16].

PM can be collected by a filter-based sampler in which air is forced to flow through a membrane filter by a pump system for a defined period. The particles are then collected on the filter. The soluble fraction of particulate matter can be obtained extracting a defined amount of PM collected in ultrapure water. The sample is placed in a sonic bath for 30 min, checking that the temperature of the bath did not exceed 27 °C [17]. The extract is left to rest for 24 h before filtration with filters with pore size 0.45 μm, and the filtered solution is then prepared for the instrumental analysis.

The soluble fraction of PM samples are multicomponent aqueous solutions; therefore the definition of the species that can be formed in these solutions and of their concentrations can be obtained by the modelling process reported before. To reach this goal, it is necessary to identify and quantify the total concentration of components in the mixture. For this purpose, the researchers can use conventional analytical techniques. Metal cations can be quantified by atomic spectroscopy/spectrometry or ion chromatography (IC), with the first useful for cations present in low concentration, whereas IC can quantify major species. Both inorganic and organic anions can be quantified by IC, too. Liquid chromatography is instead useful to characterize organic molecules.

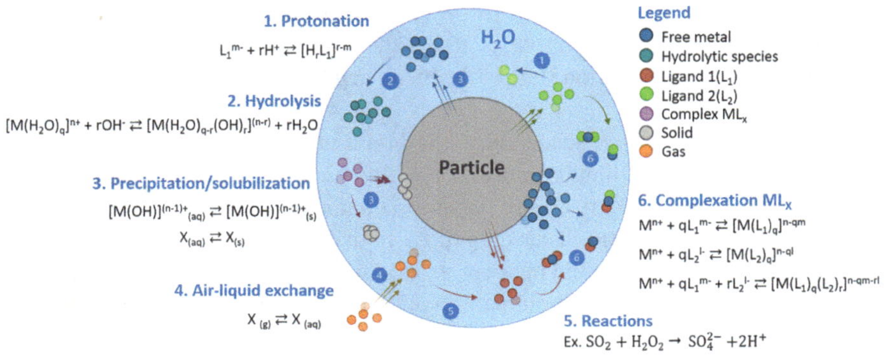

Fig. 2 Graphical representation of possible chemical reactions involved in an aqueous chemical system in equilibrium with particulate matter

In the case of soluble fraction of atmospheric particulate samples, it is reasonable suppose the presence of metal cations, inorganic anions, and some low mass carboxylic acids such as Na^+, K^+, NH_4^+, Ca^{2+}, Mg^{2+}, Mn^{2+}, Cu^{2+}, Zn^{2+}, $Fe^{3+/2+}$, Al^{3+}, Cl^-, NO_2^-, NO_3^-, SO_4^{2-}, PO_4^{3-}, formate, acetate, malonate, succinate, and oxalate. Metal cations can derive from marine, crustal, and anthropogenic sources. NO_3^- and SO_4^{2-} derive from nitric and sulfuric acids that are the most important acidic species usually present in the atmosphere, respectively produced upon the oxidation of SO_2 and NO_x. They may be neutralized by ammonia, which is the main basic species in the atmosphere [18]. Low-molecular-weight organic acids are of both primary and secondary origin. Monocarboxylic acids (formic, acetic) derive mainly from the early oxidation stages of volatile organic compounds (VOCs), while dicarboxylic acids (oxalic, malonic, succinic) are predominantly formed from aqueous processing [33]. Experimental studies have found high correlations between oxalate and formate, supporting the hypothesis that oxalic acid originates mainly from secondary processes [19].

On the basis of components present in the aqueous solution, it is possible to define a chemical model that best represents the chemistry of the multicomponent mixture. This means to prepare a list of the species that can be formed by the combination of the solution components. In order to avoid to forget relevant species, it is advisable to follow the suggestions reported in paragraph 1.1. A useful graphical representation of chemical reactions involved in an aqueous chemical system in equilibrium with a solid particle is represented in Fig. 2. Reactions of precipitation/solubilization, protonation, hydrolysis, and complexation can be concurrently considered in the chemical model, and for each species a formation equilibrium can be written, associated with the corresponding formation constant. As reported before, the species can be represented in a Morel's table, in order to describe the system with a notation suitable for computation, and the formation constants can be found in databases and scientific literature. Since the components are numerous, the chemical model can contain a lot of species. As an example, a chemical model proposed in a study devoted to the characterization of the soluble fraction of Artic PM_{10} is composed by 20 components that participate to 12 protonation processes, 32 hydrolytic equilibria, and 138 complex species or ionic pair formation [20]. In this contest, the mixed species, i.e., those complexes that can be formed by more than one different ligand or polynuclear species, were mostly neglected. Only the mixed complexes with protons or hydroxide anion were considered.

2.2 Species Formed in the Soluble Fraction of PM

The speciation model was applied to define the main species present in both Antarctic atmospheric depositions and the soluble fraction of Arctic PM_{10} [16, 20]. The formation constants used were those reported in the literature and databases at 298.15 K. This temperature is far away from those found at the Arctic/Antarctic environment, but unfortunately the most of chemical equilibria involved in the system have not yet been studied at low temperature, and the dependence of the stability constants on the temperature has not yet been investigated. However, the results collected were useful to define which were the main species that could form in the aqueous solution in contact with PM. The low concentration of the ions detected in solution makes the estimated ionic strength very low, allowing for the use of the thermodynamic equilibrium constants in calculations ($T = 298.15$ K, $I = 0$ mol L^{-1}), and, whenever possible, the selected constants were those obtained with non-interacting cations. In fact, in the experimental definition of the equilibrium constants, the ionic species used for the buffering of the ionic strengths can affect the interactions between the components under investigation. For this reason, when it is possible, ionic species with very low ligand tendency are chosen. In particular, perchlorate and nitrate were considered non-interacting anions, while among the cations, tetra-alkylammonium salts were considered inert. The stability constant values were then adjusted for the real ionic strength of the solution applying the extended Debye-Hückel equation implemented in PyES calculations.

The models highlight that at the concentration level of the extraction solution, (i) the main cations such as Na^+, K^+, Mg^{2+}, and Ca^{2+} occur almost completely in the form of aquoions that are not interacting with no one of the other component considered; (ii) among minor metals, only Al, Cu, Fe, Mn, and Zn were present in significant concentrations and the speciation study shows that most cations occurred as aquoions over the whole pH range, except for Fe^{3+} and Al^{3+} that occurred predominantly as hydrolytic forms; (iii) the transition metals and Al^{3+} show interesting interactions with oxalate.

The modelling process allows not only to describe the solutions under study but also lets to foresee what happens for different solution compositions, just changing the information as input of the software. This allows highlighting the roles of inorganic anions, such as chloride, sulfate and hydrogen phosphate, or organic acids, such as acetate, formate, malonate, succinate, and oxalate, on the speciation of metal cations or observing what are the species formed in more diluted or concentrated solutions. It is in fact possible to draw species distribution diagrams by varying the total concentration of a single component or by changing the total

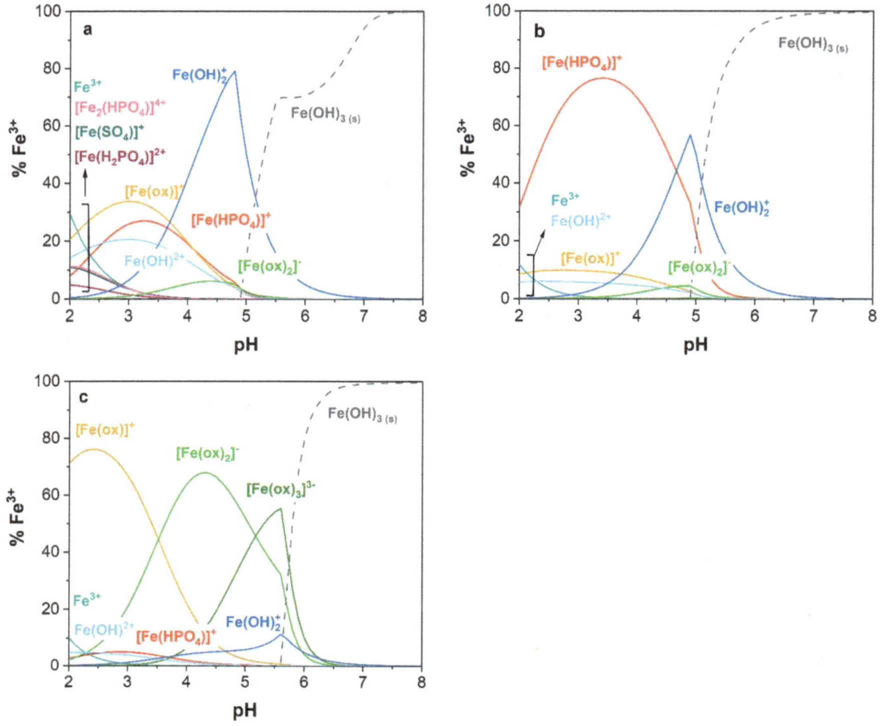

Fig. 3 Species distribution diagrams of Fe(III) in the soluble fraction of a sample of Arctic PM_{10} with $c_{Fe^{3+}} = 2.6 \times 10^{-7}$ mol dm^{-3} ($T = 298.2$ K) and (**a**) $c_{PO_4^{-3}} = 1.1 \times 10^{-5}$ mol dm^{-3}, $c_{ox^{2-}} = 2.5 \times 10^{-6}$ mol dm^{-3}; (**b**) $c_{PO_4^{-3}} = 1.1 \times 10^{-4}$ mol dm^{-3}, $c_{ox^{2-}} = 2.5 \times 10^{-6}$ mol dm^{-3}; (**c**) $c_{PO_4^{-3}} = 1.1 \times 10^{-5}$ mol dm^{-3}, $c_{ox^{2-}} = 2.5 \times 10^{-5}$ mol dm^{-3}

concentrations of all components and observe the variation of the estimated species concentration. As an example, Fig. 3 reports the species distribution diagrams of Fe^{3+} in the soluble fraction of a sample of artic PM_{10} and of the same sample with phosphate and oxalate total concentrations increased by a factor of 10. It is evident how the increasing of the anion concentration changes the chemical form of Fe^{3+}.

Questions

- What are the chemical-physic laws involved in calculation species concentration at chemical equilibrium conditions?
- How can we describe a multicomponent chemical system, in which several chemical species exist, in a matrix format?
- What information is needed to plot a species distribution diagram?
- Where can we find the values of equilibrium constants?
- What parameters influence the values of the equilibrium constants?

Glossary

- Speciation: the distribution of an element among defined chemical species in a system.
- Species distribution diagram: a graph representing the concentration of the species as a function of an independent variable.
- Linear combination: a mathematical expression in which a series of terms are added after multiplying each term by a constant.
- Nonlinear equations are mathematical equations where variables have exponents greater than one and can have multiple solutions, no solutions, or complex number solutions. Solving them involves techniques like substitution, elimination, or graphing intersections, but they are generally harder to solve than linear systems and sometimes require the application of approximation methods, such as Newton-Raphson method.
- Water dissociation constant (K_w): equilibrium constant for water's self-ionization.
  $H_2O \rightleftarrows H^+ + OH^-$
- Thermodynamic equilibrium constant (or standard equilibrium constant): quantity defined by $K = e^{-\Delta_r G^\circ / RT}$ where $\Delta_r G^\circ$ is the standard reaction Gibbs energy, R the gas constant, and T the thermodynamic temperature [37].
- Equilibrium constant: quantity characterizing the equilibrium of a chemical reaction and defined by an expression of the type $K = \prod_B x_B^{\nu_B}$ where ν_B is the stoichiometric coefficient of a reactant (negative) or a product (positive) of the reaction and x stands for the concentration [37].

Further Reading

- W. Stumm, J.J. Morgan, Aquatic Chemistry Chemical Equilibria and Rates in Natural Waters, John Wiley & Sons, New York, 3rd Ed. 1995.
- B. J. Finlayson-Pitts, J. P. Pitts, Jr., Chemistry of the Upper and Lower Atmosphere Theory, Experiments, and Applications, Eds., San Diego: Academic Press, 2000.

Bibliography

1. Templeton DM, Ariese F, Cornelis R, Danielsson L-G, Muntau H, Van Leeuwen HP, Łobiński R, Clinical Chemistry Section, Commission on Toxicology (2000) Guidelines for Terms Related To Chemical Speciation and Fractionation of Elements. Definitions, Structural Aspects, and Methodological Approaches. Pure Appl Chem 72:1453–1470. https://doi.org/10.1351/pac200072081453
2. Hartley FR, Burgess C, Alcock R (1980) Solution equilibria.1st edn. Ellis Horwood Limited, Chirchester, UK

3. Morel F, Morgan J (1972) Numerical method for computing equilibriums in aqueous chemical systems. Environ Sci Technol 6:58–67. https://doi.org/10.1021/es60060a006
4. Brown PL, Ekberg C (2016) Hydrolysis of metal ions. Wiley-VCH Verlag GmbH & Co., Weinheim, Germany
5. Baes FC, Mesmer RE (1976) The hydrolysis of cations. John Wiley & Sons, New York
6. Castellino L, Alladio E, Bertinetti S, Lando G, Gama S, Berto S, De Stefano C, Blasco S, García-España E, Milea D (2023) PyES—an open-source software for the computation of solution and precipitation equilibria. Chemom Intell Lab Syst 239:104860. https://doi.org/10.1016/j.chemolab.2023.104860
7. ReactLab EQSIM (2023). https://jplusconsulting.com/EQSIM/. Accessed 6 Nov 2025
8. Alderighi L, Gans P, Ienco A, Peters D, Sabatini A, Vacca A (1999) Hyperquad simulation and speciation (HySS): a utility program for the investigation of equilibria involving soluble and partially soluble species. Coord Chem Rev 184:311–318. https://doi.org/10.1016/S0010-8545(98)00260-4
9. De Robertis A, De Stefano C, Sammartano S, Rigano C (1986) The calculation of equilibrium concentrations in large multimetal/multiligand systems. Anal Chim Acta 191:385–398. https://doi.org/10.1016/S0003-2670(00)86324-2
10. Medusa webpage (2020) https://www.kth.se/che/medusa/. Accessed 6 Nov 2025
11. PHREEQC webpage (2025) https://www.usgs.gov/software/phreeqc-version-3. Accessed 6 Nov 2025
12. MINEQL+webpage (2023) https://mineql.com/. Accessed 6 Nov 2025
13. Pettit L, Sukhno I, Buzko V, Gans P, Puigdomenech I, Wanner H, Polushin A (2025) Aqu Solut. https://www.acadsoft.co.uk/aq_solutions.htm
14. Debye P, Hückel (1923) The theory of electrolytes. I. Lowering of freezing point and related phenomena, Phys. Z. 24:185–206
15. Davies CW (1962) Ion Association, Butterworths, London
16. Bromley LA (1973) Thermodynamic properties of strong electrolytes in aqueous solutions, AIChE J. 19:313–320. https://doi.org/10.1002/aic.690190216
17. Pitzer KS (1973) Thermodynamics of Electrolytes. I. Theoretical basis and general equations, J. Phys. Chem. 77:268–277
18. Pitzer KS, Raton B, Arbor A, London B (1991) Activity coefficients in electrolyte solutions, in: Molecular Structure and Statistical Thermodynamics, second ed., World Scientific Publishing, Singapore, pp. 435–485. https://doi.org/10.1142/9789812795960_0057
19. NEA, Modelling in Aquatic Chemistry, OECD Publishing, Paris, (2020). https://www.oecd-nea.org/jcms/pl_37657/modelling-in-aquatic-chemistry
20. Guggenheim EA (1935) The specific thermodynamic properties of aqueous solutions of strong electrolytes, Dublin Philos. Mag. J. Sci. 19:588–643. London, Edinburgh. https://doi.org/10.1080/14786443508561403
21. De Stefano C, Mineo P, Rigano C, Sammartano S (1993) Ionic stength dependence of formation constants. XVII. The calculation of equilibrium concentration and formation constants, Ann. Di Chim. 83:243–262
22. N. Hatada, Stability constant explorer, (2023). https://n-hatada.github.io/stability-constant-explorer/english.html
23. IUPAC, NIST, IUPAC-NIST Solubility Database, Version 1.1 (2012) https://srdata.nist.gov/solubility/index.aspx
24. Nuclear Energy Agency (NEA) of the Organization for Economic Co-operation and Development (OECD), Chemical Thermodynamics Series, (2023) https://www.oecd-nea.org/dbtdb/info/publications
25. Nuclear Energy Agency (NEA) of the Organization for Economic co-operation and Development (OECD), TDB electronic database (2023) https://www.oecd-nea.org/jcms/pl_37223/electronic-database-of-the-tdb-project
26. Network for Equilibria and Chemical Thermodynamics Advanced Research COST ACTION 18202, Hydrolysis (and associated equilibria) constants (2024) https://www.cost-nectar.eu/pages/wg1_period.html

27. Lambert F, Kug J-S, Park RJ, Mahowald N, Winckler G, Abe-Ouchi A, R. O'ishi T, Takemura, Lee J-H (2013) The role of mineral-dust aerosols in polar temperature amplification. Nat Clim Chang 3:487–491. https://doi.org/10.1038/nclimate1785
28. Petters MD, Kreidenweis SM (2013) A single parameter representation of hygroscopic growth and cloud condensation nucleus activity-part 3: including surfactant partitioning. Atmos Chem Phys 13:1081–1091. https://doi.org/10.5194/acp-13-1081-2013
29. Mahowald NM, Baker AR, Bergametti G, Brooks N, Duce RA, Jickells TD, Kubilay N, Prospero JM, Tegen I (2005) Atmospheric global dust cycle and iron inputs to the ocean. Global Biogeochem Cycles 19, GB4025. https://doi.org/10.1029/2004gb002402
30. Fountoukis C, Nenes A (2007) ISORROPIAII: A computationally efficient thermodynamic equilibrium model for K^+-Ca^{2+}-Mg^{2+}-NH_4^+-Na^+-SO_4^{2-}-NO_3^--Cl^--H_2O aerosols. Atmos Chem Phys 7:4639–4659. https://doi.org/10.5194/acp-7-4639-2007
31. Anderson JO, Thundiyil JG (2012) Clearing the air : a review of the effects of particulate matter air pollution on human health. J Med Toxicol, 8:166–175. https://doi.org/10.1007/s13181-011-0203-1
32. Ito A (2015) Atmospheric processing of combustion aerosols as a source of bioavailable iron. Environ Sci Technol Lett 2:70–75. https://doi.org/10.1021/acs.estlett.5b00007
33. Zhou Y, Zhang Y, Griffith SM, Wu G, Li L, Zhao Y, Li M, Zhou Z, Yu J (2020) Field evidence of Fe-mediated photochemical degradation of oxalate and subsequent sulfate formation observed by single particle mass spectrometry. Environ Sci Technol 54:6562–6574. https://doi.org/10.1021/acs.est.0c00443
34. Rimauro J, Napolitano G, Fasciolo G, Venditti P, Riccio A, Chianese E (2025) Study on toxicity and bioavailability of metals from urban PM2.5 and PM10 extracted in simulated biological fluids : in vitro and in vivo assessment. Ecotoxicol Environ Saf 302:118707. https://doi.org/10.1016/j.ecoenv.2025.118707
35. Adamson IYR, Prieditis H, Vincent R, Adamson F, Pharmacol RTA (1999) Pulmonary toxicity of an atmospheric particulate sample is due to the soluble fraction. Toxicol Appl Pharmacol 157:43–50. https://doi.org/10.1006/taap.1999.8658
36. Haswani D, Raman RS (2025) Understanding PM-bound trace element solubilities : Controls, anthropogenic and natural sources. Environ Pollut 383:126823. https://doi.org/10.1016/j.envpol.2025.126823
37. 'standard equilibrium constant' in IUPAC Compendium of Chemical Terminology, 5th ed. International Union of Pure and Applied Chemistry; 2025. Online version 5.0.0, 2025. https://goldbook.iupac.org/terms/view/S05915

Sunlight Can Fight Pollution: Photochemical Self-Depuration of Surface Waters and Effects of Climate Change

Davide Vione

Abstract

This chapter provides a brief introduction to the photochemical reactions that occur in sunlit surface waters and account for the attenuation of biorecalcitrant contaminants. Some examples of the possible effects of climate change on these light-induced reactions are also described and discussed. Because of the complexity of the natural environment, these phenomena are typically addressed by a combination of field measurements, laboratory experiments, and modeling.

Keywords

Photochemical transformation · Direct photolysis and indirect photochemistry · Contaminants of emerging concern · Abiotic attenuation of pollutants · Photochemistry of rivers and lakes · Effects of climate change

Chapter Learning Goals

- The importance of direct photolysis and indirect photochemistry in the natural attenuation of contaminants.
- What happens to photochemical reactions if the organic matter dissolved in surface waters increases.
- Photochemical implications of water evaporation in arid environments.
- Effects of ice cover and ice melting on photochemical reactions.
- Implications of vegetation cover and treeline shifts for photochemical reactions.

D. Vione (✉)
Department of Chemistry, University of Turin, Torino, Italy
e-mail: davide.vione@unito.it

M. Minella (ed.), *Complexity in Environmental Chemistry*,
https://doi.org/10.1007/978-3-032-31774-2_11

1 Introduction

Photochemical transformation reactions are important pathways that account for the degradation of many of the so-called contaminants of emerging concern (CECs) in sunlit surface waters. CECs are man-made compounds that belong to several different classes, including pharmaceuticals and personal care products, household chemicals, some pesticides, and some compounds manufactured by industry. CECs occur at trace levels in natural waters but can harm wildlife and human health through their endocrine disruption, toxic, and/or mutagenic actions. Moreover, antibiotic residues can also contribute to the spread of the phenomenon of anti-microbial resistance in surface waters.

Several CECs are biorecalcitrant and polar, for which reason they are hardly removed by conventional wastewater treatment technologies. They are, for instance, able to survive biological treatment steps because biorecalcitrance preserves them from biodegradation, while polarity prevents adsorption on biosolids that would cause physical removal of these compounds from the water phase. Therefore, CECs often reach natural waters through wastewater discharge. Other routes of CECs to water bodies include agriculture, animal husbandry and fisheries (which make wide use of antibiotics, for instance), recreational activities (e.g., UV filters in sunscreens), and industrial discharges (e.g., pharmaceutical residues) [1].

The same properties that shield CECs from removal during wastewater treatment are also operational in the natural environment; thus biodegradation and sorption to sediments are often scarcely effective attenuation pathways for CECs in water. In this framework, photodegradation may be important in the removal of many CECs from surface waters [2].

The photoreactions that take place in natural surface waters are usually divided into direct photolysis and indirect photochemistry. In the case of direct photolysis, the CEC absorbs sunlight and undergoes transformation, for instance, through bond breaking, light-induced ionization, or excited-state reactivity. As far as indirect photochemistry is concerned, sunlight is absorbed by natural compounds called photosensitizers such as nitrate, nitrite, and the chromophoric dissolved organic matter (CDOM). Sunlight absorption by the photosensitizers triggers the formation of several photochemically produced reactive intermediates (PPRIs), which are reactive transient species that can react with CECs and induce their transformation. Important PPRIs involved in indirect photochemistry processes are the hydroxyl ($^{\bullet}OH$) and carbonate ($CO_3^{\bullet -}$) radicals, singlet oxygen ($^1O_2$), and the triplet excited states of CDOM ($^3CDOM^*$). The main pathways that account for the formation of the mentioned PPRIs in sunlit natural waters are outlined below [3] (ISC = inter-system crossing).

$$NO_3^- + h\upsilon + H^+ \rightarrow {}^{\bullet}OH + {}^{\bullet}NO_2 \quad (1)$$

$$NO_2^- + h\upsilon + H^+ \rightarrow {}^{\bullet}OH + {}^{\bullet}NO \quad (2)$$

$$CDOM + h\upsilon - [\ldots] \rightarrow CDOM(OH)_n \xrightarrow{h\upsilon} {}^{\bullet}CDOM(OH)_{n-1} + {}^{\bullet}OH \quad (3)$$

$$CDOM + h\upsilon \rightarrow {}^1CDOM^* \xrightarrow{ISC} {}^3CDOM^* \quad (4)$$

$${}^3CDOM^* + O_2 \rightarrow CDOM + {}^1O_2 \quad (5)$$

$${}^{\bullet}OH + HCO_3^- \rightarrow H_2O + CO_3^{\bullet-} \quad (6)$$

$${}^{\bullet}OH + CO_3^{2-} \rightarrow OH^- + CO_3^{\bullet-} \quad (7)$$

$${}^3CDOM^* + CO_3^{2-} \rightarrow CDOM^{\bullet-} + CO_3^{\bullet-} \quad (8)$$

In the case of reaction (3), CDOM irradiation produces excited species that undergo hydroxylation upon reaction with water. Afterward, the photolysis of some CDOM hydroxyderivatives yields $^{\bullet}OH$ [3].

Other than being photogenerated, the PPRIs are also scavenged or quenched by several natural water components. Therefore, $^{\bullet}OH$ in freshwater is scavenged by dissolved organic matter (DOM, not necessarily chromophoric), carbonate, bicarbonate, and, to a lesser extent, nitrite. In the case of seawater, the main $^{\bullet}OH$ scavenger is bromide. The radical $CO_3^{\bullet-}$ is mostly scavenged by DOM, $^3CDOM^*$ is quenched by dissolved oxygen to yield $^1O_2$ (reaction 5) or simply ground-state O_2, while $^1O_2$ is deactivated by collision with the water solvent. This budget between photogeneration and scavenging causes PPRIs to reach very low steady-state concentrations in sunlit surface waters, which typically range between 10^{-18}–10^{-14} M [4]. A graphical outline of the mentioned processes is provided in Fig. 1.

Fig. 1 Outline of the processes that drive the production and scavenging of PPRIs (green ovals) upon sunlight irradiation of photosensitizers (yellow rectangles) and interaction with other water components (blue rectangles)

The occurrence of PPRIs in natural waters is largely affected by sunlight irradiance and water depth. In the case of irradiance, overall and especially under fair weather conditions, photoreaction kinetics are faster in spring or summer than in autumn or winter. As far as depth is concerned, shallow waters are usually better illuminated by sunlight compared to deep ones, because the lower depths of deep waters are often in the dark [3]. Therefore, indirect photochemical reactions are slower in deeper waters and their reaction rates are more or less inversely proportional to the water depth. A similar behavior with depth is also observed for the direct photolysis processes.

The chemical composition of water is another important environmental factor, because it can determine which photoreaction pathways are enhanced and which are inhibited. For instance, $^{\bullet}OH$ reactions are enhanced both in the presence of low values of dissolved organic carbon (DOC, which quantifies DOM), because organic matter is a major $^{\bullet}OH$ sink, and in the presence of high levels of the photochemical $^{\bullet}OH$ sources nitrate and nitrite. Of course, opposite conditions decrease the importance of the $^{\bullet}OH$-triggered processes. Other conditions that can be detrimental to $^{\bullet}OH$ reactions involve high bromide (e.g., seawater) and high carbonate levels. The latter can be found in the presence of elevated inorganic carbon and high pH [4].

The reactions induced by $CO_3^{\bullet-}$ are favored by low DOC values and by high levels of carbonate and bicarbonate. In these conditions, high nitrate and nitrite enhance the $CO_3^{\bullet-}$ processes, because $^{\bullet}OH$ is a major PPRI that is involved in the generation of $CO_3^{\bullet-}$ (reactions 6 and 7) [5]. In contrast, the processes induced by $^3CDOM^*$ and $^1O_2$ are enhanced in the presence of high DOC values, in which the amount of CDOM is also expected to be high [4]. In this framework, Fig. 2 reports the steady-state concentrations of $^{\bullet}OH$, $CO_3^{\bullet-}$, $^3CDOM^*$, and $^1O_2$ as a function of the DOC, showing a decrease with increasing DOC for $^{\bullet}OH$ and $CO_3^{\bullet-}$ and an increase for $^3CDOM^*$ and $^1O_2$.

Note that the trends shown in Fig. 2 were obtained by means of modeling tools. The experimental measurement of PPRIs is a very challenging task, and sometimes it cannot properly take the variability of environmental conditions into account. For instance, experimental measurements struggle to properly include the effects of water-column depths, which are typically much larger in the natural environment than in laboratory conditions. Moreover, the large variability in the chemistry and irradiance values of natural waters can be reproduced in the laboratory only by means of extremely time-consuming procedures. In these circumstances, photochemical modeling is very useful to bridge the gap between laboratory experiments and the natural environment.

In the case of the direct photolysis, the main inhibitor is CDOM that is a major sunlight absorber in natural waters and competes with pollutants for the absorption of sunlight. Therefore, direct photolysis reactions are fastest in clear waters with low DOC and low CDOM. At the same time, however, high DOC levels inhibit the $^{\bullet}OH$ and $CO_3^{\bullet-}$ reactions to a higher extent than they inhibit the direct photolysis. Therefore, for contaminants that are degraded by all photochemical processes, the direct photolysis may reach higher relative importance with intermediate values of the DOC [3] (at high DOC, $^3CDOM^*$ and $^1O_2$ would often prevail).

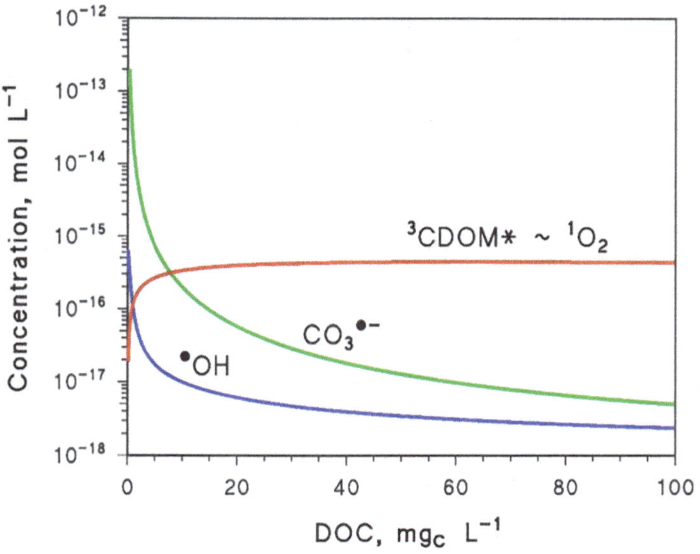

Fig. 2 Trend of the modeled steady-state concentrations of PPRIs as a function of the DOC, under fair weather summertime irradiation conditions [6]. Note the increase in [^{3}CDOM*] and [$^1O_2$] and the decrease in [$^{\bullet}$OH] and [$CO_3^{\bullet-}$]

An example of how solar irradiance and the DOC may interact in the photodegradation of contaminants on a global scale is provided in Fig. 3. Here, a DOC map of the global lakes is shown (Fig. 3a), which clearly suggests that the highest DOC values are found at elevated Nordic latitudes in the boreal hemisphere. On the other hand, the average yearly irradiance is maximum at the equator and in the tropical belt, and it gets lower as one moves northward or southward from there. In the case of a contaminant that is mainly degraded by the direct photolysis (diclofenac, Fig. 3b), it is shown that photodegradation kinetics follow irradiance; thus photodegradation is the fastest at the tropics and the slowest at more elevated latitudes in both hemispheres. In the case of Nordic lakes in the boreal hemisphere, the combination of relatively low irradiance and high DOC values accounts for quite slow photodegradation of diclofenac. In contrast, for a contaminant that mainly reacts with ^{3}CDOM* (clofibric acid, Fig. 3c), the kinetics of photodegradation largely follow the global distribution of lake-water DOC. Actually, the fastest kinetics of clofibric acid photodegradation are observed at elevated Nordic latitudes, where the highest DOC values are also found, despite lower sunlight irradiance compared to the tropical belt. On a global scale, the DOC would thus be more important than irradiance to define the kinetics of triplet sensitization (reaction with ^{3}CDOM*) [7].

The described processes are involved in the photodegradation of contaminants, but they are also affected by modifications of the natural environment. In particular, climate change can affect water chemistry and depth, which may have important consequences on photoreaction kinetics [3]. Important phenomena that are at least partially affected by climate are the browning of boreal water bodies, the duration

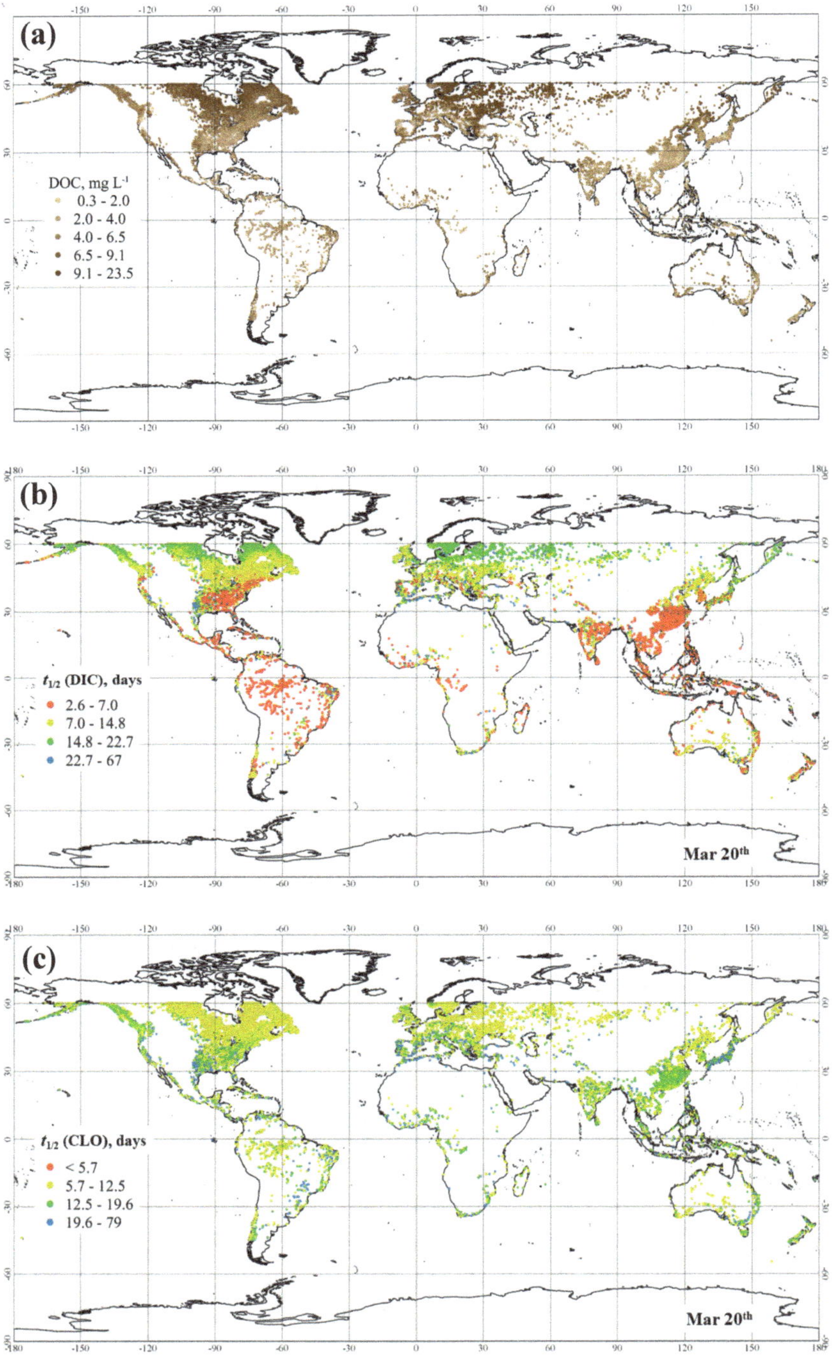

Fig. 3 Quasi-global maps (60°S–60°N) of (**a**) lake-water DOC values; (**b**) modeled half-life times ($t_{½}$) of the anti-pain drug Diclofenac, and (**c**) modeled half-life times of Clofibric acid (metabolite of the lipid regulating drugs Clofibrates) [7]

of summertime (or otherwise dry-season) evaporative concentration of water, shortening of the ice-cover season in mountain and/or high-latitude lakes, and the gradual increase in altitude of the treeline in mountain environments. The photochemical implications of these phenomena will be discussed in the following paragraphs of this chapter.

2 Browning of Boreal Lake Waters

One of the most important environmental changes that could have considerable impact on the photochemistry of surface freshwaters is the phenomenon of browning or brownification. It consists in the progressive increase in the UV-visible absorbance and DOC values of water, currently observed in many surface-water bodies located in the Scandinavian peninsula and in the Eastern US, due to increased export of organic matter from soils to water [8]. The phenomenon of browning has several causes, among which:

(i) An increase in the pH of rainwater, as a consequence of the desulfurization of fuels and the subsequent decline in rainwater acidity. This is an unexpected side effect of pollution decrease, and the issue is that acid rain is a poor solvent for the humic and fulvic acids contained in soil. In contrast, less acidic rainwater allows for more extensive deprotonation of such soil substances, which thus become more water-soluble and can undergo leaching from soil into water bodies, where they increase both the absorbance and the DOC values of surface waters. Changes in soil use (e.g., forest vs. agriculture) can also alter the export of humic substances from the watersheds. The fact that humic and fulvic acids absorb visible light and are brown in color accounts for the name of the phenomenon.
(ii) Higher precipitation in the relevant regions, as a consequence of climate change. A wetter climate can increase both the content of organic matter in soil and its export to surface waters, the latter especially in the presence of heavy rain events.

With reference to Fig. 2, an increase in the (C)DOM content of surface waters is expected to inhibit the reactions triggered by $^{\bullet}OH$ and $CO_3^{\bullet-}$ while favoring the processes induced by $^3CDOM^*$ and $^1O_2$. As far as the direct photolysis is concerned, brownified water is darker due to CDOM, which competes with the contaminants for the absorption of sunlight. Therefore, the kinetics of the direct photolysis of contaminants would slow down as a consequence of browning [9].

A further effect of water browning is a shallower thermocline in lakes during the summer stratification period. Summer stratification in lakes is due to the fact that the near-surface water is heated by sunlight, which decreases water density as the temperature increases. If the temperature gradient between surface and deep water is high enough, stratification may occur where the surface-water layer (epilimnion) floats over bottom water (hypolimnion), and mass exchange between the two layers is inhibited. Water in the lower depths of a lake is poorly illuminated by sunlight, and, below a certain irradiance threshold, the inhibition of photosynthesis in deep water may cause the latter to become anoxic and thus unsuitable for the life of higher organisms. Browning decreases the penetration depth of sunlight in lake water, which decreases the depth of the surface layer that can be heated by sunlight during summer. As a consequence of browning, the epilimnion becomes shallower during stratification and a larger fraction of the lake loses contact with the atmosphere. Under suitable conditions, the hypolimnion may become anoxic. In these circumstances, the available lake volume for the life of, e.g., fish, during summer may be considerably decreased [10].

A shallower epilimnion together with the browning of water has photochemical implications as well. Interestingly, direct photolysis processes in the epilimnion would be poorly affected, because of the opposing effects of shallower water, which accelerates the direct photolysis kinetics, and higher CDOM that inhibits the direct photolysis. In contrast, the reactions triggered by $^{\bullet}OH$ and $CO_3^{\bullet-}$ would be slower in a shallower but (C)DOM-richer epilimnion. In this case, the positive effect of a shallower water column on the photogeneration rates of the radicals $^{\bullet}OH$ and $CO_3^{\bullet-}$ would be more than offset by the combination of higher DOM that would enhance radical scavenging and higher CDOM that would compete for sunlight absorption and thus inhibit the photolysis of nitrate and nitrite (note that $^{\bullet}OH$ production in reaction (3) is less effective than in reactions (1) and (2)). Reactions induced by $^3CDOM^*$ and $^1O_2$ would instead be enhanced in the epilimnion of a brownified lake [9]. Possible effects on the photodegradation of contaminants in a lake epilimnion are shown in Fig. 4. Three compounds

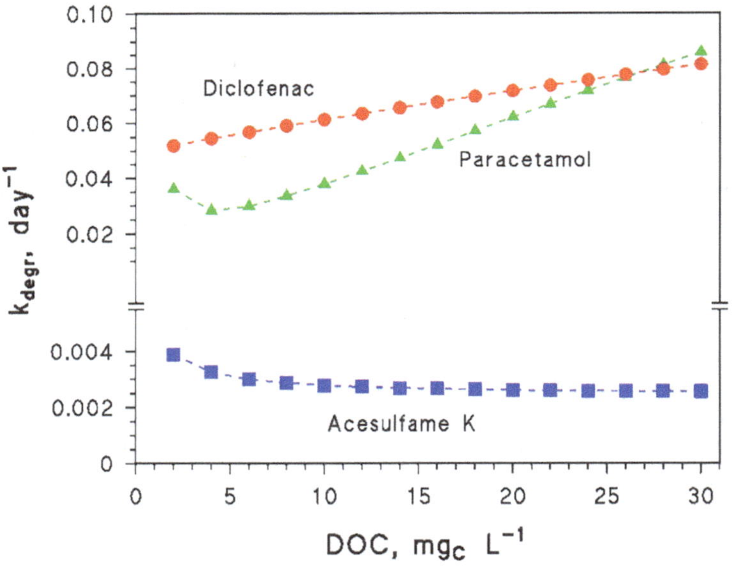

Fig. 4 First-order photodegradation rate constants of three pollutants in epilimnion water, modeled as a function of the DOC. Note that water browning would cause the DOC to increase and the epilimnion to become shallower. Here it is hypothesized that the depth of the epilimnion would vary as d_{epi} [m] = 30 (DOC [mg_C L^{-1}])$^{-1}$. Also note the break in the Y-axis, due to the fact that the photodegradation of acesulfame K is much slower than those of the other two compounds

are taken into account, namely, the antipyretic drug paracetamol (which is mainly degraded by $CO_3^{•-}$ and $^3CDOM^*$), the pain killer diclofenac (which mainly undergoes direct photolysis, but with an increasing importance of $^3CDOM^*$ as the browning process proceeds), and the artificial sweetener acesulfame K, which only reacts with $^•OH$.

Note that, in Fig. 4, the progression of the browning process is indicated by an increase of the DOC. At the same time, the epilimnion depth decreases. It can be seen that:

(i) The degradation of paracetamol first slows down and then accelerates as the DOC increases, because of the combined inhibition of $CO_3^{•-}$ photodegradation and enhancement of paracetamol transformation by $^3CDOM^*$.
(ii) The degradation of acesulfame K slows down because the $^•OH$-mediated processes are inhibited with increasing DOC.
(iii) The degradation of diclofenac accelerates with increasing DOC, because of the growing importance of the $^3CDOM^*$ reaction that is enhanced, while the kinetics of the direct photolysis are about constant.

3 Dry-Season Evaporative Concentration of Lake Water

In warm regions during the dry season, the scarcity of precipitation coupled with elevated temperatures can cause lakes to lose important amounts of water due to evaporation. In these circumstances water evaporates, but non-volatile solutes do not, causing evaporative concentration of these solutes [11]. As far as photoreactions are concerned, both sources and scavengers of PPRIs typically increase in concentration because of evaporation, with the exceptions of O_2 ($^3CDOM^*$ scavenger) and H_2O ($^1O_2$ quencher). An example of the trends of the steady-state concentrations of $^•OH$, $CO_3^{•-}$, and $^1O_2$ ~ $^3CDOM^*$ upon water evaporation is provided in Fig. 5, as a function of water depth. In this framework, a depth decrease implies evaporative concentration of the photochemically significant solutes nitrate, nitrite, inorganic carbon (HCO_3^- + CO_3^{2-}), and organic matter, the latter quantified as the DOC. The result is an increase in the steady-state concentrations of all PPRIs as depth decreases (i.e., as evaporation progresses), but [$^3CDOM^*$] and [$^1O_2$] increase faster than [$^•OH$] or [$CO_3^{•-}$]. The reason is that while all the PPRI sources (NO_3^-, NO_2^-, and CDOM) get more concentrated, not all the quenchers/scavengers behave the same. In particular, the scavengers of $^•OH$ and $CO_3^{•-}$ (DOM, HCO_3^-, and CO_3^{2-}) undergo concentration, thereby partially offsetting the effect on the sources, while the quenchers of $^3CDOM^*$ and $^1O_2$ (O_2 and H_2O, respectively) not. Note that non-volatile pollutants would also get more concentrated, which means that the pollution phenomenon would be exacerbated by evaporative water concentration [12]. In this framework, an increase in PPRI levels would partially offset the increased pollution by triggering faster pollutant attenuation, but it would also shift the photoreaction pathways from $^•OH$ and $CO_3^{•-}$ to $^3CDOM^*$ and $^1O_2$. The outcome could be

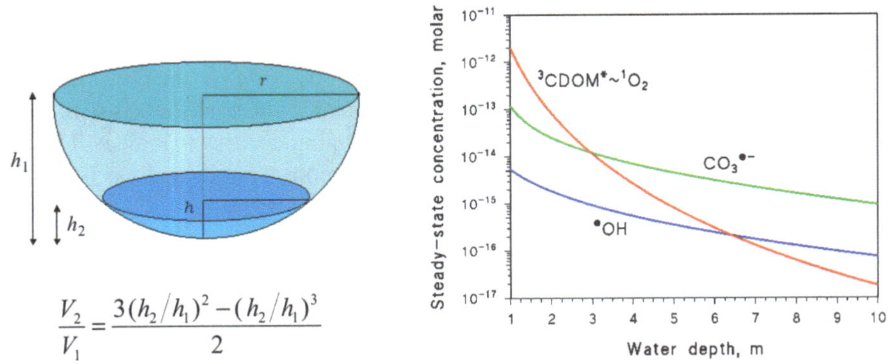

Fig. 5 Simplified schematic of water evaporation in a lake of arbitrary shape (left) and modeled trends of the steady-state concentrations of PPRIs [12]. Note that water evaporation causes depth to decrease; thus evaporative concentration proceeds from right to left on the X-axis. When d = 10 m, water conditions were assumed as 10^{-6} M NO_3^-, 10^{-8} M NO_2^-, 10^{-5} M HCO_3^-, 10^{-7} M CO_3^{2-}, and 0.1 mg_C L^{-1} DOC

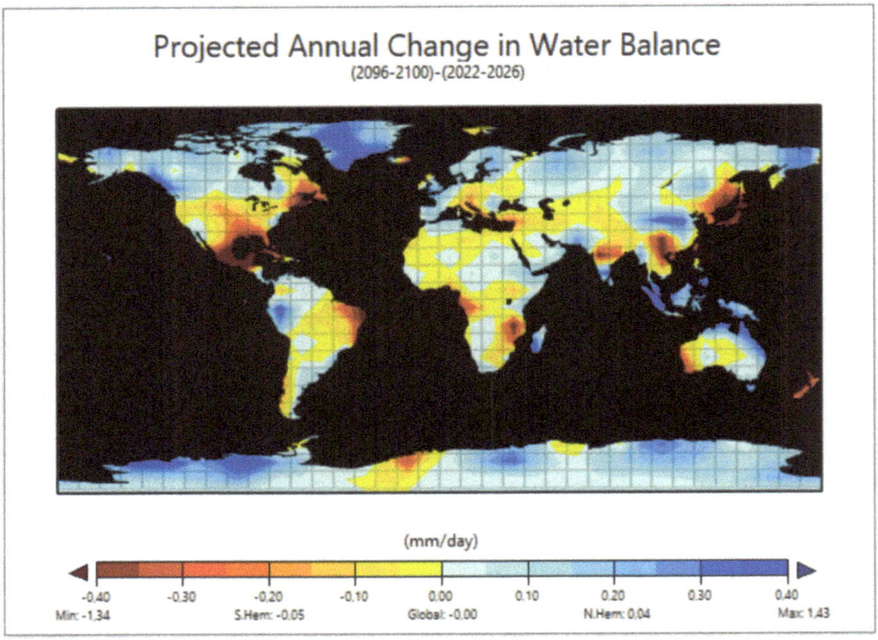

Fig. 6 Comparison between water balance (precipitation minus evaporation) at the end of the 21st century and at present [12]. The EdCGM climate model was run by assuming a linear increase of atmospheric CO_2, extrapolating the behavior of the first quarter of the century

variable, mainly depending on the transformation products of each pollutant and on their effects on the natural environment and on human health.

In this framework, Fig. 6 shows the predictions of a climate model concerning the budget in water balance between precipitation and evaporation, with a comparison between the present and the end of the 21st century. The regions that are highlighted in orange or red are those where the water balance is most negative, i.e., where evaporation is expected to most strongly overcome precipitation. In such circumstances, the phenomenon of evaporative water concentration and its above-described consequences are most likely to take place.

4 Shortening of the Ice Cover Period

A very likely event in a warming world is the occurrence of warmer summers and winters, which would produce faster ice melting both in mountain regions and at elevated latitudes. For instance, gradual disappearance of glaciers below 3000 m asl altitude is foreseen during this century in the European Alps [13].

Ice cover in a lake has important photochemical implications, because the ice layer and especially the occurrence of snow on the ice can effectively prevent the penetration of sunlight in the water column. In these circumstances, photochemical reactions in the underlying water would be effectively inhibited and, sometimes, completely switched off. Global warming will cause winter ice cover on lakes to last for shorter periods, with two main consequences on lake ecosystems [3, 14, 15]:

(i) Higher penetration of UV and, especially, harmful UV-B radiation within the water column for a longer fraction of the year, with potential damage to aquatic organisms and/or the need for the latter to recur to radiation protection or anti-oxidant defenses. The typical outcome is an increased level of stress or even the death of the organisms involved.
(ii) A higher degree of lake-water irradiation during the year, which enhances photochemical reactions and their ability to attenuate pollution. Note that, in such mostly remote environments, many contaminants would arise from atmospheric depositions.

An interesting issue with lake environments in the European Alps is that the ice-melting period in late Spring-early Summer and the ice-reforming period in late Summer-Autumn are not characterized by the same level of incident sunlight irradiance. Actually, ice melting occurs in a period when the clear-sky irradiance is next to its maximum. In such circumstances it makes a huge difference, from a photochemical point of view, to have water covered by an ice layer or to have it free of ice. In contrast, ice reforms in a period when irradiance is almost an order of magnitude lower compared to the melting season. Therefore, the photochemical reactions that would be impeded by reformed ice would be slow anyway [3]. This state of things is shown in Fig. 7 for the steady-state concentrations of major PPRIs, which are proportional to the irradiance and are correspondingly higher during ice-melt than during ice reforming.

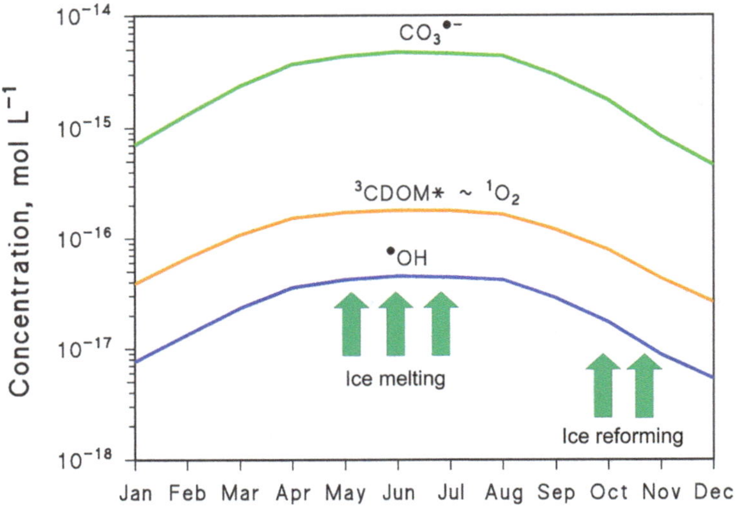

Fig. 7 Fair-weather steady-state concentrations of PPRIs in different months of the year, modeled under mid-latitude conditions. The typical period in which ice melts or reforms in lake water in the European Alps is also highlighted. Water conditions: 10^{-4} M NO_3^-, 10^{-6} M NO_2^-, 10^{-3} M HCO_3^-, 10^{-5} M CO_3^{2-}, 1 mgC L^{-1} DOC, and 3 m depth [3]. (Copyright 2019 American Chemical Society, reproduced with permission)

Therefore, if the melting of ice is anticipated by 1–2 weeks in late Spring-early Summer due to global warming, this event has far higher photochemical implications than a corresponding delay in ice reforming in late Summer or Autumn.

5 Increase in the Altitude of the Treeline

In a mountain environment, the treeline is defined by the maximum altitude at which trees still manage to grow. Above that line, a combination of low average temperatures, strong winds, and water scarcity in summer prevents the growth of trees, so that vegetation is only made up of grass. Trees and especially their fallen leaves play a major role in providing soil with a reservoir of organic matter. Most notably, the microbial degradation of vegetal debris in woods and forests enriches the soil with humic material. This material, and especially water-soluble humic and fulvic acids, can be leached by precipitation and reach surface-water bodies, where they give an important contribution to radiation absorption and photochemical reactivity [16].

In particular, humic and fulvic acids are largely aromatic in nature and absorb sunlight, in both the UV and visible regions, to a higher extent than the organic matter that is released from dead algae. While the ability of algal-derived material to produce PPRIs per unit of photons absorbed (in other words, its PPRI photogeneration quantum yield) can be higher compared to humic and fulvic acids, this does not make up for the fact that the latter absorb sunlight to a much higher extent. As a consequence, the potential to photochemically produce PPRIs is considerably higher in water from lakes surrounded by forests, than in water from lakes surrounded by meadows or bare rocks. In the latter case, the soil, if any, is rather poor of organic material and of humic substances that reach the lake through runoff, and most of the lake-water CDOM is made up of less photoactive material originating from algal lysis or atmospheric depositions [17].

Climate change and the related increase in temperature are going to allow trees to grow at higher altitude than before, thereby shifting the treeline upward and providing potential for high-mountain soil and surface waters to become enriched with humic substances. In such a scenario, an increase in the photochemical reactivity of CDOM can be expected, with a corresponding increase in the ability of organic matter to photochemically generate PPRIs [17]. This phenomenon would likely enhance processes triggered by PPRIs that are directly photogenerated by CDOM and are not scavenged or quenched by DOM, such as $^3CDOM^*$ and $^1O_2$. Differently, the impact on reactions induced by $^{\bullet}OH$ and $CO_3^{\bullet-}$ would be less straightforward. On the one side, more photoreactive CDOM would produce such radicals or their precursors to a higher extent, but much less is currently known of the ability of soil-leached DOM to act as radical scavenger. Therefore, processes induced by $^{\bullet}OH$ and $CO_3^{\bullet-}$ would not necessarily undergo an enhancement. Moreover, an increase in light-absorbing substances in lake water would inhibit the direct photolysis of contaminants, due to competition for irradiance. As a consequence, the effect of an upward shift of the treeline on photochemical reactions in mountain lakes is likely to be substrate-dependent and maybe also environment-dependent, on the basis of the photochemical reaction pathways that a contaminant would mostly undergo.

6 Conclusions and Outlook

Photochemical reactions (direct photolysis and indirect photochemistry) play an important role in the attenuation of biologically recalcitrant contaminants in sunlit surface waters. The kinetics and pathways of pollutant attenuation largely depend on environmental factors such as sunlight irradiance, water depth, and water chemistry. All these factors can be modified, to at least some extent, by phenomena that are connected with the ongoing climate change. As a consequence, climate change has an impact on photochemical reactions, which is however variable depending on the affected environment and the involved phenomenon. Furthermore, consequences of climate change are not the same in all the affected environments, but they may show considerable variations.

Here, some possible scenarios and their potential photochemical implications are discussed. The brownification of water in boreal environments has the potential to enhance $^3CDOM^*$ and $^1O_2$ photoreactions at the expense of $^{\bullet}OH$, $CO_3^{\bullet-}$, and the direct photolysis. The outcome can be very variable, depending on the prevailing photoreaction pathway(s) that different contaminants may follow. In some cases, opposite effects may be observed in different environments if a contaminant undergoes phototransformation pathways that are affected to different extents. A typical example is with phenols and anilines, which undergo degradation by $CO_3^{\bullet-}$ at low

DOC and by $^3CDOM^*$ at high DOC. In this case, water browning could inhibit photodegradation in low-DOC environments and enhance it in high-DOC ones.

Another phenomenon that could produce different effects in different environments or with different contaminants is the change in CDOM, connected with upward shifts of the treeline that would enrich lake water with humic and fulvic acids leached from soil.

While browning is partly due to increased precipitations in some environments, other environments could become more arid and prone to evaporative concentration of water during summer or otherwise during a dry season. Photochemical reactions would in most cases be enhanced by a decrease in water depth, although some processes (especially $^3CDOM^*$ and $^1O_2$) would be more favored compared to $^{\bullet}OH$, $CO_3^{\bullet-}$, and the direct photolysis.

A shorter period of ice cover is another event that would enhance all photochemical reactions and all of them to similar extents in this case. However, earlier ice melting in late Spring-early Summer may have higher photochemical implications than delayed ice reforming in late Summer and Autumn.

Questions

- Which is the difference between direct photolysis and indirect photochemistry?
- What is water browning, and how could it affect photochemical reactions?
- If water is concentrated by evaporation, how do the concentration trends of the different photosensitizers and scavenger/quenchers affect the processes triggered by PPRIs?
- Why does ice melting in Alpine lakes in late Spring/early Summer have different photochemical implications than ice reforming in Autumn?
- How can an upward shift in the treeline modify water (C)DOM, and how could this phenomenon affect photochemical reactions?

Glossary

- **Acid rain**: rain that has lower pH than usual (the natural pH of rainwater is around 5.5), because of the presence of atmospheric pollutants such as HNO_3 and H_2SO_4. These compounds originate from the atmospheric oxidation of nitrogen oxides and sulfur dioxide, respectively. Acid rain can damage plants, water bodies, and even monuments made up of marble or limestone.
- **Browning**: a phenomenon for which surface waters get enriched with sunlight-absorbing organic matter, due to enhanced leaching of organic compounds from soil. The increased content of organic matter causes water to assume a yellowish or brownish color.

(continued)

- **Contaminants of emerging concern** (CECs): they are pollutants that occur at trace levels (usually, μg L^{-1} or lower) in natural waters and can be harmful to both aquatic life forms and humans. The monitoring of CECs requires rather sophisticated techniques, which explains why attention to the occurrence of these compounds in environmental waters has been raised only recently.
- **Desulfurization**: it is a procedure used to eliminate sulfur and sulfur-containing compounds from fuels such as coal and oil derivatives, in order to reduce the emissions of sulfur dioxide and control the phenomenon of acid rain.
- **Dissolved organic matter** (DOM): it is the pool of organic compounds that occur in the dissolved phase in natural waters. Operationally, DOM is defined as the fraction of organic matter that passes through a filter with pores having diameter of 0.45 μm. The fraction of DOM that can absorb light is defined as the chromophoric DOM (CDOM).
- **Evaporative concentration**: it is the phenomenon for which water and volatile solutes evaporate at relatively high temperature, while non-volatile solutes do not and get concentrated as the consequence.
- **Half-life time** ($t_{½}$): it is the time it takes for a reaction to halve the initial concentration of a given compound.
- **Humic substances**: they are complex aromatic compounds of natural origin, produced by microbial decay of plant and animal spoils. In water, humic substances absorb sunlight and trigger photochemical reactions. In order of increasing molecular weight and decreasing water solubility, humic substances are divided into fulvic acids, humic acids, and humin.
- **Inter-system crossing**: it is a process by which a singlet state (in a molecule with an even number of electrons, all electrons in each pair have antiparallel spin) is transformed into a triplet state (in the same molecule, two electrons occur in different orbitals with parallel spin), or vice versa.
- **Irradiance**: it is the intensity of a radiation source (e.g., a lamp or sunlight), typically expressed in units of W m^{-2}.
- **Photochemically produced reactive intermediates** (PPRIs): they are short-lived species that are generated when photosensitizers absorb sunlight. PPRIs can react with pollutants and cause their degradation.
- **Photolysis**: as a narrow definition, it is the breaking of a chemical bond caused by the absorption of radiation. This definition is often extended to encompass all types of transformation induced by radiation absorption.
- **Photosensitizer**: it is a compound that can absorb radiation and generate PPRIs. Natural photosensitizers include nitrate, nitrite, and chromophoric dissolved organic matter.
- **Quantum yield** (Φ): it defines the efficiency of a photoinduced process. It is the ratio between a quantity produced as an effect of radiation absorption (e.g., the rate of a chemical transformation reaction) and the rate of radiation absorption by the relevant compound/system. Typically (with few exceptions), it is $0 \le \Phi \le 1$.

(continued)

- **Quencher**: it is a compound that deactivates an excited state through energy transfer.
- **Scavenger**: it is a compound that consumes a radical or an excited state through a chemical reaction.
- **Steady-state concentration**: it is a constant concentration with time, which can be reached by a species because of the interplay between formation and (usually fast) transformation. Typically, all PPRIs reach steady-state concentrations in water.
- **Thermocline**: in a thermally stratified lake, it is the zone where the variation of water temperature with depth is the highest. The thermocline divides the surface layer of the lake (epilimnion) from the bottom layer (hypolimnion).
- **Treeline altitude**: it is the highest altitude in a mountain where trees still manage to grow. Above the treeline, the vegetation is made up of grass only.

Further Reading
- P. Boule (ed.) *Environmental Photochemistry*. Springer-Verlag Berlin Heidelberg, 1999.
- P. Calza, D. Vione (eds.) *Surface Water Photochemistry*. RSC Publishing, 2016.

Bibliography

1. Richardson SD, Ternes TA (2022) Water analysis: emerging contaminants and current issues. Anal Chem 94(1):382–416. https://doi.org/10.1021/acs.analchem.1c04640
2. Guo Z, Kodikara D, Albi LS, Hatano Y, Chen G, Yoshimura C, Wang J (2023) Photodegradation of organic micropollutants in aquatic environment: importance, factors and processes. Water Res 231:118236. https://doi.org/10.1016/j.watres.2022.118236
3. Vione D, Scozzaro A (2019) Photochemistry of surface fresh waters in the framework of climate change. Environ Sci Technol 53(14):7945–7963. https://doi.org/10.1021/acs.est.9b00968
4. Xu Y, Zhang Y, Qiu L, Zhang M, Yang J, Ji R, Vione D, Chen Z, Gu C (2024) Photochemical behavior of dissolved organic matter in environmental surface waters: a review. Eco-Environ Health 3(4):529–542. https://doi.org/10.1016/j.eehl.2024.06.002
5. Cai Z, Zuo G, Huang L, Lou X, Zhang G, He H, Zhang S (2026) Carbonate radicals in environmental systems: mechanistic insights and engineering applications. Water Res 288(Pt B):124703. https://doi.org/10.1016/j.watres.2025.124703
6. Vione D (2023) Photochemical reactions and the self-depuration of sunlit freshwaters. Environ Res Ecol 2:012001. https://doi.org/10.1088/2752-664X/ac8eb8
7. Carena L, García-Gil Á, Marugán J, Vione D (2024) Global modeling of photochemical reactions in lake water: a comparison between triplet sensitization and direct photolysis. Eco-Environ Health 4(1):100123. https://doi.org/10.1016/j.eehl.2024.09.001

8. Blanchet CC, Arzel C, Davranche A, Kahilainen KK, Secondi J, Taipale S, Lindberg H, Loehr J, Manninen-Johansen S, Sundell J, Maanan M, Nummi P (2022) Ecology and extent of freshwater browning—what we know and what should be studied next in the context of global change. Sci Total Environ 812:152420. https://doi.org/10.1016/j.scitotenv.2021.152420
9. Calderaro F, Vione D (2020) Possible effect of climate change on surface-water photochemistry: a model assessment of the impact of browning on the photodegradation of pollutants in lakes during summer stratification. Epilimnion vs whole-lake phototransformation. Molecules 25(12):2795. https://doi.org/10.3390/molecules25122795
10. Williamson CE, Overholt EP, Pilla RM, Leach TH, Brentrup JA, Knoll LB, Mette EM, Moeller RE (2015) Ecological consequences of long-term browning in lakes. Sci Rep 5:18666. https://doi.org/10.1038/srep18666
11. Achite M, Kumar M, Elshaboury N, Srivastava A, Elbeltagi A, Salem A (2025) Comparative assessment of standalone and hybrid deep neural networks for modeling daily pan evaporation in a semi-arid environment. Sci Rep 15(1):20179. https://doi.org/10.1038/s41598-025-05985-z
12. Rosso A, Vione D (2024) Pollutant photodegradation affected by evaporative water concentration in a climate change scenario. Molecules 29(11):2655. https://doi.org/10.3390/molecules29112655
13. Cannone N, Diolaiuti G, Guglielmin M, Smiraglia C (2008) Accelerating climate change impacts on alpine glacier forefield ecosystems in the European Alps. Ecol Appl 18(3):637–648. https://doi.org/10.1890/07-1188.1
14. Hörtnagl P, Pérez MT, Sommaruga R (2011) Contrasting effects of ultraviolet radiation on the growth efficiency of freshwater bacteria. Aquat Ecol 45(1):125–136. https://doi.org/10.1007/s10452-010-9341-9
15. Neale PJ, Hylander S, Banaszak AT, Häder DP, Rose KC, Vione D, Wängberg SÅ, Jansen MAK, Busquets R, Andersen MPS, Madronich S, Hanson ML, Schikowski T, Solomon KR, Sulzberger B, Wallington TJ, Heikkilä AM, Pandey KK, Andrady AL, Bruckman LS, White CC, Zhu L, Bernhard GH, Bais A, Aucamp PJ, Chiodo G, Cordero RR, Petropavlovskikh I, Neale RE, Olsen CM, Hales S, Lal A, Lingham G, Rhodes LE, Young AR, Robson TM, Robinson SA, Barnes PW, Bornman JF, Harper AB, Lee H, Calderón RM, Ossola R, Paul ND, Revell LE, Wang QW, Zepp RG (2025) Environmental consequences of interacting effects of changes in stratospheric ozone, ultraviolet radiation, and climate: UNEP environmental effects assessment panel, update 2024. Photochem Photobiol Sci 24(3):357–392. https://doi.org/10.1007/s43630-025-00687-x
16. Schreder S, Sommaruga R, Psenner R, Chimani B, Ganekind M, Koinig KA (2023) Changes in air temperature, but not in precipitation, determine long-term trends in water chemistry of high mountain lakes of the Alps with and without rock glacier influence. Sci Total Environ 905:167750. https://doi.org/10.1016/j.scitotenv.2023.167750
17. De Laurentiis E, Minella M, Maurino V, Minero C, Brigante M, Mailhot G, Vione D (2012) Photochemical production of organic matter triplet states in water samples from mountain lakes, located below or above the tree line. Chemosphere 88(10):1208–1213. https://doi.org/10.1016/j.chemosphere.2012.03.071

The Complex (Photo)Chemistry in a Snowflake

Luca Carena

Abstract

Environmental ice and snow contain a variety of inorganic and organic compounds, some of which are from natural sources while others are anthropogenic. For instance, contaminants usually released at lower latitudes by anthropogenic activities are found in ice and snow of remote polar regions because of long-range transport, showing that the presence of these (potentially harmful) contaminants is a worldwide problem. Therefore, it is necessary to study the fate of these compounds in the Earth's cryosphere. Although a low temperature is generally related to slow reaction kinetics, ice and snowpack are chemically active environments. Indeed, compounds present in ice and snow can undergo chemical transformations that affect their environmental fate. Interestingly, some of these reactions are significantly accelerated compared to liquid-water conditions. Water freezing and/or the interaction with sunlight (as far as ice/snowpack exposed to solar radiation is concerned) trigger these reactions, making chemical reactivity in ice and snow distinctive. In this brief chapter, the main features and some examples of (photo)chemical reactions that occur in ice and snow are described and discussed.

Keywords

Ice · Snow · Freeze-concentration effect · Freezing potential · Direct photolysis · Contaminants

---

L. Carena (✉)
Department of Chemistry, University of Torino, Torino, Italy
e-mail: luca.carena@unito.it

M. Minella (ed.), *Complexity in Environmental Chemistry*,
https://doi.org/10.1007/978-3-032-31774-2_12

Chapter Learning Goals
In this brief chapter, you will learn key features of ice and snow that make chemical and photochemical reactions occurring in these matrices unique. First of all, the possible location of solutes in ice and snow will be addressed, due to the considerable role that solute location in ice/snow plays in (photo) chemical reactions. This will allow you to understand why some reactions are significantly accelerated in ice/snow compared to the corresponding reactions occurring in liquid solutions, despite the lower temperature.

1 Introduction

Ice and snow are the components of the Earth's cryosphere, the environmental compartment where water is present in solid form and that includes sea ice, lake ice, river ice, glaciers, and snowpacks. Frozen water is also present in the atmosphere, for instance, in high-altitude clouds (e.g., *cirrus*), in tropospheric mixed clouds, and as solid precipitation (e.g., a snowfall). Besides its key role in the energy balance of the Earth's system (e.g., because it reflects solar radiation back into the space), the cryosphere is also important in the context of the environmental fate of both natural compounds and anthropogenic contaminants. After emission and depending on their volatility, these chemical compounds can be transported far from their sources by long-range atmospheric transport and/or by ocean currents, and they eventually enter areas covered by ice and snow, such as the polar regions. In the atmosphere, natural and anthropogenic compounds can partition in ice particles and be then deposited on the ground during solid precipitation events [1]. Therefore, the presence of anthropogenic contaminants in remote polar regions is not a surprise, and these contaminants include compounds such as bisphenol A (an additive used in the plastic industry) and benzophenone-3 (a UV filter used in sunscreens and cosmetics) [2]. Furthermore, some compounds can undergo chemical and photochemical reactions in ice and snow. For instance, both inorganic and organic molecules, formerly dissolved in liquid water, can be transformed during water freezing; in addition, sunlight induces photochemical reactions in ice and snow exposed to solar radiation, which can alter the chemical structure of solutes present in these matrices.

Although the title refers to "a snowflake," this chapter briefly addresses the chemical and photochemical reactions that take place in ice and snow, which are different matrices. Differently from ice, snowpack is a porous medium that is mainly made up of snowflakes (i.e., ice crystals) and air. Firstly, the possible location of solutes in ice and snow is described (Sect. 2), since (photo)chemical reactions strongly depend on this feature. Freezing-induced reactions are then covered in Sect. 3, describing the processes that trigger chemical reactions during water freezing and giving some examples of these reactions. Finally, Sect. 4 concerns chemical reactions that are induced by light (i.e., photochemical reactions) in ice and snow.

The study of ice and snow (photo)chemistry is important to understand the fate of both natural compounds and anthropogenic contaminants in cold environments. However, (photo)chemical reactions in ice and snow, and their kinetics, are extremely difficult to model, because they are affected by several variables and their dependence is often unknown, making the modeling of the reactive system highly complex [3].

2 Location of Solutes and Chemical Reactivity in Ice and Snow

Low temperatures are usually related to slow kinetics of chemical reactions. However, some chemical processes can surprisingly occur faster in ice than in liquid water, although this is not always true because other reactions in ice may be as fast as or even slower than the corresponding reactions in liquid water.

The reasons for an increase in reaction rates in ice can be often related to the exclusion of water solutes from the ice bulk and their confinement in so-called liquid-like regions (LLRs; compartments of water with liquid-like properties). Indeed, during water freezing, solutes are rejected, as impurities, from the growing ice lattice into LLRs of ice where they get concentrated, because of the low volume of LLRs [4]. These LLRs can be found on the ice surface, at the ice-grain boundaries, and as micropockets within the ice bulk. As far as the ice-air interface is concerned, the ice surface shows a disordered thin layer of quasi-liquid water, called the quasi-liquid layer (QLL). The thickness of the QLL varies from a few nm to tens of nm depending on temperature, and the QLL becomes thinner as temperature decreases [5]. For laboratory ice samples, the location and the size of LLRs have been shown to be dependent on the freezing procedure (e.g., freezing rate, direction of the growing ice front, type of solutes) [6]. Furthermore, when ice is produced by simple freezing of aqueous solutions, solutes seem to be mainly present in compartments of liquid-like water in ice, such as micropockets. In contrast, when artificial snow is prepared through nebulization of aqueous solutions at low temperatures, solutes would mainly occur in/on the QLL at the surface of ice crystals [7]. Since the chemical reactivity of solutes is affected by their location in ice (*vide infra*), the freezing procedure is an important variable that should be taken into account in laboratory experiments.

The concentration of solutes sequestered in LLRs of ice increases compared to original liquid-water conditions (i.e., the pre-frozen solution). It is possible to determine a concentration factor (F) for a solute S that is completely rejected from the ice phase into LLRs. It is $F = C_{S,LLRs} (C_{S,lw})^{-1}$, where $C_{S,LLRs}$ is the molal concentration of S in LLRs of ice, while $C_{S,lw}$ is the molal concentration of S in the corresponding pre-frozen solution [4]. By assuming that S does not precipitate in LLRs (i.e., temperature is above the eutectic of the system S-H_2O) and that no chemical reactions or physical processes decrease the amount of S, the concentration factor F is inversely proportional to the mass fraction of water contained in LLRs ($\phi_{H2O,LLRs}$), so that $F = (\phi_{H2O,LLRs})^{-1}$. This fraction can be assessed through the freezing-point depression model, as follows [4, 8]:

$$\phi_{H_2O,LLRs} \approx \frac{MM_{H_2O} RT_f}{10^3 \Delta_f H^0} \left(\frac{T}{T_f - T} \right) C_{TS,lw} \quad (1)$$

where MM_{H2O} is the molar mass of water (g mol^{-1}), R is the ideal gas constant (J mol^{-1} K^{-1}), T_f is the freezing temperature of pure water (K), $\Delta_f H^0$ is the fusion enthalpy of water (J mol^{-1}), and $C_{TS,lw}$ is the total solute concentration of the pre-frozen solution (mol kg^{-1}). Therefore, the higher the fraction of liquid-like water is, the lower will be F (indeed, a higher volume of LLRs makes solutes more diluted). As reported in Eq. 1, the concentration factor depends on both temperature and total solute concentration. Below the freezing point of pure water, F increases as temperature decreases, because the fraction of liquid-like water in ice is progressively lower, and solutes are consequently more concentrated in the remaining LLRs. On the contrary, F is inversely proportional to the total solute concentration. Therefore, ice formed by freezing of a liquid solution with high concentrations of solutes will have a significant fraction (i.e., mass or volume) of liquid-like water where solutes will be confined. For instance, at the same temperature, the F value of a probe solute S would be 10 times higher in LLRs of ice with total solute concentration of ~ 10^{-4} mol kg^{-1} (e.g., snow from continental regions) than in LLRs of ice with total solute concentration of ~ 10^{-3} mol kg^{-1} (e.g., snow from coastal areas). Therefore, at the same value of $C_{S,lw}$ in the pre-frozen solution, the concentration of S in LLRs will be 10 times higher in ice formed from a liquid solution with $C_{TS,lw} \sim 10^{-4}$ mol kg^{-1} compared to ice from liquid water with $C_{TS,lw} \sim 10^{-3}$ mol kg^{-1}.

However, besides rejection into LLRs, it must be pointed out that the ice lattice can incorporate small molecules, such as H_2O_2, HNO_3, and formaldehyde, as well as ions [9, 10]. In particular, ion incorporation in the ice phase is related to interesting physical-chemical phenomena. During the freezing of aqueous solutions of diluted salts, a transient potential difference is produced at the interface between the growing ice front and liquid water [11]. This potential difference, called the Workman-Reynolds Freezing Potential (WRFP; from E. J. Workman and S. E. Reynolds, who firstly reported this phenomenon in 1950 [12]), is due to a temporary charge separation that develops at the freezing front [13]. Most of the ions are in fact rejected from the ice bulk during water freezing and form brine compartments, as described above, but a small fraction can be trapped in the ice lattice. This ion incorporation is preferential. For instance, in the case of NaCl, the distribution coefficient of Cl^-, namely, the ratio $[Cl^-]_{ice}$ $([Cl^-]_{LLRs})^{-1}$ (where $[Cl^-]_{ice}$ is the chloride concentration trapped in the ice phase, while $[Cl^-]_{LLRs}$ is the chloride concentration in LLRs), has been determined to be $2.7 \cdot 10^{-3}$ [14]. In contrast, the distribution coefficient of Na^+ has been estimated to be slightly lower ($2.6 \cdot 10^{-3}$) [15, 16]. This means that Cl^- is preferentially embedded in the ice lattice compared to Na^+ [17], and a transient accumulation of negative charges in the ice phase is produced when an aqueous solution of NaCl is frozen. Negative ice is thus counterbalanced by an excess of Na^+ in LLRs. On the contrary, in the case of ammonium chloride, NH_4^+ ions are more incorporated in the ice bulk than Cl^-, making ice positively charged [13, 18]. The charge separation that develops at the ice-LLRs interface has been modelled as a capacitor, with the WRFP representing the potential difference between the plates of the capacitor [19]. The freezing potential is a transient phenomenon, because neutralization of the charge imbalance in the two phases would occur through the migration of H^+ and HO^- ions at the interface between ice and liquid-like water. Diffusion of ionic species partitioned between ice and LLRs has also been invoked to account for neutralization of the WRFP [13]. Interestingly, the measured WRFP can span from tens to 200 V in frozen aqueous solutions with salt concentration ranging from 10^{-6} to 10^{-4} mol L^{-1} [11, 12, 18].

The differential incorporation of ions in the ice lattice depends on their electronegativity. Among halogen anions, trapping of F^- in the ice lattice is more favored than that of Cl^- or Br^-, because F^- is more electronegative. In contrast, in the case of the ammonium cation, a molecular structure similar to that of water allows NH_4^+ to substitute a H_2O molecule in the ice lattice. Therefore, the identity of the ions affects both the sign and the value of the WRFP [20]. However, the WRFP value also depends on salt concentration, ice growth rate, as well as the surface of the growing ice (single ice crystal or polycrystalline ice) [21]. As far as NaCl solutions are concerned, a parabolic dependence of the WRFP on [NaCl] (logarithmic scale) has been observed, with WRFP values peaking between 10^{-4} and 10^{-3} mol L^{-1} NaCl while being lower at around 10^{-5} and 10^{-2} mol L^{-1} NaCl [11].

Organic compounds are rejected from the ice lattice and segregated in LLRs, as well as many inorganic species. For instance, freezing of aqueous solutions containing sodium acetate or ammonium acetate forms a transient charge separation at the ice-LLRs interface, as already seen for NaCl and NH_4Cl [18]. In these instances, acetate is mainly excluded from the ice phase, which makes LLRs negatively charged. Rose Bengal, a large organic dye which is commonly used in photochemical experiments as singlet oxygen ($^1O_2$) photosensitizer in both liquid water and ice, is confined in LLRs of ice where it can produce significant $^1O_2$ concentrations [22]. Furthermore, volatile and semi-volatile organic molecules can partition between gas and ice phases, through adsorption on the QLL at the ice surface. For instance, gaseous valerophenone adsorbs on the surface of the ice crystals of snow [7]. Moreover, anthracene has been found to form clusters on the QLL after adsorption, which can affect its photoreactivity [23].

Solute location in ice and snow can significantly affect (photo)chemical reactions. Kurková and co-workers have shown that the photolysis of 4-methyldibenzyl ketone (4MeDBK) taking place in liquid-like micropockets of ice can proceed differently from that observed in the QLL on ice crystals of snow [7]. Briefly, 4MeDBK undergoes direct photolysis under UV light irradiation (λ > 290 nm), yielding 4-methylbenzyl and benzyl radicals as intermediates (hereafter, I_A and I_B, respectively), while CO is lost by decarbonylation. The radicals I_A and I_B can then react to form the final products 1-methyl-4-(2-phenylethyl)benzene ($P_{A\text{-}B}$, which is produced through combination of I_A and I_B), 1,2-bis(4-methylphenyl)ethane ($P_{A\text{-}A}$, formed from the reaction $I_A + I_A$), and 1,2-diphenylethane ($P_{B\text{-}B}$, produced from $I_B + I_B$) (Fig. 1). In the absence of cage effects (i.e., when I_A and I_B are able to diffuse and

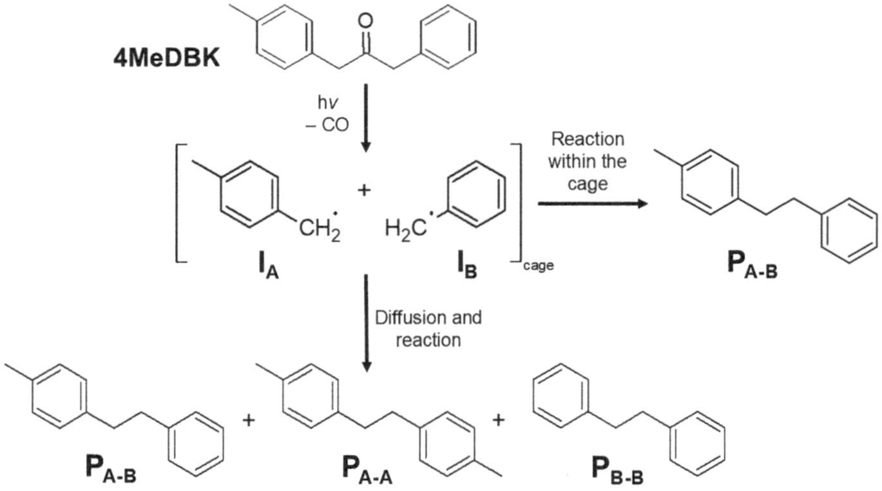

Fig. 1 Reaction scheme of 4-methyldibenzyl ketone (4MeDBK) phototransformation showing the production of radical intermediates (I_A and I_B) and their reactions to yield the final products $P_{A\text{-}B}$, $P_{A\text{-}A}$, and $P_{B\text{-}B}$. This reaction was studied by Kurková and co-workers to investigate the different photoreactivity of small organic molecules occurring at the surface of the ice grains of snow and in micropockets within ice [7]. Diffusion phenomena are definitely more common in ice micropockets

react), the statistical distribution of these compounds is $P_{A\text{-}B}$: $P_{A\text{-}A}$: $P_{B\text{-}B}$ = 2:1:1, as expected in a liquid solution. However, this product distribution is not observed when 4MeDBK is present in the QLL at the surface of ice crystals of snow at 243 K, because 4MeDBK is mainly phototransformed into $P_{A\text{-}B}$ [7]. The 2:1:1 statistical distribution of 4MeDBK photoproducts in the QLL of snow is sharply obtained at higher temperature values, approaching the melting point of water [7]. On the contrary, at the same temperature of 243 K but in liquid-like micropockets within ice, the distribution $P_{A\text{-}B}$: $P_{A\text{-}A}$: $P_{B\text{-}B}$ is close to its statistical value, and the main formation of $P_{A\text{-}B}$ occurs at significantly lower temperature values in ice (around 200 K) than in snow [24].

These findings have shown that the cage effect on photolysis of molecules can be important in the QLL, and they further suggest that, at the same temperature, diffusion of molecules partitioned in the QLL is limited compared to diffusion of molecules dissolved in ice micropockets. Furthermore, some compounds can form clusters by self-association on the QLL as well, which would favor the recombination reactions of radical intermediates [7].

3 Chemical Reactions Induced by Freezing

Some chemical reactions of environmental relevance are induced or significantly accelerated by water freezing. Here, some of these reactions will be described, while further examples can be found in previous reviews [25–27].

Higher reaction rates in ice than in liquid water are somehow counterintuitive, because a low temperature typically slows down the kinetics of chemical reactions. However, some processes taking place during water freezing can affect the rate of chemical reactions. These processes are known as the *freeze-concentration effect* and the *freeze-potential effect* [27]. The freeze-concentration effect refers to the rejection of solutes from the ice lattice into LLRs, where they get concentrated as discussed above. Higher concentration of reagents, as is common in LLRs, can increase the rate of some reactions. Second- or higher-order reactions with small activation energy could be thus accelerated by freezing, and they could consequently occur faster in LLRs of ice than in liquid water, despite the lower temperature of the LLRs. However, the total solute concentration in the pre-frozen solution ($C_{TS,lw}$) should be low enough to allow for a significant concentration of reagents in the LLRs of ice (i.e., the concentration factor F should be high; see Eq. 1 and related discussion) [28].

The freeze-potential effect is related to the formation of the WRFP. The transient potential difference at the interface between ice and LLRs (i.e., the WRFP), produced by the differential inclusion of ions in the ice lattice during water freezing, can affect the rate of chemical reactions as well. The concentration of ionic compounds in LLRs can increase at the interface with the ice phase, because ions are attracted by the electrostatic forces due to the WRFP. Furthermore, the neutralization of the charge separation predominantly occurs through variation of H^+ and HO^- concentrations in the ice phase and LLRs. For instance, when an aqueous solution of diluted NaCl is frozen, the preferential inclusion of Cl^- in the ice lattice and the higher rejection of Na^+ in LLRs produce a positive charge in LLRs. To restore the charge balance, the $[HO^-]$ in LLRs increases. Therefore, the pH variation in LLRs can potentially affect the rate of chemical reactions in these microenvironments.

Although solute segregation in LLRs and charge separation between the ice phase and LLRs usually play a major role, further effects can alter the rate of chemical reactions upon freezing. The QLL present at the ice surface may act as a catalytic surface, because of its particular physical-chemical characteristics. These are different from the bulk phase and can affect the bonding forces among molecules in/on the QLL (this is the so-called *catalytic effect* of the ice surface). Moreover, the direction of the growing ice front, e.g., from bottom toward the top or from wall sides toward the central part of a solution, can affect the concentration gradient of solutes in the frozen solution and, consequently, the reaction rates (*convection effect*) [26, 27]. Finally, temperature is an important parameter to take into account during lab experiments on reactions affected by freezing (*temperature effect*). When an aqueous solution is frozen, ice and LLRs are formed when the temperature of the liquid solution drops below its freezing temperature. Further cooling increases the solute concentration in LLRs (see the temperature dependence of F), until solutes precipitate when the temperature reaches the eutectic of the system solute-H_2O. Below the eutectic temperature, solutes are not dissolved in LLRs anymore, and their reactions are inhibited as a consequence. Therefore, chemical reactions in LLRs can take place mainly between the freezing temperature and the eutectic temperature of the solution. The latter depends on the type of solutes. For instance, the eutectic temperature of an aqueous solution of NaCl is ~ 252 K, while that of NaBr is ~ 245 K. However, experimental results showed that brine LLRs should still exist below the eutectic temperature of the system salt-H_2O [5]. Finally, the freezing rate can affect the rate of chemical reactions as well, because LLRs properties, solute concentrations in LLRs, and the WRFP can all depend on the freezing rate [6, 21, 29].

3.1 Nitrite Oxidation Accelerated by Freezing

Takenaka and co-workers have shown that nitrite (NO_2^-; initial concentration of 100 μmol L^{-1}) is significantly oxidized to nitrate (NO_3^-) during freezing of slightly acidic aqueous solutions at 252–263 K (the pH value of the pre-frozen solution was ~ 4 by addition of H_2SO_4). Moreover, the reaction in ice can be 10^5 times faster than the corresponding reaction taking place in liquid water [18, 30]. This acceleration is mainly accounted for by the freeze-concentration effect.

During freezing, Na^+, NO_2^-, and SO_4^{2-} are rejected from the ice lattice into LLRs, where they get concentrated. Increased local concentrations of acids (like H_2SO_4) in LLRs have also been observed in ice at 253 K [31]. Consequently, because of the rejection of H_2SO_4 from the ice phase, the pH of LLRs is expected to be lower than that of the pre-frozen solution. An additional source of acidity may be the neutralization of the charge separation between ice and LLRs, because Na^+ should be more incorporated in the ice lattice than NO_2^- and SO_4^{2-} [32]. This issue would increase the [H^+] in LLRs to restore the charge balance (*vide infra* for further details). The higher [H^+] favors the protonation of nitrite in LLRs, yielding nitrous acid (HNO_2). Note that (*i*) HNO_2 prevails over NO_2^- below the pK_a of nitrous acid (~ 3.3 at 298 K and ~ 3.6 at 258 K [32]) and (*ii*) HNO_2 is able to react with O_2 dissolved in LLRs through reaction **R1**. Therefore, the concentration of solutes and the decrease of pH accelerate the reaction between HNO_2 and O_2 in LLRs, producing NO_3^- [18, 30].

$$2HNO_2 + O_2 \rightarrow 2NO_3^- + 2H^+ \qquad \text{(R1)}$$

Interestingly, this oxidation would also be affected by the freeze-potential effect. Takenaka and co-workers have shown that when an aqueous solution of $NaNO_2$ (100 μmol L^{-1}) at pH = 4.5 is frozen at 252 K in the presence of a salt (1 mmol L^{-1}), the conversion of nitrite to nitrate is affected by the identity of the salt added to the solution [18]. For instance, the presence of NH_4Cl or Na_2SO_4 does not influence the oxidation extent of N(III). On the contrary, the presence of NaCl or KF completely inhibits the oxidation of nitrite. This observation has been accounted for by the role of the WRFP sign that develops through the interface between ice and LLRs. Both NH_4Cl and Na_2SO_4 make ice positively charged, because NH_4^+ and Na^+ are preferentially trapped in the ice lattice than, respectively, Cl^- and SO_4^{2-}. The consequence of this phenomenon is a negatively charged LLR, and, as described above, neutralization of this charge imbalance takes place through the increase of [H^+] in LLRs.

Therefore, the formation of HNO_2 (the N(III) species which is actually oxidized) in these compartments and its reaction with O_2 are not inhibited. Conversely, in the case of NaCl and KF, the anions are more incorporated in the ice phase, while Na^+ and K^+ accumulate in LLRs. The positive charge in LLRs is neutralized by the increase of $[HO^-]$, which buffers the pH decrease below the initial value of 4.5 caused by acid concentration, and thus prevents the conversion of NO_2^- into HNO_2 and the consequent oxidation of N(III) by O_2 [18].

In the natural environment, NO_2^- oxidation by freezing could be relevant to acidic-water droplets in the atmosphere (e.g., acid rains and acid fogs that can show pH values between 2 and 4) under cold conditions, since nitrite concentration in atmospheric liquid water (e.g., fog) can reach some tens micromolar.

3.2 Nitrite-Mediated Transformation of Organic Compounds Induced by Freezing

Freezing of slightly acidic aqueous solutions containing NO_2^- can trigger the transformation of organic compounds. For example, dimethylamine is transformed into harmful N-nitrosodimethylamine upon freezing of an aqueous solution (initial pH = 4.6) at 258 K and in the presence of $NaNO_2$ (1 mmol L^{-1}) [32] (Fig. 2a). This transformation has been accounted for by the increase of $[H^+]$ in LLRs because of both freeze-concentration and freeze-potential effects. Indeed, NO_2^- accumulates in LLRs of ice more than Na^+, because of their different distribution coefficients [32]. Then, neutralization of the negative-charge excess in LLRs through an $[H^+]$ increase shifts the equilibrium between HNO_2 and NO_2^- toward HNO_2. Finally, the

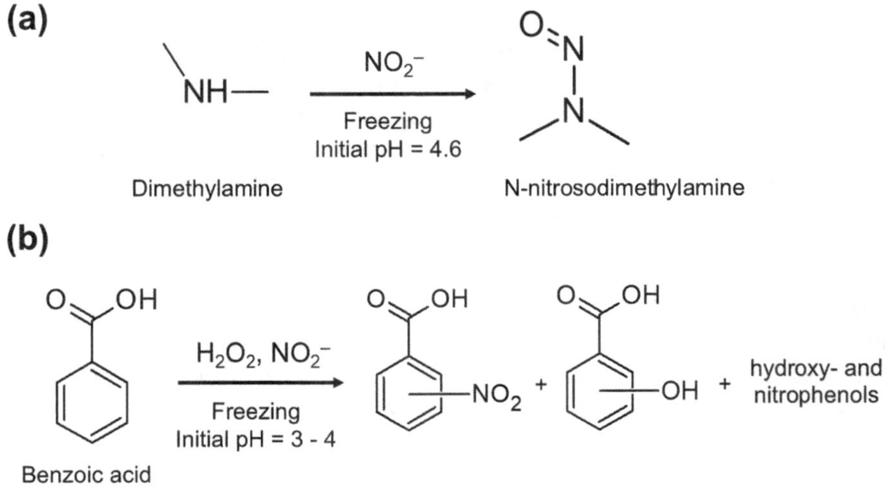

Fig. 2 Reaction scheme of the freezing-induced, nitrite-mediated degradation of (**a**) dimethylamine and (**b**) benzoic acid, as discussed in the text

formation of HNO_2 triggers the conversion of dimethylamine into N-nitrosodimethylamine [32]. However, it must be pointed out that concomitant N(III) oxidation to NO_3^- can take place as well, as previously described.

Another example is that, in the presence of hydrogen peroxide (H_2O_2; 1 mmol L^{-1}) and $NaNO_2$ (0.5 mmol L^{-1}), benzoic acid can be degraded during water freezing to yield hydroxylated and nitrated compounds (Fig. 2b). This reaction proceeds faster in ice (253–263 K) than in liquid water (278–308 K). Furthermore, the pH of the pre-frozen solution plays a key role, because benzoic acid is degraded when the initial pH is 3–4 but no degradation occurs at pH 5 [33]. Both H_2O_2 and NO_2^- are rejected from the ice phase into LLRs during freezing, where they get concentrated. Here, NO_2^- forms HNO_2 due to increased [H^+] (see previous discussion). Ahn and co-workers [33] have proposed that peroxynitrous acid (HOONO) is produced by the reaction between H_2O_2 and HNO_2 in LLRs, and HOONO reacts with benzoic acid to yield both hydroxy- and nitro-benzoic acids as well as hydroxy- and nitrophenols. Other benzene derivatives (e.g., phenol, hydroquinone, 4-chlorophenol, 4-nitrophenol, and nitrobenzene) have been found to be significantly degraded under similar freezing conditions [33]. These reactions are of environmental relevance because, for instance, H_2O_2, nitrite, and aromatic compounds are found in the atmospheric aqueous phase. Furthermore, H_2O_2 can be produced at the ice-LLRs interface during freezing of dilute salt solutions, because of the freeze-potential effect [34].

3.3 Degradation of Gallic Acid Promoted by Freezing

At initial water pH = 7, gallic acid has been found to be partially degraded by freezing at 253 K, in the presence of NaCl at concentration between 10^{-4} and 10^{-2} mol L^{-1} [15]. This phenomenon can be accounted for by the segregation and concentration of gallic acid in LLRs and by the freeze-potential effect due to the presence of NaCl. As previously described, an excess of Na^+ in LLRs during water freezing is neutralized by the increase of the pH in LLRs. Alkaline conditions promote the degradation of gallic acid in these compartments, and this reaction is even more important when the pH of the pre-frozen solution is >7 [15]. In this case, besides the increase of [HO^-] in LLRs due to neutralization of the Cl^- and Na^+ imbalance, additional HO^- would be concentrated in LLRs by the freeze-concentration effect. That would increase even more the pH value and promote a higher degradation of gallic acid.

The rise in pH of LLRs can be really significant. Ice formed from a solution containing NaCl (5 mmol L^{-1}) and in equilibrium with atmospheric CO_2 (i.e., initial pH of liquid water ~ 5.6) would contain LLRs at pH ~ 8.6, while the pH of LLRs can be as high as 11.4 in the absence of dissolved CO_2 at 263 K (because of the lack of buffering capacity from the carbonate system) [15].

4 Photochemical Reactions in Ice and Snow

Photochemical reactions taking place in different environmental compartments play an important role in the transformation of both natural compounds and anthropogenic contaminants. Despite the high albedo, sunlight can penetrate snowpacks and trigger photochemical reactions. Indeed, several molecules present in environmental ice and snow are able to absorb the energy of solar photons in the UV-B, UV-A, or visible range and undergo chemical transformation as a consequence. The transmission of UV and visible light in the snow column depends on the scattering and absorption properties of snow, which are in turn related to, e.g., the size and the shape of the ice crystals, and the type and concentration of solutes present in the ice grains. The *e*-folding depth is defined as the snow depth over which the light intensity incident on the snowpack surface decreases by a factor of *e*; it shows how deep sunlight can penetrate downward the snow column. For typical ice-crystal types found in the upper layer of snowpacks, the *e*-folding depth varies from a few centimeters (2–5 cm) to 25–30 cm [35]. This makes a big difference with liquid-water bodies, such as lakes and oceans, where sunlight can penetrate the water column for several meters and thus trigger photochemical reactions at higher depths.

In general, photochemical reactions can be divided into direct and indirect photolysis. Direct photolysis occurs when a molecule absorbs solar light and the acquired energy induces a chemical transformation of the molecule itself. Indirect photolysis, instead, refers to the reaction of a molecule with a transient reactive intermediate. The latter is photochemically produced by a third chemical species involved in the process, which is called photosensitizer. These transient intermediates, also named *Photochemically Produced Reactive Intermediates* (PPRIs), have short lifetimes (hence "transient") because of their high reactivity and deactivation rates. PPRIs can react with both natural compounds and anthropogenic contaminants, thereby inducing their photodegradation. In environmental ice and snow, relevant photosensitizers are nitrite, nitrate, hydrogen peroxide, and organic matter (OM), which photoproduce PPRIs such as the highly reactive hydroxyl radical ($HO^\bullet$), singlet oxygen ($^1O_2$), and excited triplet states of OM ($^3OM^*$).

$$NO_3^- \xrightarrow{h\nu, H^+} {}^\bullet NO_2 + HO^\bullet \tag{R2}$$

$$NO_3^- \xrightarrow{h\nu} NO_2^- + O \tag{R3}$$

$$NO_2^- \xrightarrow{h\nu, H^+} {}^\bullet NO + HO^\bullet \tag{R4}$$

$$H_2O_2 \xrightarrow{h\nu} 2HO^\bullet \tag{R5}$$

$$OM \xrightarrow{h\nu} {}^1OM^* \xrightarrow{ISC} {}^3OM^* \tag{R6}$$

$${}^3OM^* + O_2 \rightarrow OM + {}^1O_2 \tag{R7}$$

In particular, nitrate photolysis in the UV-B wavelength range forms both $HO^{\bullet}$ and NO_2^- in ice (reactions **R2** and **R3**, respectively), while nitrite undergoes photolysis mainly by UV-A light to yield $HO^{\bullet}$ (**R4**) [36–38]. Both UV-B and UV-A photons trigger H_2O_2 photolysis in ice, which results in the production of $HO^{\bullet}$ (**R5**). Organic matter is able to absorb over a wider wavelength range that extends well into the visible region, and it is firstly excited to a singlet state ($^1OM^*$) that then produces an excited triplet state ($^3OM^*$) through inter-system crossing (ISC; **R6**). Organic matter is the main photosensitizer for $^1O_2$, because one of the several deactivation pathways of $^3OM^*$ takes place through energy transfer to ground-state molecular oxygen, yielding $^1O_2$ (**R7**). Finally, OM can also photosensitize the production of $HO^{\bullet}$ radicals in ice [39].

Once they have been photoproduced, PPRIs are rapidly deactivated or they react. For example, $HO^{\bullet}$ radicals react with OM, NO_2^- and H_2O_2, and these reactions produce other radicals that are potentially reactive toward natural compounds and contaminants. Nitrite is oxidized by $HO^{\bullet}$ to nitrogen dioxide ($^{\bullet}NO_2$), while reaction between $HO^{\bullet}$ and H_2O_2 forms the hydroperoxyl radical ($HO_2^{\bullet}$). $^3OM^*$ can be deactivated to ground-state OM, yielding $^1O_2$ when molecular oxygen acts as a quencher, and can react with chemical compounds by energy or electron transfer. Finally, $^1O_2$ is mainly quenched by water molecules, although it is slightly reactive toward electron-rich molecules. Therefore, both quenching and scavenging processes consume PPRIs and keep their lifetimes and concentrations in environmental waters low.

Both direct and indirect photolysis of natural and anthropogenic compounds have been observed in ice and snow. Interestingly, these reactions sometimes occur faster in ice/snow than in liquid water, showing that ice photochemistry is significant. However, other reactions take place in ice/snow as fast as or even more slowly than in liquid water. Furthermore, since ice and snow are different matrices, the same photochemical process can proceed differently in ice than in snow [40]. In the following sections, some examples of these reactions will be discussed, describing the possible reasons for the rate increase (or decrease) observed in ice and snow compared to liquid water.

4.1 Direct Photolysis in Ice and Snow

When a molecule (S) undergoes direct photolysis in water, the reaction rate is a function of (i) the intensity of light; (ii) the ability of S to absorb that light; and (iii) the efficiency of the absorbed photons to trigger a chemical transformation of S. All these conditions must be satisfied for the direct photolysis of S to take place. For instance, if the light is intense and it is efficiently absorbed by S, but the absorbed photons are unable to induce a chemical reaction, the direct photolysis of S is expected to be slow or even not to occur.

$$R_{\mathrm{d.p.}} = \int p^0(\lambda) f_S(\lambda) \phi_S(\lambda) \mathrm{d}\lambda \qquad (2)$$

Equation 2 explains these considerations about the rate of S direct photolysis ($R_{d.p.}$, units of mol L^{-1} s^{-1}), where $p^0(\lambda)$ is the spectral photon flux density in the irradiated water sample (it has units of E L^{-1} s^{-1} nm^{-1}, where E means "Einstein" and 1 E = 1 mol of photons), $f_S(\lambda)$ is the fraction of light absorbed by S (unitless), and $\phi_S(\lambda)$ is the quantum yield of S direct photolysis (unitless). In general, the quantum yield ϕ of a molecule is defined as the number of events that the molecule undergoes per photon absorbed, and it ranges between 0 and 1. These "events" can be photophysical processes (e.g., fluorescence, phosphorescence, or ISC) or photochemical reactions, such as the direct photolysis [41]. The fraction of light absorbed by S, $f_S(\lambda)$, is directly proportional to both S concentration, [S] (mol L^{-1}), and the ability of S to absorb a photon of wavelength λ. This ability is quantified by the molar absorption coefficient of S at the wavelength λ, $\varepsilon_S(\lambda)$, which has units of L mol^{-1} cm^{-1}. Therefore, more light will be absorbed by S when $\varepsilon_S(\lambda)$ and/or [S] is high.

In a low-light-absorbing water sample, with optical path l (cm), Eq. 2 can be expressed as:

$$R_{d.p.} \cong 2.303 l[S] \int p^0(\lambda)\varepsilon_S(\lambda)\phi_S(\lambda)d\lambda \qquad (3)$$

Note that $p^0(\lambda)$, $\varepsilon_S(\lambda)$, and $\phi_S(\lambda)$ are all functions of light wavelength, and for this reason the integrals of Eqs. 2 and 3 are defined over the wavelength range of $p^0(\lambda)$.

The direct photolysis of a molecule can be faster in ice and snow than in liquid water, but sometimes significantly slower reactions can occur as well. Here, the direct photolysis reactions of guaiacol and vanillin are reported as examples. Guaiacol (2-methoxyphenol) and vanillin (4-hydroxy-3-methoxybenzaldehyde) are two semi-volatile methoxyphenols that are emitted into the atmosphere from biomass burning. They can undergo deposition on ice/snow, and, interestingly, vanillin is used as a tracer of biomass burning in environmental ice.

Firstly, values of $p^0(\lambda)$ can be higher in frozen samples than in liquid aqueous solutions [42] because of light scattering, resulting in potentially positive effects on the rates of the direct photolysis of compounds present in LLRs, as expected by Eq. 3. Variations of $R_{d.p.}$ can also be due to differences in $\varepsilon_S(\lambda)$ and $\phi_S(\lambda)$ between liquid water and ice/snow. Hullar and co-workers have shown that, under simulated solar light, guaiacol direct photolysis is ~ 6 times faster in LLRs within ice at 263 K (i.e., in micropockets) than in liquid water at 278 K. However, a 77-fold acceleration of the direct photolysis (relative to liquid water) has been observed when guaiacol was present in the QLL at the surface of snow ice grains [43]. This would be mainly accounted for by an increase in the quantum yield of guaiacol direct photolysis in LLRs in ice, which is more significant in the QLL. Additionally, on the ice-grain surface, the absorption spectrum of guaiacol shows both a bathochromic shift of ~5 nm and an increase of ~6% (i.e., $\varepsilon(\lambda)$ of guaiacol is slightly higher and shifted by 5 nm toward longer wavelengths), which ensures higher radiation absorption by guaiacol on ice than in liquid water [43].

On the contrary, under UV-A light, the direct photolysis of vanillin in snow at 243 K is slower than in liquid water [44]. Although vanillin efficiently absorbs UV-A light and its quantum yield of direct photolysis is not negligible, the direct

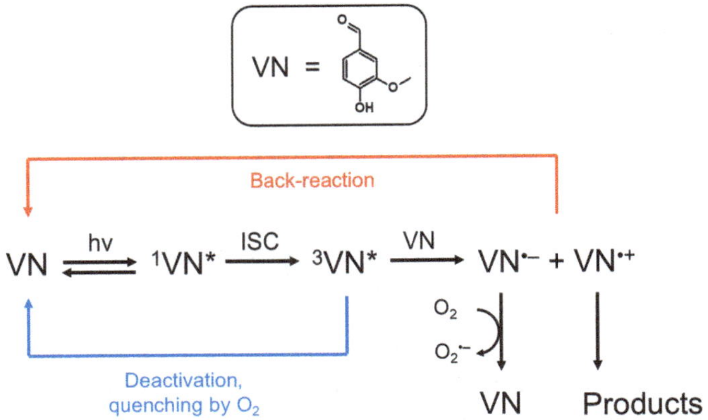

Fig. 3 Reaction scheme of vanillin (VN) direct photolysis. The reaction between $VN^{\bullet-}$ and $VN^{\bullet+}$ (back-reaction) is highlighted by the red pathway. The vanillin chemical structure is reported in the inset

photolysis mechanism (Fig. 3) would limit its transformation in snow. Indeed, vanillin (VN) forms an excited triplet state ($^3VN^*$) upon light absorption and inter-system crossing, which can either deactivate (e.g., by internal conversion or quenching by O_2; blue pathway in Fig. 3) or react with a ground-state molecule of VN. In the latter case, two radical intermediates are produced ($VN^{\bullet-}$ and $VN^{\bullet+}$). $VN^{\bullet-}$ reduces molecular oxygen, yielding VN and $O_2^{\bullet-}$, whereas $VN^{\bullet+}$ undergoes further degradation to the final products. However, when the concentration of vanillin is high enough, the reaction between $VN^{\bullet-}$ and $VN^{\bullet+}$ can take place as well, yielding two molecules of VN (red pathway in Fig. 3). When this reaction prevails over $VN^{\bullet+}$ degradation, the result is a limited transformation of vanillin. In liquid solution, the reaction between $VN^{\bullet-}$ and $VN^{\bullet+}$ is slow at low concentrations of vanillin, because the interaction between the two radical intermediates is unlikely. On the contrary, upon freezing, the freeze-concentration effect makes the VN concentration in the QLL high; thus the reaction $VN^{\bullet-} + VN^{\bullet+}$ becomes more important and results in 98% inhibition of the direct photolysis of VN [44].

Finally, compounds such as pyrene and anthracene do not show any enhancement or inhibition of their direct photolysis in the LLRs of ice, compared to liquid water [45]. These findings suggest that variations in direct photolysis rates between ice and liquid water strictly depend on the identity of the compounds that are photodegraded.

4.2 Photoproduction of $^1O_2$ Is Significantly Higher in Ice Compared to Liquid Water

The concentration of singlet oxygen photoproduced in ice has been found to be 10^2–10^4 times higher than in liquid water, under the same irradiation conditions [22, 46], because of freeze-concentration of the $^1O_2$ photosensitizers (PS) in the LLRs of ice. According to the steady-state approximation, the formation rate of singlet oxygen ($R_{f,{}^1O_2}$) in LLRs is equal to the overall rate of $^1O_2$ consumption (Eq. 4); thus the steady-state concentration of $^1O_2$ can be assessed through Eq. 5. The consumption of singlet oxygen is due to (i) quenching (deactivation) by water molecules, the first-order rate constant of which is k'_{d,H_2O} , and (ii) scavenging by solutes S_i, with reaction (pseudo-first order) rate coefficients $k'_{S_i,+{}^1O_2}$.

$$R_{f,{}^1O_2} = k'_{d,H_2O}\left[{}^1O_2\right] + \sum_i k'_{S_i+{}^1O_2}\left[{}^1O_2\right] \tag{4}$$

$$\left[{}^1O_2\right] = \frac{R_{f,{}^1O_2}}{k'_{d,H_2O} + \sum_i k'_{S_i+{}^1O_2}} \tag{5}$$

Quenching by water molecules is often the main deactivation pathway of singlet oxygen, which means that $k'_{d,H_2O} >> \Sigma_i\ k'_{S_i,+{}^1O_2}$. Therefore, it is also $[{}^1O_2] \sim R_{f,{}^1O_2}\ (k'_{d,H_2O})^{-1}$.

The formation rate of $^1O_2$ by the photosensitizer PS can be expressed as $R_{f,{}^1O_2} = k'_{f,{}^1O_2}\ [PS]_{LLRs}$, where $k'_{f,{}^1O_2}$ is the rate coefficient of $^1O_2$ photoproduction by PS, which depends on both the photon flux in irradiated ice and the efficiency of light absorption by PS. PS is usually organic matter that is not soluble in the ice lattice and is thus rejected into LLRs during water freezing. Therefore, $[PS]_{LLRs}$ is the concentration of PS in LLRs that can be estimated through the concentration factor F, as $[PS]_{LLRs} = F\ [PS]_{lw}$, where $[PS]_{lw}$ is the PS concentration in the completely unfrozen solution. The steady-state $[{}^1O_2]$ is thus directly proportional to F [22].

$$\left[{}^1O_2\right] \approx \frac{k'_{f,{}^1O_2}}{k'_{d,H_2O}} F[PS]_{lw} \tag{6}$$

According to Eq. 1 (which is based on the freezing point depression model) and Eq. 6, when the system is above the eutectic temperature, the concentration of $^1O_2$ in LLRs is expected: (i) to increase by decreasing the ice temperature at constant total solute concentration (because PS gets more concentrated in a lower volume of LLRs) and (ii) to decrease by increasing the total solute concentration at constant temperature (because higher solute concentrations increase the volume of LLRs, producing more diluted PS as a consequence). In general, experimental evidence showed that this is qualitatively true [8, 46], but the freezing point depression model overestimates the enhancement of $[{}^1O_2]$ in ice at total solute concentration < 1 mmol kg^{-1} [8]. Finally, the dependence of both $k'_{f,{}^1O_2}$ and k'_{d,H_2O} on temperature might also affect $[{}^1O_2]$ in LLRs, although their values have been usually considered to approximate those observed in liquid water [22].

Singlet oxygen is not the most reactive among the PPRIs produced in environmental waters, but it shows good reactivity with electron-rich compounds (both natural and anthropogenic). Amino acids (e.g., tyrosine and tryptophan) and contaminants, such as chlorophenolates and the basic forms of bisphenols, are typical compounds that are degraded by $^1O_2$. Despite its low and selective reactivity, the significantly high concentrations reached in LLRs can make $^1O_2$ an important sink for selected compounds in environmental ice and snow.

Questions

1. Why does a transient potential difference develop at the interface between ice and liquid-like water during freezing of diluted salt solutions?
2. What are the main processes that can accelerate a chemical reaction in water by freezing?
3. Describe how the freeze-concentration effect can depend on temperature and total solute concentration.
4. Is the oxidation of nitrite to nitrate expected to occur when an aqueous solution (initial water pH = 7) containing 5 mmol L^{-1} NaCl and 0.1 mmol L^{-1} $NaNO_2$ is frozen at 253 K?
5. What are the variables that can affect the direct photolysis of a chemical compound in ice/snow?

Glossary

- *Liquid-like regions (LLRs)*: they are compartments of water with liquid-like properties that occur in ice, at the ice-grain borders, and at the surface of ice grains. The fraction of water in LLRs can be assessed through the freezing point depression model.
- *Quasi-liquid layer (QLL)*: it is a thin layer of quasi-liquid water occurring at the surface of ice grains.
- *Freeze-concentration effect*: during water freezing, solutes are rejected from the growing ice lattice into liquid-like regions. Here, solutes are segregated, and their concentration is higher compared to initial liquid conditions, because of the lower solvent mass in which they partition.
- *Workman-Reynolds Freezing Potential (WRFP)*: it is the potential difference that develops, during water freezing, at the interface between the growing ice front and liquid-like regions because of charge imbalance in these two phases. This potential difference is related to the different solubility of ions in the ice lattice.
- *Direct photolysis*: it is a chemical reaction that takes place when a compound absorbs light and gets transformed as a consequence.

(continued)

- *Photochemically Produced Reactive Intermediates (PPRIs)*: they are transient chemical species of significant reactivity (they are often radicals) which are produced by photochemical reactions of compounds that absorb light (the latter are called "photosensitizers").
- *Indirect photolysis*: it refers to reactions between a chemical compound and PPRIs.

Bibliography

1. Grannas AM, Bogdal C, Hageman KJ, Halsall C, Harner T, Hung H, Kallenborn R, Klán P, Klánová J, Macdonald RW, Meyer T, Wania F (2013) The role of the global cryosphere in the fate of organic contaminants. Atmos Chem Phys 13:3271–3305. https://doi.org/10.5194/acp-13-3271-2013
2. D'Amico M, Kallenborn R, Scoto F, Gambaro A, Gallet JC, Spolaor A, Vecchiato M (2024) Chemicals of emerging Arctic Concern in North-Western Spitsbergen snow: distribution and sources. Sci Total Environ 908:168401. https://doi.org/10.1016/j.scitotenv.2023.168401
3. Domine F, Bock J, Voisin D, Donaldson DJ (2013) Can we model snow photochemistry? Problems with the current approaches. J Phys Chem A 117:4733–4749. https://doi.org/10.1021/jp3123314
4. Cho H, Shepson PB, Barrie LA, Cowin JP, Zaveri R (2002) NMR investigation of the quasi-brine layer in ice/brine mixtures. J Phys Chem B 106:11226–11232. https://doi.org/10.1021/jp020449+
5. Bartels-Rausch T, Jacobi H-W, Kahan TF, Thomas JL, Thomson ES, Abbatt JPD, Ammann M, Blackford JR, Bluhm H, Boxe C, Domine F, Frey MM, Gladich I, Guzmán MI, Heger D, Huthwelker T, Klán P, Kuhs WF, Kuo MH, Maus S, Moussa SG, McNeill VF, Newberg JT, Pettersson JBC, Roeselová M, Sodeau JR (2014) A review of air–ice chemical and physical interactions (AICI): liquids, quasi-liquids, and solids in snow. Atmos Chem Phys 14:1587–1633. https://doi.org/10.5194/acp-14-1587-2014
6. Hullar T, Anastasio C (2016) Direct visualization of solute locations in laboratory ice samples. Cryosph 10:2057–2068. https://doi.org/10.5194/tc-10-2057-2016
7. Kurková R, Ray D, Nachtigallová D, Klan P (2011) Chemistry of small organic molecules on snow grains: the applicability of artificial snow for environmental studies. Environ Sci Technol 45:3430–3436. https://doi.org/10.1021/es104095g
8. Bower JP, Anastasio C (2013) Using singlet molecular oxygen to probe the solute and temperature dependence of liquid-like regions in/on ice. J Phys Chem A 117:6612–6621. https://doi.org/10.1021/jp404071y
9. Perrier S, Houdier S, Dominé F, Cabanes A, Legagneux L, Sumner AL, Shepson PB, Formaldehyde in Arctic snow. (2002) Incorporation into ice particles and evolution in the snowpack. Atmos Environ 36:2695–2705. https://doi.org/10.1016/S1352-2310(02)00110-3
10. Dominé F, Thibert E (1996) Mechanism of incorporation of trace gases in ice grown from the gas phase. Geophys Res Lett 23:3627–3630. https://doi.org/10.1029/96GL03290
11. Wilson PW, Haymet ADJ (2008) Workman–Reynolds freezing potential measurements between ice and dilute salt solutions for single ice crystal faces. J Phys Chem B 112:11750–11755. https://doi.org/10.1021/jp804047x
12. Workman EJ, Reynolds SE (1950) Electrical phenomena occurring during the freezing of dilute aqueous solutions and their possible relationship to thunderstorm electricity. Phys Rev 78:254–259. https://doi.org/10.1103/PhysRev.78.254

13. Haymet ADJ, Wilson PW (2017) The Workman–Reynolds “Freezing Potential”: a new look at the inherent physical process. J Mol Liq 228:243–246. https://doi.org/10.1016/j.molliq.2016.07.081
14. Gross GW, Wong PM, Humes K (1977) Concentration dependent solute redistribution at the ice–water phase boundary. III. Spontaneous convection. Chloride solutions. J Chem Phys 67:5264–5274. https://doi.org/10.1063/1.434704
15. Takenaka N, Tanaka M, Okitsu K, Bandow H (2006) Rise in the pH of an unfrozen solution in ice due to the presence of NaCl and promotion of decomposition of gallic acids owing to a change in the pH. J Phys Chem A 110:10628–10632. https://doi.org/10.1021/jp0634095
16. Watanabe H, Otsuka T, Harada M, Okada T (2014) Imbalance between anion and cation distribution at ice Interface with liquid phase in frozen electrolyte as evaluated by fluorometric measurements of pH. J Phys Chem C 118:15723–15731. https://doi.org/10.1021/jp5031653
17. Conde MM, Rovere M, Gallo P (2017) Spontaneous NaCl-doped ice at seawater conditions: focus on the mechanisms of ion inclusion. Phys Chem Chem Phys 19:9566–9574. https://doi.org/10.1039/C7CP00665A
18. Takenaka N, Ueda A, Daimon T, Bandow H, Dohmaru T, Maeda Y (1996) Acceleration mechanism of chemical reaction by freezing: the reaction of nitrous acid with dissolved oxygen. J Phys Chem 100:13874–13884. https://doi.org/10.1021/jp9525806
19. Bronshteyn VL, Chernov AA (1991) Freezing potentials arising on solidification of dilute aqueous solutions of electrolytes. J Cryst Growth 112:129–145. https://doi.org/10.1016/0022-0248(91)90918-U
20. Pruppacher HR, Steinberger EH, Wang TL (1968) On the electrical effects that accompany the spontaneous growth of ice in supercooled aqueous solutions. J Geophys Res 73:571–584. https://doi.org/10.1029/JB073i002p00571
21. Wilson PW, Haymet ADJ (2010) Effect of ice growth rate on the measured Workman–Reynolds freezing potential between ice and dilute NaCl solutions. J Phys Chem B 114:12585–12588. https://doi.org/10.1021/jp105001c
22. Bower JP, Anastasio C (2013) Measuring a 10,000-fold enhancement of singlet molecular oxygen (1O2*) concentration on illuminated ice relative to the corresponding liquid solution. Atmos Environ 75:188–195. https://doi.org/10.1016/j.atmosenv.2013.04.054
23. Chakraborty S, Stubbs AD, Kahan TF (2022) Direct observation of anthracene clusters at ice surfaces. J Am Chem Soc 144:751–756. https://doi.org/10.1021/jacs.1c09220
24. Ruzicka R, Baráková L, Klán P (2005) Photodecarbonylation of dibenzyl ketones and trapping of radical intermediates by copper(II) chloride in frozen aqueous solutions. J Phys Chem B 109:9346–9353. https://doi.org/10.1021/jp044661k
25. Lv J, Zuo W, Tian C, Wang M, Liao Q, Lin Z (2023) Freeze-accelerated reactions on environmental relevant processes. Cell Rep Phys Sci 4. https://doi.org/10.1016/j.xcrp.2023.101456
26. Sun Y, Li J, Li S, Li J, Jing B, Yang T, Yang Y, Ao Z (2025) Inorganic substrates in frozen solutions: transformation mechanisms and interactions with organic compounds—a review. Water Res 273:123068. https://doi.org/10.1016/j.watres.2024.123068
27. O’Concubhair R, Sodeau JR (2013) The effect of freezing on reactions with environmental impact. Acc Chem Res 46:2716–2724. https://doi.org/10.1021/ar400114e
28. Takenaka N, Bandow H (2007) Chemical kinetics of reactions in the unfrozen solution of ice. J Phys Chem A 111:8780–8786. https://doi.org/10.1021/jp0738356
29. Heger D, Jirkovský J, Klán P (2005) Aggregation of methylene blue in frozen aqueous solutions studied by absorption spectroscopy. J Phys Chem A 109:6702–6709. https://doi.org/10.1021/jp050439j
30. Takenaka N, Ueda A, Maeda Y (1992) Acceleration of the rate of nitrite oxidation by freezing in aqueous solution. Nature 358:736–738. https://doi.org/10.1038/358736a0
31. Heger D, Klánová J, Klán P (2006) Enhanced protonation of cresol red in acidic aqueous solutions caused by freezing. J Phys Chem B 110:1277–1287. https://doi.org/10.1021/jp0553683
32. Kitada K, Suda Y, Takenaka N (2017) Acceleration and reaction mechanism of the N-Nitrosation reaction of dimethylamine with nitrite in ice. J Phys Chem A 121:5383–5388. https://doi.org/10.1021/acs.jpca.7b03246

33. Ahn YY, Kim J, Kim K (2022) Frozen hydrogen peroxide and nitrite solution: the acceleration of benzoic acid oxidation via the decreased pH in ice. Environ Sci Technol 56:2323–2333. https://doi.org/10.1021/acs.est.1c05705
34. Song J, Werner L, Da Silva YCM, Theis A, Donaldson DJ, George C (2024) Spontaneous production of H2O2 at the liquid–ice interface: a potential source of atmospheric oxidants. Environ Sci Technol. https://doi.org/10.1021/acs.est.4c07546
35. Domine F, Albert M, Huthwelker T, Jacobi H-W, Kokhanovsky AA, Lehning M, Picard G, Simpson WR (2008) Snow physics as relevant to snow photochemistry. Atmos Chem Phys 8:171–208. https://doi.org/10.5194/acp-8-171-2008
36. Chu L, Anastasio C (2007) Temperature and wavelength dependence of nitrite photolysis in frozen and aqueous solutions. Environ Sci Technol 41:3626–3632. https://doi.org/10.1021/es062731q
37. Benedict KB, Anastasio C (2017) Quantum yields of nitrite (NO2–) from the photolysis of nitrate (NO3–) in ice at 313 nm. J Phys Chem A 121:8474–8483. https://doi.org/10.1021/acs.jpca.7b08839
38. Jacobi H-W, Kleffmann J, Villena G, Wiesen P, King M, France J, Anastasio C, Staebler R (2014) Role of nitrite in the photochemical formation of radicals in the snow. Environ Sci Technol 48:165–172. https://doi.org/10.1021/es404002c
39. Grannas AM, Pagano LP, Pierce BC, Bobby R, Fede A (2014) Role of dissolved organic matter in ice photochemistry. Environ Sci Technol 48:10725–10733. https://doi.org/10.1021/es5023834
40. Frassati S, Carena L, Barbaro E, Roman M, Feltracco M, Minella M, Sordello F, Minero C, Spolaor A, Scalabrin E, Barbante C, Gambaro A (2025) Photodegradation of bisphenol a and identification of photoproducts in artificial snow under UVA radiation. Environ Pollut:126503. https://doi.org/10.1016/j.envpol.2025.126503
41. Braslavsky SE (2007) Glossary of terms used in photochemistry 3rd edition. Pure Appl Chem 79:293–465. https://doi.org/10.1351/pac200779030293
42. McFall AS, Anastasio C (2016) Photon flux dependence on solute environment in water ices. Environ Chem 13:682–687. https://doi.org/10.1071/EN15199
43. Hullar T, Bononi FC, Chen Z, Magadia D, Palmer O, Tran T, Rocca D, Andreussi O, Donadio D, Anastasio C (2020) Photodecay of guaiacol is faster in ice, and even more rapid on ice, than in aqueous solution. Environ Sci Process Impacts 22:1666–1677. https://doi.org/10.1039/D0EM00242A
44. Carena L, Zoppi B, Sordello F, Fabbri D, Minella M, Minero C (2023) Phototransformation of vanillin in artificial snow by direct photolysis and mediated by nitrite. Environ Sci Technol 57:8785–8795. https://doi.org/10.1021/acs.est.3c01931
45. Hullar T, Magadia D, Anastasio C (2018) Photodegradation rate constants for anthracene and pyrene are similar in/on ice and in aqueous solution. Environ Sci Technol 52:12225–12234. https://doi.org/10.1021/acs.est.8b02350
46. Fede A, Grannas AM (2015) Photochemical production of singlet oxygen from dissolved organic matter in ice. Environ Sci Technol 49:12808–12815. https://doi.org/10.1021/acs.est.5b03600

Long-Term Data in Support of the Study of Aquatic Ecosystems: Unravelling the Role of Internal and External Drivers in the Long-Term Evolution of Lake Chemistry Through Experimental and Modeled Data

Michela Rogora, Claudia Dresti, Paola Giacomotti, Arianna Orru', Gabriele Tartari, and Andrea Fenocchi

Abstract

Lakes are complex ecosystems, with varying chemical composition, defined and regulated by both internal and external drivers, including morphological and land cover features, biological processes, meteo-climatic factors, and anthropogenic pressures. Lake chemistry is highly variable among different lakes and within the same lake, both in space and time. Dissolved oxygen and algal nutrients, particularly phosphorus and nitrogen compounds, are among the most important variables used in the assessment of lake water quality within research and monitoring programs. Trends in these parameters indeed affect biological communities and the overall chemical and ecological status of lakes.

Long-term monitoring of lake physical and chemical properties, performed with standard sampling and analytical methods, is an invaluable tool to assess how climate change and other pressures are impacting lake ecosystems. However, data collection and analysis are not enough to capture the complexity of lake dynamics: to better understand the potential impacts of climate change and other stressors on lake processes and make reliable forecast, the use of mathematical models is essential. Alongside the analysis of multi-decadal data series, numerical modeling may indeed contribute to the evaluation of lake evolution, provid-

M. Rogora (✉) · C. Dresti · P. Giacomotti · A. Orru' · G. Tartari
National Research Council of Italy (CNR) – Water Research Institute (IRSA), Verbania, Italy
e-mail: michela.rogora@cnr.it

A. Fenocchi
Department of Civil Engineering and Architecture, University of Pavia, Pavia, Italy

M. Minella (ed.), *Complexity in Environmental Chemistry*,
https://doi.org/10.1007/978-3-032-31774-2_13

ing complementary information to those addressed by monitoring campaigns and developing forecasts of the effects of natural and anthropogenic drivers.

In this contribution we focus on a category of lakes, i.e., deep lakes, and on selected water quality parameters providing some examples of long-term monitoring programs and modeling approaches. We use as a case study Lake Maggiore, a trans-boundary deep subalpine lake for which long-term data exist, dating back to the 1980s. In the last few decades, Lake Maggiore has been steadily affected by climate change which is causing important modifications in the oxygenation status and in the distribution of nutrients. The primary objective of this chapter is to establish a comprehensive framework for the design and implementation of robust long-term research on lake chemistry under increasing anthropogenic pressures and ongoing climate change. In addition, the chapter identifies key limitations of current monitoring programs and proposes strategies for their improvement and integration, with the aim of supporting effective management and long-term conservation of lake ecosystems.

Keywords

Deep lakes · Hydrochemistry · Water quality · Monitoring · Modeling

Chapter Learning Goals

- How long-term data may contribute to a better understanding of deep lake functioning and evolution under climate change.
- Why quality control of chemical data deriving from lake monitoring is essential for a proper evaluation of lake water quality.
- Why and how numerical modeling may complement the information on lake evolution obtained from long-term data.
- What are the key functional differences between modeling approaches applied to lakes and the fundamental criteria leading to the choice for a specific case study.
- How Lake Maggiore is evolving, according to long-term data assessments and numerical modeling predictions.

1 Introduction

Lakes are complex and varied ecosystems, covering a wide range of physical, chemical, and biological features, mutually interacting and shaped and regulated by both internal processes and external drivers [1]. Lake chemistry in particular is highly variable among lakes and within the same lake, both in space and time. Catchment and lake morphological and land cover features, meteo-climatic drivers, and anthropogenic pressures affect lake chemical composition and its variability [2]. Besides external factors, in-lake processes also affect the chemical properties of lake water, through complex and variable interactions between abiotic and biotic factors: as an example, primary producers regulate the amount and ratios of algal nutrients through both uptake and release processes. These exchanges in turn depend on physical and physico-chemical conditions, such as water temperature, water column stability, and oxygen availability [1, 3].

Research on lakes as water resources and key providers of ecosystem services is constantly evolving, also in support of stakeholders responsible for their protection and management. Alongside the analysis of multi-decadal data series, a key element for understanding their evolution, approaches based on innovative technologies, such as molecular barcoding techniques, satellite imagery, numerical modeling, and high-frequency monitoring, gained ground in recent years [4, 5]. Micropollutants and emerging contaminants, including microplastics, also acquired increasing attention and started to be included in ongoing monitoring programs [6, 7].

Climate change has been recognized as a primary driver of lake ecosystem shift, affecting water quality and security, productivity, and biodiversity [8–10]. Long-term ecological data, including thermal and chemical properties of water, collected with standard methodologies, are an invaluable tool to assess how climate change is impacting lake ecosystems, also through the interaction with other pressures. As an example, climate change has been shown to foster the effects of eutrophication on already impacted water bodies, enhancing stratification, oxygen deficiency at the bottom, and internal nutrient loads: these effects may counteract the efforts to restore the original trophic status through the abatement of external loads and other remediation actions [11, 12].

Time series of data from both research projects and institutional monitoring programs contribute to a better understanding of climate change effects on terrestrial and aquatic ecosystems [13]. The European (eLTER; https://elter-ri.eu/) and the International (ILTER; https://www.ilter.network/) Long Term Ecological Research networks are examples of collaborative efforts to help understanding environmental change across the globe focusing on site-based research and monitoring: lakes included in these networks are thoroughly studied sites where pluri-annual data series provide valuable information also for management and restoration plans [14, 15].

Collaborative networks, sharing protocols for data collection and analysis, and providing opportunities for joint elaboration are also extremely important for a better understanding of lake functioning at different scales: as an example, the Global Lake Ecological Observatory Network (GLEON; https://gleon.org/) promotes international team science and collaboration from different disciplines to advance the understanding of lake ecosystems [16]. These initiatives have led to the development of several joint papers, modeling frameworks and teaching and training materials [17].

In this contribution we will focus on a subset of physico-chemical and chemical parameters which are routinely monitored in lakes worldwide and commonly adopted in the assessment of lake ecological status and in long-term monitoring and research programs. These parameters include dissolved oxygen, algal nutrients, and major ions. Trends in these parameters, both in absolute values and as relative ratios, indeed affect biological communities and the overall lake food webs [18, 19]. We also focus on a specific class of lakes, namely, deep lakes, commonly defined as systems in which the maximum mixing depth remains shallower than the mean depth. In these environments, the thermal regime and associated hydrodynamic processes exert a primary control on the vertical distribution of oxygen and nutrients throughout the water column [20]. We will present examples from the deep subalpine lakes (DSL) in Northern Italy [4] and specifically for Lake Maggiore, a transboundary lake for which long-term high frequency data exist and date back to the 1980s thanks to long-term research programs jointly funded by Italy and Switzerland [21]. We will discuss how the collection of high-quality and time-consistent data is a fundamental requirement in long-term ecological research on aquatic ecosystems.

Besides data collection and analysis, building and testing numerical models may contribute to a better understanding of lake dynamics and processes: models provide a formal framework to test hypotheses on lake functioning that cannot be directly verified through observations alone. By simulating system behavior under controlled and repeatable conditions and by identifying and linking causal sets of variables in a coherent way, modeling enables the attribution of observed changes to specific drivers, such as climate forcing, nutrient loading, or internal feedback mechanisms, thereby strengthening the interpretative power of long-term monitoring programs [22, 23]. To this aim, we present selected applications that demonstrate the integration of long-term limnological datasets into hydrodynamic and ecological modelling frameworks, with the aim of forecasting lake responses to projected climate change and varying catchment loading scenarios [24, 25]. The ultimate aim of this contribution is to provide a general framework for robust long-term research on lake chemistry in the context of anthropogenic pressures and ongoing climate change, providing examples of water chemistry data collection and analysis and of applications of these data in limnological studies.

2 Long-Term Research on Deep Lakes

2.1 Characteristics and Monitoring of Deep Lakes

Large, deep lakes are a well-defined typology of lake ecosystems in which internal processes, especially vertical mixing, play a fundamental role in lake functioning. In this contribution, deep lakes are defined as basins in which a substantial fraction of the deep-water mass remains isolated from the surface layer for most of the year, becoming hydraulically and thermally connected to the upper water column only during late winter or early spring, when thermal stratification breaks down [1]: at this time of the year, lakes may undergo a complete or at least a deep partial circulation, depending on the meteo-climatic conditions and the morphometric characteristics of the lake, allowed by the thermal and density homogenization. This leads to a redistribution of oxygen and nutrients along the mixed water column [26].

The late winter and spring overturn is extremely important both for the reoxygenation of hypoxic or anoxic deep waters and for the nutrient replenishment of the trophogenic layers [27, 28]. Therefore, the trophic status of deep lakes depends not only on external nutrient loads but also on the occurrence and vertical extent of spring turnover [26]. Vertical mixing and renewal time are also crucial in the distribution of possible chemical pollutants, as well as in the recovery pattern following reduction of external pollution sources [29]. In very deep lakes in warm temperate climates, such as the DSL, full overturn does not occur every year but only at the end of very cold and windy winters. These lakes, classified as oligomictic, are therefore not subject to a regular homogenization of oxygen and nutrients along the water column [30, 31].

Mixing depth is considered one of the most critical variables controlling DO and nutrient vertical distribution, as well as nutrient replenishment of the trophogenic layers [32]. In the DSL extended vertical mixing (i.e., to depths of 150 m or more) rarely occurred after the last complete overturn of 2006, several years being characterized by very limited mixing depths, rarely extended beyond 30% of the water column (or 50% of lake volume) [33]. In Lake Maggiore mixing depth showed a clear negative trend, with values at spring overturn mostly below 75 m in recent years [34]. For the abovementioned reasons, in the monitoring of deep lakes, the assessment of lake water chemical composition at different depths is fundamental: as an example, guidance on surface water ecological monitoring under the EU Water Framework Directive (WFD) states that "in stratified water bodies such as lakes, waters samples may be taken in different depths to give a better representation of the water column compared to a single sampling depth," with the number of samples to be collected depending on lake size, depth, and ecological quality classification [35]. Institutional monitoring schemes, such as those set at national level under the WFD, identify sampling depths according to specific criteria and to represent the main layers (e.g., surface water, epilimnion, hypolimnion, deep hypolimnion). However, these depths may vary in time according to the existing condition at the time of the sampling. Instead, in long-term research programs aiming to assess the response of a lake ecosystem to natural and anthropogenic drivers, a fixed number of samples collected at time-invariant depths are usually adopted, allowing an easier organization and especially analysis of the data: as an example, multi-decadal time series of limnological data exist for the DSL, with the main variables collected at fixed depths for each lake, possibly at monthly frequency [4, 34].

To characterize the vertical pattern of water chemistry, sampling is typically conducted from a research boat at the deepest point of the lake, using Niskin bottles or comparable devices specifically designed for deep-water sampling. In most cases, a single sampling point is selected and maintained as long-term research station, assuming that data regularly collected at the deepest point may properly describe the temporal evolution of lake chemistry [36]. However, this approach may be less reliable in large lakes with relevant spatial heterogeneity: ecological parameters used for water quality assessment (e.g., chlorophyll-a) may exhibit strong spatial variability, making the representativeness of data collected in one single point questionable [37]. In lakes with spatial heterogeneity, for example, due to the presence of inflows or to multi-basin morphological characteristics, more sampling stations might be implemented, as recommended, for instance, by the implementation strategy for the WFD. In addition, remote sensing data along with modeling tools may provide additional information to in situ time series and help in optimizing sampling

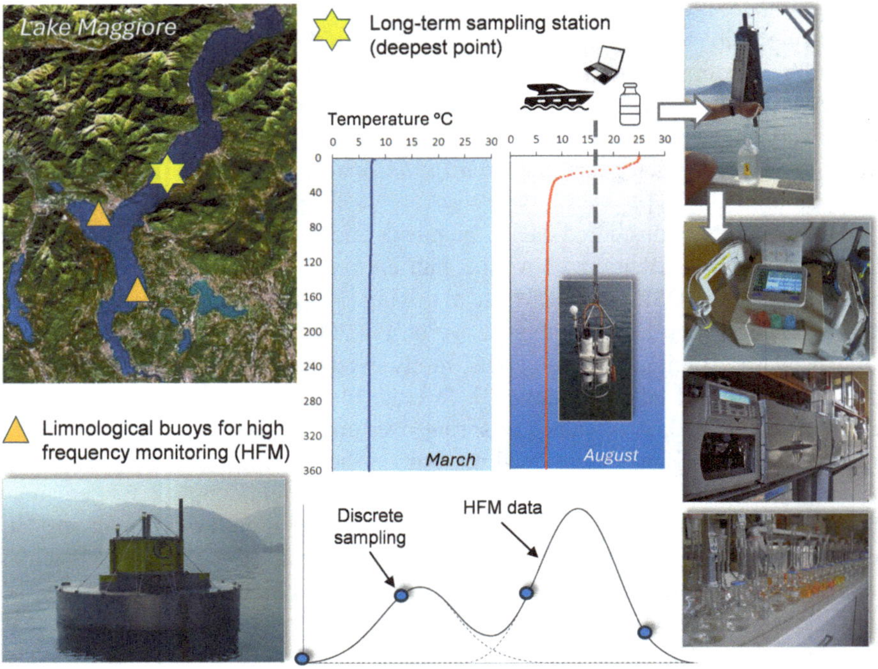

Fig. 1 Scheme of the monitoring program of the deep Lake Maggiore: discrete CTD profiles and collection of water samples for chemical analyses are performed monthly at the deepest point; two limnological buoys with in situ sensors collect high-frequency monitoring (HFM) data at littoral/semi-pelagic stations

efforts [38]. Besides the collection of discrete water samples for laboratory analysis, vertical profiles of temperature and other basic chemical variables (typically pH, electrical conductivity, redox potential, dissolved oxygen) are normally collected with multiparameter (CTD, conductivity-temperature-depth) probes (Fig. 1). The frequency of field measurements and sampling may vary from one lake to another, depending on lake characteristics, parameters of interest, and objectives of monitoring programs: as an example, a monthly or fortnightly frequency has been assumed as adequate to detect change in nutrient levels in lakes [39]. The adopted monitoring plan is necessarily a compromise between the cost of additional, more frequent measures and the connected benefits in terms of lake status evaluation [37].

Variables that are commonly analyzed in deep lake monitoring programs include base parameters (pH, conductivity, alkalinity, dissolved oxygen) and algal nutrients (phosphorus (P), both as P-PO_4 and total P; the main inorganic forms of nitrogen (N), i.e., N-NO_3, N-NO_2, N-NH_4, and total N; reactive silica (Si); total organic carbon (TOC)). This set of variables may be complemented by major ions, trace elements, and organic pollutants, also in relation to specific environmental problems.

Dissolved oxygen (DO) availability in lakes is fundamental for their functioning because DO levels and variability control important ecosystem processes [40].

Therefore, DO is a fundamental parameter in the ecological assessment of water bodies. In lake trophic layers, it represents a proxy of primary productivity, and, in deep layers, it provides information on the redox status and the potential release of P (i.e., internal loading) and toxic substances [41, 42]. Deep-water DO is also one of the key criteria used to classify lake ecological status under the WFD [35]. Recently, several studies have documented a widespread decrease of DO in lakes, with climate change playing a significant role in the DO decline in deep-water layers [28, 43]. Therefore, monitoring DO levels, with reliable and high-resolution methods, has become a priority to detect the impact of climate on lake oxygenation and identify critical thresholds [44].

The monitoring of algal nutrients is of critical importance, given their central role in eutrophication processes and their contribution to the definition of water quality standards and ecological status classes within national assessment frameworks. Nutrients are indeed key elements in lake functioning, limiting primary production and therefore regulating the whole lake food web [45]. Phosphorus has been recognized as the pivotal factor regulating phytoplankton abundance and therefore the main target for restoration measures in eutrophic lakes [46]. However, according to recent findings, both N and P may be co-limiting in different seasons and in different layers of the same lake, requiring a good temporal and vertical resolution of the measurements for identification of the actual limiting nutrient [47, 48]. In addition, an increasing number of studies demonstrated the critical role of N availability in the development of phytoplankton and cyanobacteria [49], and whole-lake experiments indicated that harmful algal blooms (HAB) are fostered by combined P and N enrichment [50]. This evidence suggests that both nutrients are important for HAB control and more generally that recovery plans should be targeted to reduce both N and P inputs to lakes [51].

To assess changes in time in nutrient levels and ratios, as well as in other chemical parameters, the first step is to rely on high-quality data. The validity of any assessment based on monitoring data depends on the accuracy of the sampling and analytical procedures and of the analysis: therefore, it is recommended that standard protocols and methods are adopted and that laboratories implement effective quality control procedures and participate in interlaboratory tests [52]. When dealing with time series, the consistency and comparability of data through time should be guaranteed by quality assurance/quality control (QA/QC) procedures, such as the analysis of reference materials and the regular use of control charts. In addition, any change in the analytical methods must be carefully evaluated, to ensure comparability of the data before and after the introduction of a new technique [52, 53].

The long-term evolution of deep-lake thermal, physical, and chemical properties is usually assessed through annual average data or based on the data collected at spring overturn, considered as a crucial period for deep lakes, as previously described. To calculate average data along the water column (e.g., epilimnetic and hypolimnetic averages) in stratified lakes, volumes of the water layers obtained by lake bathymetry (through the derived hypsometric curve) are considered, leading to more meaningful volume-weighted values. However, diverse approaches exist, for example, to define and calculate epilimnion depth [54]. Therefore, a time-consistent method to calculate average data must be used and clearly described, also to avoid potential problems that may arise for multi-lake comparisons or long-term time series analysis.

High-frequency data, both spatially and temporally, may improve the estimate of average values and provide a greater bulk of information on the variability of lake water physical and chemical properties (Fig. 1): as an example, highly resolved vertical profiles, as those obtained with CTD profiling systems, provide better insights into stratification and circulation patterns with respect to correspondent data collected at a few discrete depths [55]. Similarly, data collected by in situ sensors, at hourly or sub-hourly frequency, enable a deeper understating of short-term processes taking place in lakes, also in response to extreme weather events [56]; high-frequency data may also support the development of alert systems with respect to potentially harmful conditions (e.g., algal blooms; [5]). However, due to logistic constraints, these systems are rarely deployed in the pelagic area of deep lakes, being more common in the littoral/semi-pelagic zone or eventually deployed for short-term studies [57, 58].

In the following section, we will present examples of long-term series of physico-chemical and chemical data from discrete monitoring for a case study, discussing the main outcomes that can be derived from these data for limnological research and water resource management.

2.2 Lake Maggiore as a Case Study

2.2.1 Data Collection

The DSL, namely, lakes Lugano, Maggiore, Como, Garda, and Iseo, represent a resource of enormous environmental and economic value, facing increasing pressures, including global warming, the introduction and establishment of allochthonous species, and the presence of emerging chemical pollutants [4]. Lake Maggiore (surface area 212.5 km^2; maximum depth 370 m) is located between Italy and Switzerland and drains the upstream part of the watershed of River Ticino, which is also the only outflow of the lake, in addition to the whole River Toce catchment. The lake level is regulated to allow for a 0.5 m difference between summer and winter [59]. From a mixing regime point of view, Lake Maggiore can be classified as oligomictic, i.e., a full thermal and chemical homogenization only occurs following exceptionally cold and windy winters [31].

Lake Maggiore underwent eutrophication since the 1960s. Subsequently, because of multiple remediation measures aimed at reducing external nutrient inputs, the lake progressively recovered from eutrophication and attained a stable oligotrophic state by the late 1990s, with average total P concentrations in the water column around 10 $\mu g\ L^{-1}$ [34]. Studies on the lake have been performed continuously since the early 1980s within the framework of the limnological campaigns funded by the International Commission for the Protection of Italian Swiss Waters (CIPAIS; www.cipais.org). These sampling campaigns are conducted on a regular monthly basis at the Ghiffa station, which is located at the deepest point of the lake (Fig. 1). These pluriannual data series have allowed the inclusion of Lake Maggiore in the eLTER

Table 1 Chemical variables analyzed in the water chemistry laboratory of CNR IRSA Verbania on freshwater (lakes and rivers) samples

Variable	Unit	Range	Method	Calibration frequency	Control charts
pH		3–10	Pot.	Weekly	x
Conductivity	$\mu S\ cm^{-1}$	0.5–800	Cond.	Yearly	x
Alkalinity	$meq\ L^{-1}$	0–10	Pot. Tit.	Yearly	x
Chloride	$mg\ L^{-1}$	0.02–20	IC	Daily	x
Sulfate	$mg\ L^{-1}$	0.1–100	IC	Daily	x
Nitrate	$mg\ N\ L^{-1}$	0.01–6	IC	Daily	x
Ammonium	$\mu g\ N\ L^{-1}$	5–3000	Spec.	Yearly	x
Calcium	$mg\ L^{-1}$	0.03–100	IC	Daily	x
Magnesium	$mg\ L^{-1}$	0.02–20	IC	Daily	x
Sodium	$mg\ L^{-1}$	0.02–20	IC	Daily	x
Potassium	$mg\ L^{-1}$	0.02–10	IC	Daily	x
P-PO4	$\mu g\ P\ L^{-1}$	3–400	Spec.	Yearly	x
Total P	$\mu g\ P\ L^{-1}$	3–400	Spec.	Yearly	x
Total N	$mg\ N\ L^{-1}$	0.1–10	Spec.	Yearly	x
Silica	$mg\ Si\ L^{-1}$	0.02–10	Spec.	Yearly	x
TOC	$mg\ L^{-1}$	0.5–10	HTCO NDIR	Daily	x
Dissolved oxygen	$mg\ L^{-1}$	0.2–15	Vol. tit.	Monthly	

Methods: Pot., potentiometric. Cond., conductometric. Pot. Tit., potentiometric titration (Gran's method). IC, ion chromatography. Spec., spectrophotometry. HTCO NDIR, high-temperature catalytic oxidation with infrared detector. Vol. Tit., volumetric titration (Winkler's method)

network, as well as in other international networks and joint data analysis initiative [9, 14, 43]. In the last few decades, climate change has turned out to be a primary factor in the evolution of Lake Maggiore and the other DSL, modifying the timing and vertical extent of stratification and circulation and causing changes in the distribution of nutrients, both in time and along the water column [33, 34, 36].

Water samples for chemical analyses have been routinely collected in Lake Maggiore at the following depths: 0, 5, 10, 20, 30, 50, 100, 150, 200, 250, 300, and 360 m. Chemical analyses have been run since the beginning of the monitoring at the water chemistry laboratory of the CNR Water Research Institute (CNR IRSA) in Verbania, using standard methods for freshwater samples [60]. CTD probes (304 Plus and 310 Ocean Seven, Idronaut) have been adopted in recent years to get vertical profiles of water temperature and conductivity (since 2016) and dissolved oxygen (since 2022). The annual loads of P and N to the lake are estimated through a monthly monitoring of the main tributaries, considering the same variables of lake samples, as described in [34]. The variables which are routinely analyzed on Lake Maggiore water samples are detailed in Table 1, together with the analytical interval and the methodology used for each variable.

2.2.2 Data Quality Control

Regular QA/QC procedures are used in the laboratory to ensure the data quality and the comparability of the data in time. In particular, control charts, i.e., stable samples, analyzed for an extended period [60, 61], have been adopted in the laboratory as an internal quality check. Control charts are prepared starting from natural or artificial solutions and kept stable for a period of several months. Different concentration levels (at least two) are considered, within the range most frequently measured values. Analysis of control charts is done in association with each series of analyses in the laboratory, together with the determination of blanks. Their use allows verifying the correct setting of analytical conditions and the consistency of values obtained by different operators over time and between calibrations [52], which is extremely important in the assessment of trends in time series of chemical data. An example of the graphical elaboration of the data obtained from the control charts for the determination of total P is presented in Fig. 2a. For each control chart, a relative standard deviation (RSD) can be calculated and plotted against the average concentration to get at an index of repeatability dispersion (Fig. 2b).

2.2.3 High-Frequency Monitoring System

To improve the understanding of lake processes occurring at short timescales, a high-frequency monitoring (HFM) program was started in Lake Maggiore in 2020, within a cross-border cooperation project dealing with lake quality monitoring as a critical input for successful lake management [62]. The system initially consisted of a buoy deployed in one of the lake sub-basins with about 40 m depth and equipped with sensors for limnological and meteorological variables and algal pigments. A second buoy was added in 2023, increasing the spatial coverage of HFM. Both buoys were conceived as low-cost modular systems, aimed at completing the traditional monitoring based on discrete sampling and laboratory analysis (Fig. 1). Comparison between sensor data and data from traditional monitoring for some base variables (pH, conductivity, DO) and Chl-a showed the use of in situ sensors as a reliable approach to assess the short-term variability of these parameters in Lake Maggiore [58, 62]. However, the variables that can be analyzed with the HFM approach are still limited and cannot cover the range of information that is required for the assessment of the ecological status of lakes and more generally of their water quality. Furthermore, the activities performed on data quality and validation highlighted that regular laboratory analyses of grab samples are needed to calibrate HFM sensors and validate data in order to ensure the reliability of HFM systems [58].

2.2.4 Long-Term Trends

Annual average values of total P, total N, and silica obtained from the long-term monitoring were used to identify the main phases in the long-term evolution of nutrient concentrations and ratios in Lake Maggiore (Fig. 3). Each point in the data series is obtained from 144 determinations (12 samplings per year at 12 depths): each series consists of about 5700 measurements overall done over the 40-year period. Considering total P (TP), three main phases can be identified: (1) the oligotrophication process, with a steep decrease of TP; (2) a stable phase, from the

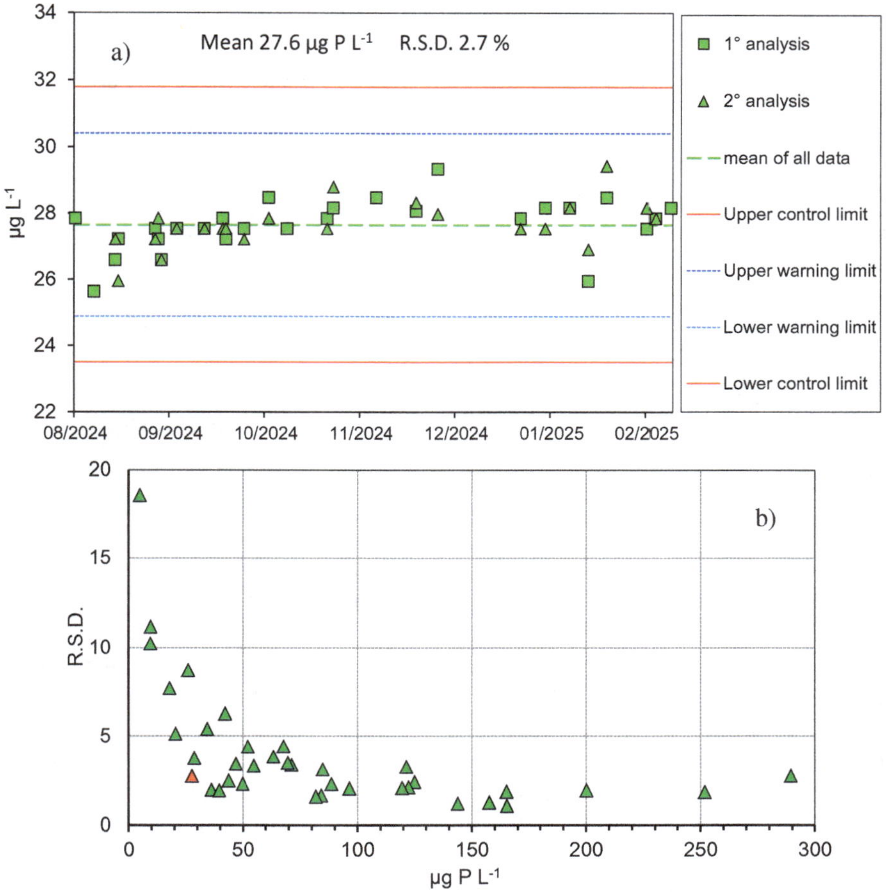

Fig. 2 Example of a control chart (**a**) and repeatability values expressed as R.S.D. (**b**) for the analysis of total P in the water chemistry laboratory of CNR IRSA in Verbania. Each value in plot (**b**) corresponds to a control chart used for a period of 2–6 months in the laboratory (e.g., the red triangle corresponds to the control chart in plot (**a**) analyzed over a 6-month period)

mid-1990s until about 2010, with TP around 9–11 μg P L^{-1}; (3) the recent period, characterized by a slight but regular increase of TP (13–14 μg P L^{-1}) until 2022 and a steeper increase afterward (Fig. 3).

Total N remained stable, with a likely slight increase, until around 2010, followed by a decline, in turn interrupted in 2022. Silica increased steadily since the 1990s. Due to these changes, the N:P ratio also significantly changed from about 200–250, as molar ratio, in the oligotrophication phase, to about 150 in the most recent years due to the concurrent increase of total P and the decrease of total N. According to the actual average concentrations of total P, Lake Maggiore should be classified as oligo-mesotrophic, and the recent steep increase could suggest a tendency toward a deterioration of lake water quality. However, the in-depth investigation based on the monthly vertical profiles of chemical data has shown that climate-driven eutrophication is affecting Lake Maggiore, as detailed in the following.

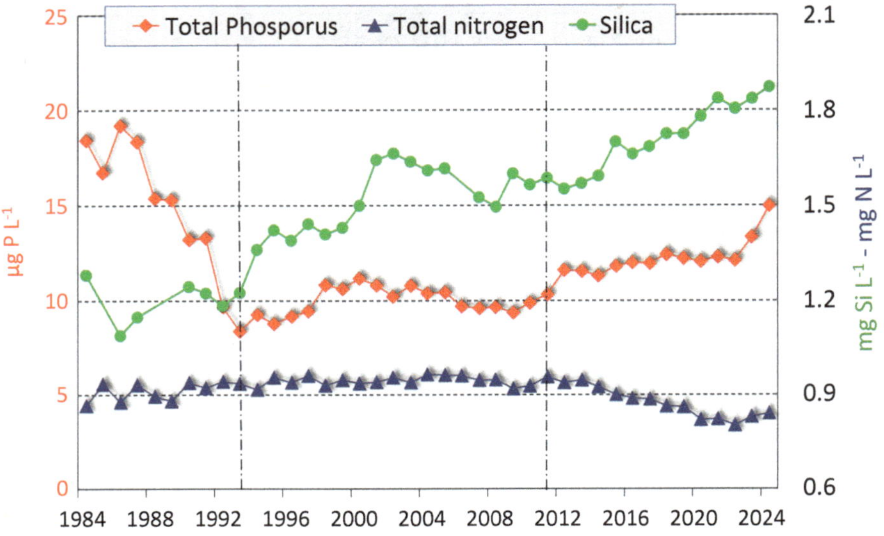

Fig. 3 Temporal evolution of total phosphorus, total nitrogen, and silica in Lake Maggiore since 1984 (volume-weighted annual average values along the water column)

Average annual data, as well as data at spring overturn, can be used for a broad classification of the lake trophic status and its temporal evolution. However, in deep lakes, it is of foremost importance to evaluate trends also in the different water layers, using the highest feasible resolution. As an example, Lake Maggiore data are usually organized as time series for epilimnetic (volume-weighted average for the 0–25 m layer), hypolimnetic (25–370 m), and, for some variable, deep-water (200–370 m) layers. The metalimnion depth may of course vary within seasons, with epilimnion and hypolimnion being not fixed-depth water layers: however, the adopted approach is particularly useful to both visualize and statistically assess long-term trends (Fig. 4). Trends of total P clearly show how concentrations are steadily increasing in the deep-water layer only, while epilimnetic concentrations are stable or even decreasing (Fig. 4a). The positive trend of hypolimnetic P is evident since 2010, thus explaining the increasing phase of total P as average values along the water column (Fig. 3). Total N shows the same pattern in surface and deep layers, with a slight increase until around 2012, followed by a decreasing phase and a new increase in 2023–2024 (Fig. 4b). As for total P, different behaviors between surface and deep layers can be seen also for reactive Si concentrations, which are rather stable in the epilimnion, despite sharp oscillations, but steeply increasing in the deep layers, particularly in the last decade (Fig. 4c). In summary, P and Si are progressively accumulating in the hypolimnion of the lake, causing an overall increase of the average concentrations in the water column in the recent period. As regards DO, a steep decrease of concentration in the hypolimnion started in 2010, with an acceleration in the last few years (Fig. 4d): the trend is even steeper for DO concentration in the deep hypolimnion (below 200 m; [63]).

These recent trends in nutrient and DO concentrations, with a decoupling of surface and bottom dynamics, have been explained by an increase of the thermal

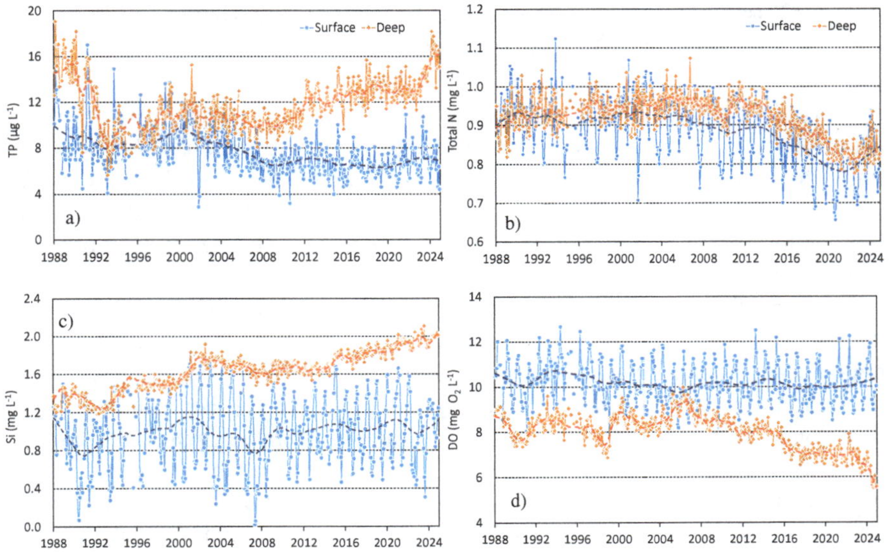

Fig. 4 Long-term trends of monthly concentrations of total phosphorus (**a**), total nitrogen (**b**), silica (**c**), and dissolved oxygen (DO) (**d**) in the surface epilimnetic layer (0–25 m) and in the deep hypolimnetic layer (25–360 m) of Lake Maggiore. The dashed lines represent the trend components obtained by seasonal decomposition through a LOESS (LOcally Estimated Scatterplot Smoothing) regression (SDL)

stability of the lake, due to different warming rates of the surface and deep-water layers [64, 65]: the increasing stability of the water column, coupled with the mild conditions of the recent winter periods, has led to a reduced frequency and intensity of deep-mixing events, promoting DO depletion and P and Si accumulation in the hypolimnion [34, 63]. A further consequence of the lack of deep-mixing events is the very limited nutrient replenishment of the upper layers at spring overturn, which can cause important effects on the whole lake food webs [27].

During the eutrophication and the following recovery, P concentrations in Lake Maggiore were tightly related to external P loads. Since 2010, there has been a decoupling between P inputs and in-lake P levels, with the latter being more controlled by processes related to thermal and hydrodynamical features in the water column. On the contrary, N temporal evolution has been controlled by external loads throughout the study period. Catchment loads of N have indeed decreased since the second half of the 2000s [34]. Studies performed on the Lake Maggiore watershed considering the role of atmospheric deposition in N inputs to the lake have shown that the atmospheric contribution may be up to 70% of the total load. Due to the reduction in the emission and deposition of N compounds in Northern Italy, as in other EU countries, the inputs of inorganic N to the lake decreased, explaining the observed trends [34]. The recent slight increase of N concentrations (Fig. 4b) may be also ascribed to atmospheric N deposition, being the end of 2023 and 2024 characterized by abundant rainfall. Despite the evident role of external inputs in regulating N in Lake Maggiore, it should be highlighted that in-lake processes also clearly control N levels, as already highlighted, with these processes in turn depending on DO and P availability [66, 67] and on meteo-hydrological conditions [68].

3 The Role of Lake Modeling in Understanding and Managing Lake Systems

Numerical modeling offers valuable possibilities to consolidate lake evolution evaluations resulting from long-term observed data on lakes. First, it allows developing future forecasts of the effects of climate change and external load variations [69, 70]. It also allows gaining complementary information to those directly addressed by monitoring campaigns [63], in addition to potentially filling voids in observed data series.

The term "modeling" applied to lakes has a wide meaning. It may refer to reproductions of the functioning of all lake spheres, i.e., the physical, (geo)chemical, and biological/ecological ones. Lentic water bodies are known to have all these spheres mutually interacting in determining ecosystem processes [1]. For example, phytoplankton primary production is shaped by water temperature; relies on inorganic nutrients, partly obtained through recycling of organic ones by bacteria, some of which consume DO in the process; and sustains the whole upper ecosystem layers, both as food and through DO production through photosynthesis, but in turn it shapes the lake thermal structure through its effects on water turbidity [1–3]. These tightly coupled interactions imply that changes occurring in one sphere may propagate across the entire lake system, often with nonlinear responses and delayed effects. Numerical modeling is therefore particularly suited to capture such cross-scale feedbacks, which are increasingly relevant under climate-driven alterations of stratification, mixing regimes and biogeochemical cycling [71, 72].

Models may either address a single sphere, and parameterize, i.e., simplify to fixed terms, all interactions with the other ones, or try to reproduce the exchanges between the spheres, in which case they are often called "coupled" models [25]. However, the most relevant distinction between lake models is between statistical (also called empirical or "black-box") and process-based models [25, 73].

Statistical models typically deal with the biological and/or ecological spheres and employ field-based inferential relationships to predict variables of interest such as nutrient or Chl-a concentration from predictor variables [74]. Some statistical models have a minimal physical basis, which serves as "conceptual model," though calibrated empirical coefficients still play the main role. Because of their low data and computational requirements, statistical models remain attractive for rapid assessments and screening-level applications. However, their reliance on historical correlations inherently limits their transferability to novel conditions, particularly under non-stationary climate forcing, where past relationships may no longer hold [75]. Examples of this are the well-known Vollenweider-OECD eutrophic-state-based models [76, 77] and all their extensions and modifications [78], which were developed to estimate lake total P concentrations at spring mixing in response to external loads of total P. For instance, the original Vollenweider model [76] expresses how lake spring-mixing total P for equal external P loading decreases with increasing residence time and increases with increasing mean depth, i.e., essentially linking it to sedimentation dynamics. Vollenweider-OECD models have been efficiently used as tools to develop lake restoration strategies from eutrophication [79], but they have multiple limitations [69]: (1) their response is steady state, i.e., they cannot reproduce the dynamic evolution of a lake with external load variations; (2) they completely neglect stratification and internal-load mechanisms, which play a relevant role in basins affected by climate change and anthropogenic pressures [80–82]; (3) model parameters were derived from regressions over families of sample basins, so that individual features of peculiar lakes cannot be included [83]. These limitations are particularly critical for deep lakes, where long water residence times, strong thermal stratification, and climate-induced changes in mixing dynamics can decouple external nutrient loads from in-lake nutrient availability over seasonal to decadal timescales [34, 84]. Furthermore, the limitations have sometimes led to overestimations of the efficiency of reduction in external nutrient loads [75, 85]. Recent developments of statistical models involve the use of artificial intelligence (AI), using machine learning (ML) and artificial neural network (ANN) techniques, to detect harmful algal blooms (HABs) in real time from field meteorological and/or limnological data, for drinking water and bathing safeguard [86, 87]. Despite the promising outlook of these techniques, it should be remembered that they are effectively black-box approaches that do not explicitly account for physical links, and as such they are prone to fail in conditions which were not addressed in the initial model training [88, 89]. This issue is especially relevant when AI-based tools are applied to long-term projections or extreme events, where extrapolation beyond the training domain may lead to physically inconsistent or ecologically implausible outcomes [90].

Process-based numerical models instead reproduce lake processes through differential equations [75, 89], which represent the cause-effect relationships at the base of physical processes, chemical reactions, and ecological variations, by linking variations between variables. By explicitly resolving underlying mechanisms, process-based models are better suited for scenario analysis and for assessing system responses to novel combinations of climatic and anthropogenic pressures, thus providing a more robust basis for long-term forecasting and management-oriented applications [91]. These models may deal with either lake physics (hydrodynamic models) or with lake biological and ecological spheres (ecological models) or address both (coupled ecological-hydrodynamic models), including the already highlighted mutual relationships between different spheres [25]. All process-based models include a time-dynamic representation but employ different schematizations of spatial variability.

Ecological-only models employ a zero-dimensional (0D) architecture, completely neglecting spatial variability, or, to represent eutrophication, they may subdivide the water column into two to three layers with fixed thickness and exchange fluxes [92, 93]. A dynamic representation of lake physics, i.e., coupling to a hydrodynamic model, is needed to avoid such extreme parameterization. Most complete ecological models represent the food chain originating and ending with inorganic nutrients after the bacterial decomposition of living organisms, reproducing DO, C, N, P, and Si cycles [25]. Representation of the food chain can range from including phytoplankton only and parameterizing effects of the higher trophic levels into its mortality rate (NPD: Nutrients-Phytoplankton-Detritus schematization [69]), to including the whole chain up to fish predators.

Hydrodynamic models, coupled or uncoupled to an ecological driver, may instead be (1) one-dimensional (1D), in which case they represent horizontally averaged processes in the water column, i.e., stratification and mixing dynamics, assuming that horizontal gradients are negligible compared to vertical ones, as is the case of deep stratified basins [94, 95]; (2) two-dimensional (2D), either in the horizontal plane, being suited to shallow mixed basins [96, 97], or in the vertical plane, apt for valley reservoirs (such as the CE-QUAL-W2 model, https://www.ce.pdx.edu/w2/); (3) three-dimensional (3D), representing both mixing and stratification processes and circulation processes [98, 99], being further possibly linked to an atmospheric model to provide the needed heterogeneous meteorological forcing, notably wind, which shape lake circulations in large basins [100, 101].

It is crucial to remind here some basic principles of modeling activity, which must guide the choice of the approach to adopt for a lake case study, addressing a specific problem. Data accuracy and consistency and availability of long-enough time series are essential not only for properly evaluating evolutionary trends, as seen before, but also for properly setting up, calibrating, and validating numerical models, according to the GIGO ("garbage in, garbage out") principle. Then, more complex models offer a more complete representation of co-occurring phenomena, but they are not necessarily better: they require more data, which may not be available, they are way more intensive to calibrate, and they require more computational resources. Furthermore, we should consider that environmental models of all sorts are mostly needed because reality is too complex to be understood: if models become too complex, interpretation of results becomes intricated, and the advantage of modeling is spoiled. The choice of the modeling approach to use should therefore be driven by the lake research and management aims and by data availability. In general, it is suggested to choose the simplest model which gets to the specific target with the desired accuracy, according to the Occam's razor principle. This explains why, for example, 3D coupled ecological-hydrodynamic models, or 0D ecological-only or 1D coupled ecological-hydrodynamic models including the whole food chain up to fish, though available, are rarely employed in actual studies [25].

3.1 Model Applications to the Lake Maggiore Case Study

As made evident in Paragraph 2.2, Lake Maggiore is suffering from reduction in spring mixing intensity due to climate warming effects, causing the intensification of the vertical (i.e., water-column) temperature gradient [81]. As an example, analysis of 1981–2022 water temperature data series obtained from CIPAIS programs highlights rising trends of +0.48 °C/decade for the 0–20 m layer and only +0.14 °C/decade for the 200–370 m layer. While water warming is observed across the DSLs, the difference between the epilimnetic and deep-hypolimnetic gradients is not uniform: for example, analysis of unpublished 1996–2022 data on Lake Garda kindly provided by Fondazione Edmund Mach highlights a practically coincident warming rate of surface and deep waters. These differences can be attributed to contrasts in basin morphology and wind forcing between the two lakes. In Lake Garda, the presence of strong periodic winds combined with an elongated, deep northern basin gives rise to secondary recirculating currents in the vertical plane, which enhance hypolimnetic mixing [102]. Such dynamics do not appear to characterize Lake Maggiore. The relevant intensity of the vertical temperature gradient intensification already observed in Lake Maggiore from long-term data, having a sensible effect on nutrient distribution in the water column (see Paragraph 2.2), led to the development of numerical modeling studies to forecast the future evolution of the lake, including its thermal structure and nutrient distribution and primary productivity features [82, 103]. The aim of these works was to develop long-term estimations of lake future evolution according to climate change scenarios of different intensities, specifically those given by CH2011 [104], further testing the effects of reductions in external nutrient loads. Due to the focus on long-term dynamics governed by vertical stratification and mixing processes, a one-dimensional coupled ecological-hydrodynamic model was adopted, namely, GLM/AED2 (General Lake Model/Aquatic EcoDynamics [22, 71]). This modeling framework is capable of reproducing the interactions among physical, chemical, and biological components [91], while not requiring spatially distributed meteorological forcing or calibration and validation datasets, and involving substantially lower computational demands than three-dimensional models. These characteristics make it particularly suitable for long-term scenario simulations and projections [105]. It should be however considered that even process-based coupled ecological-hydrodynamic models suffer from the bounds dictated by the parameterization at the biological level, which cannot be thoroughly described by differential equations, as possible for the physical sphere and, to a lesser extent, for the chemical sphere. This is due to the higher complexity of biological mechanisms: e.g., phenomena such as immigration and adaptation of phytoplankton species [91, 105] or top-down effects of parameterized upper ecosystem levels [106, 107] cannot be reproduced. Furthermore, modelling of biological populations cannot be done up to the level of detail of individual species, this leading to an unbearable list of model parameters to be calibrated [106]. Aggregation into functional groups is compulsory to obtain a reasonably manageable model. As a result of all of this, forecasts should always be critically framed, and the possible effects of model shortcomings should always be evaluated [25, 108].

Two explicit modeling choices were made in [82, 103] due to the oligomictic mixing regime of Lake Maggiore, ultimately evolving toward meromixis [31]. First, the "sawtooth" behavior of hypolimnetic water temperatures [64], increasing throughout years without complete mixing and dropping when this occurs, makes it necessary to perform continuous simulations starting from the present day up to the desired point in the future to properly reproduce the thermal structure evolution, as also done for Lake Iseo in [109]. Simulating isolated future periods, or even single years, would be meaningful for dimictic lakes only, in which the thermal structure is reset twice per year passing through the 4 °C isotherm, not carrying over the so-called climatic memory of past warm winters in the water-column heat content [31]. Second, the casual occurrence of complete-mixing events, originating from the randomness of meteorological conditions around the climatic driving force [82], and their extreme relevance on the vertical distribution of lake parameters, makes it necessary to run manifold simulations (realizations) with random meteorological data series, analyzing future forecasts from a probabilistic point of view. In the studies on Lake Maggiore, the Vector-Autoregressive Weather Generator VG [110] was employed to generate 200 random series for each considered climate scenario.

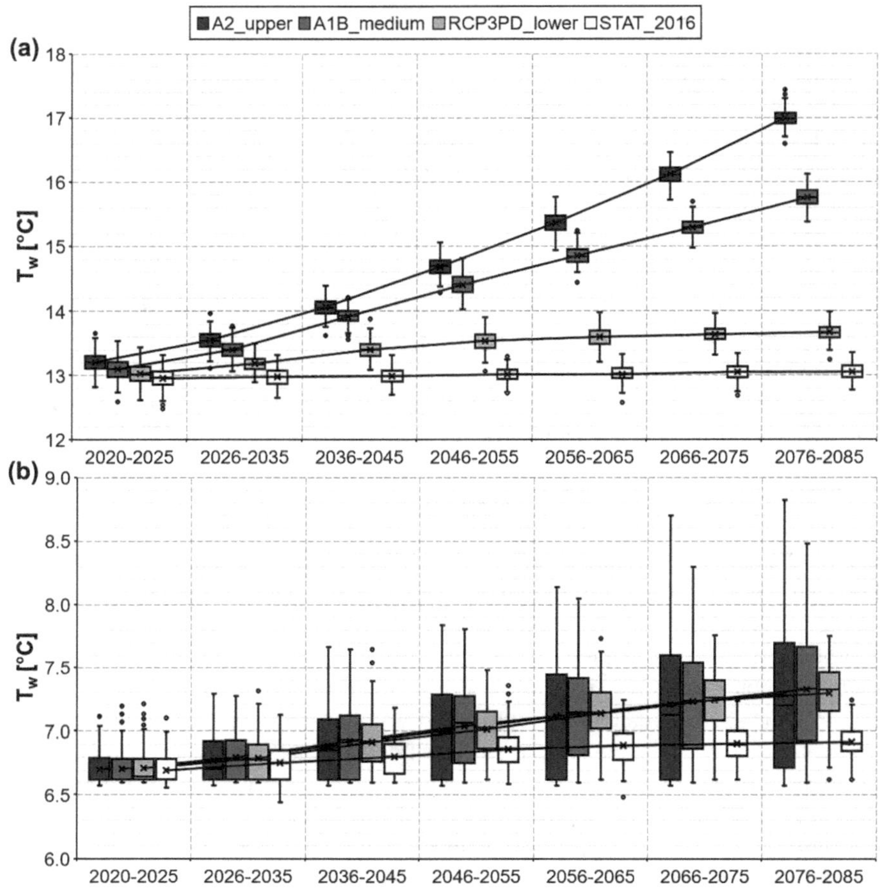

Fig. 5 Box plots of the distributions over the 200 model realizations of the volume-weighted mean annual water temperatures for 2020–2085 averaged over decades in the 0–20 m (**a**) and in the 200–370 m (**b**) layers for the A2_upper, A1B_medium, RCP3PD_lower, and STAT_2016 CH2011 climate scenarios (from [103]; crosses and solid lines identify the mean values and their trends)

Figure 5 reports the probabilistic box plots of the distributions over the 200 model realizations of the volume-weighted mean annual water temperatures for the study period 2020–2085, averaged over decades in the 0–20 m epilimnion and in the 200–370-m-deep hypolimnion of Lake Maggiore, from [103]. Results are represented for these CH2011 [104] climate scenarios: (1) A2_upper, which depicts an uncontrolled rise of global greenhouse gas (GHG) emissions; (2) A1B_medium, which describes an ordinary climate warming resulting from the reduction of global GHG releases starting from the mid-21st century; (3) RCP3PD_lower, which represents a mild heating caused by an immediate decrease of global GHG emissions; (4) STAT_2016, which considers fictitious conditions in which air temperatures are stabilized at the 2016 levels and serves as reference.

The differential increasing temperature gradient is evident for the A2_upper and A1B_medium scenarios, with 0.66 °C/decade and 0.47 °C/decade epilimnetic warming gradients, respectively, and only 0.10 °C/decade and 0.12 °C/decade in the deep hypolimnion, respectively, confirming those obtained from long-term observations. This indicates a decreasing mixing probability, effectively turning Lake Maggiore into a meromictic basin [82]. Immediate reduction of global GHG emissions, as in the RCP3PD_lower scenario, would be needed to limit the effects of climate change on the mixing regime of Lake Maggiore, stabilizing water temperatures in the epilimnion by the mid-21st century.

The corresponding box plots from simulations for concentrations of various chemical variables under invariant current external-load pressures and the different climate scenarios in the deep hypolimnion (200–370 m) of Lake Maggiore are presented in Fig. 6. The modelled forecasts all confirm the trends resulting from the

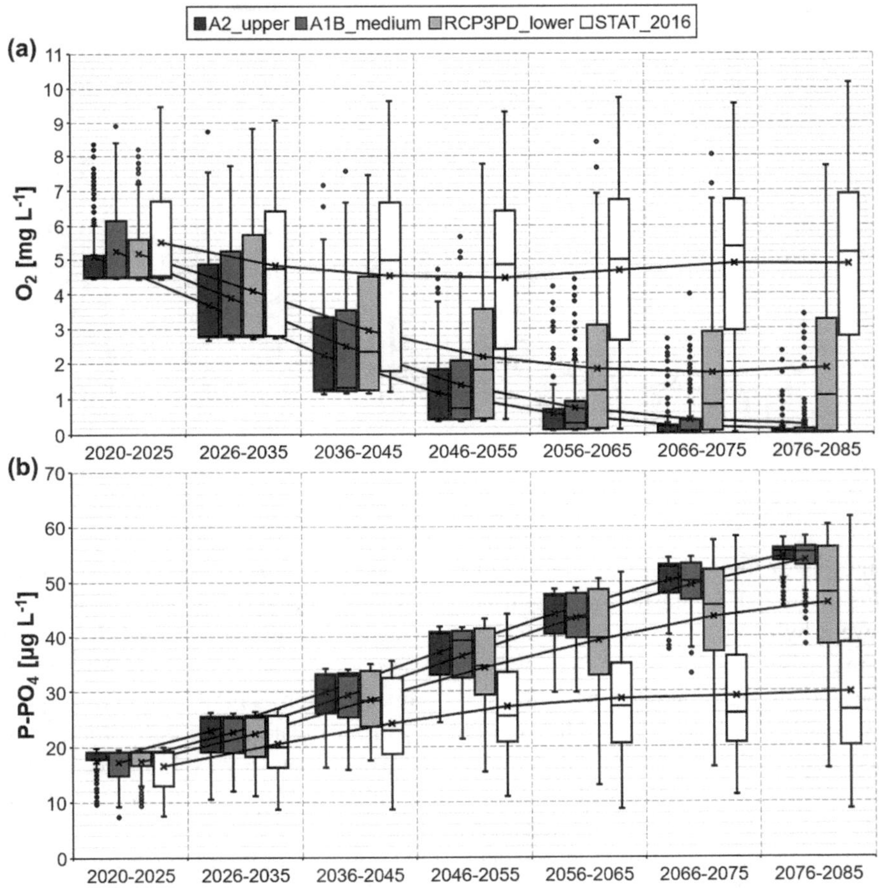

Fig. 6 Box plots of the distributions over the 200 model realizations of the volume-weighted mean annual concentrations for 2020–2085 averaged over decades of dissolved oxygen (**a**), orthophosphate (**b**), nitrates (**c**), and reactive silica (**d**) in the 200–370 m layer for the A2_upper, A1B_medium, RCP3PD_lower, and STAT_2016 CH2011 climate scenarios and invariant current external-load pressures (modified from [103]; crosses and solid lines identify the mean values and their trends)

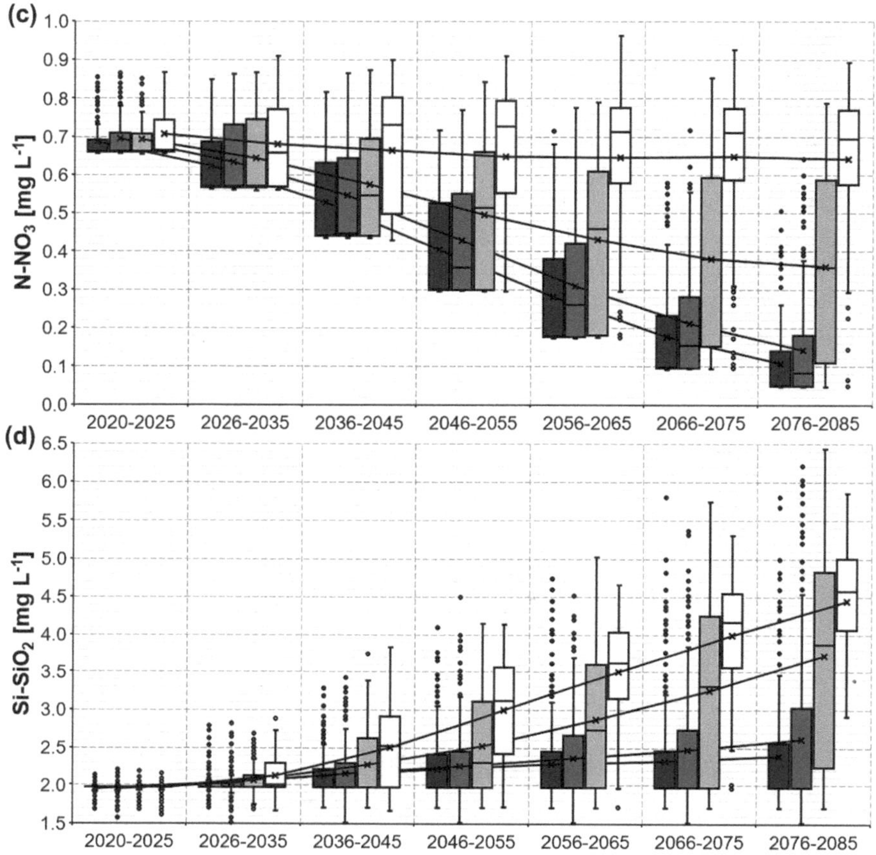

Fig. 6 (continued)

analysis of long-term observations. In the A2_upper and A1B_medium scenarios, due to simulated vanishing spring mixing, DO concentration (Fig. 6a) is set to reach severe hypoxic levels by the mid-21st century, having already dropped to around 4 mg L^{-1} in recent observations, P-PO_4 concentration (Fig. 6b) would triplicate by the end of the simulated period, N-NO_3 concentration (Fig. 6c) would drop to extremely low value, due to the impossibility of nitrification to occur without DO, and reactive Si (Fig. 6d) would also accumulate in the deep layers. In the RCP3PD_lower scenario, the effects of climate change on the deep-water chemistry of Lake Maggiore are successfully limited.

The simulated effects of climate change on phytoplankton under invariant current external-load pressures are last given in Fig. 7. A slight decrease in overall biomass, expressed in terms of Chl-a concentration (Fig. 7a), is predicted for the A2_upper and A1B_medium scenarios, due to the already observed decrease of nutrient concentrations in the photic layer with mixing decrease. Rising water temperatures [27, 42, 111] would also trigger a strong rise in cyanobacteria (Fig. 7b),

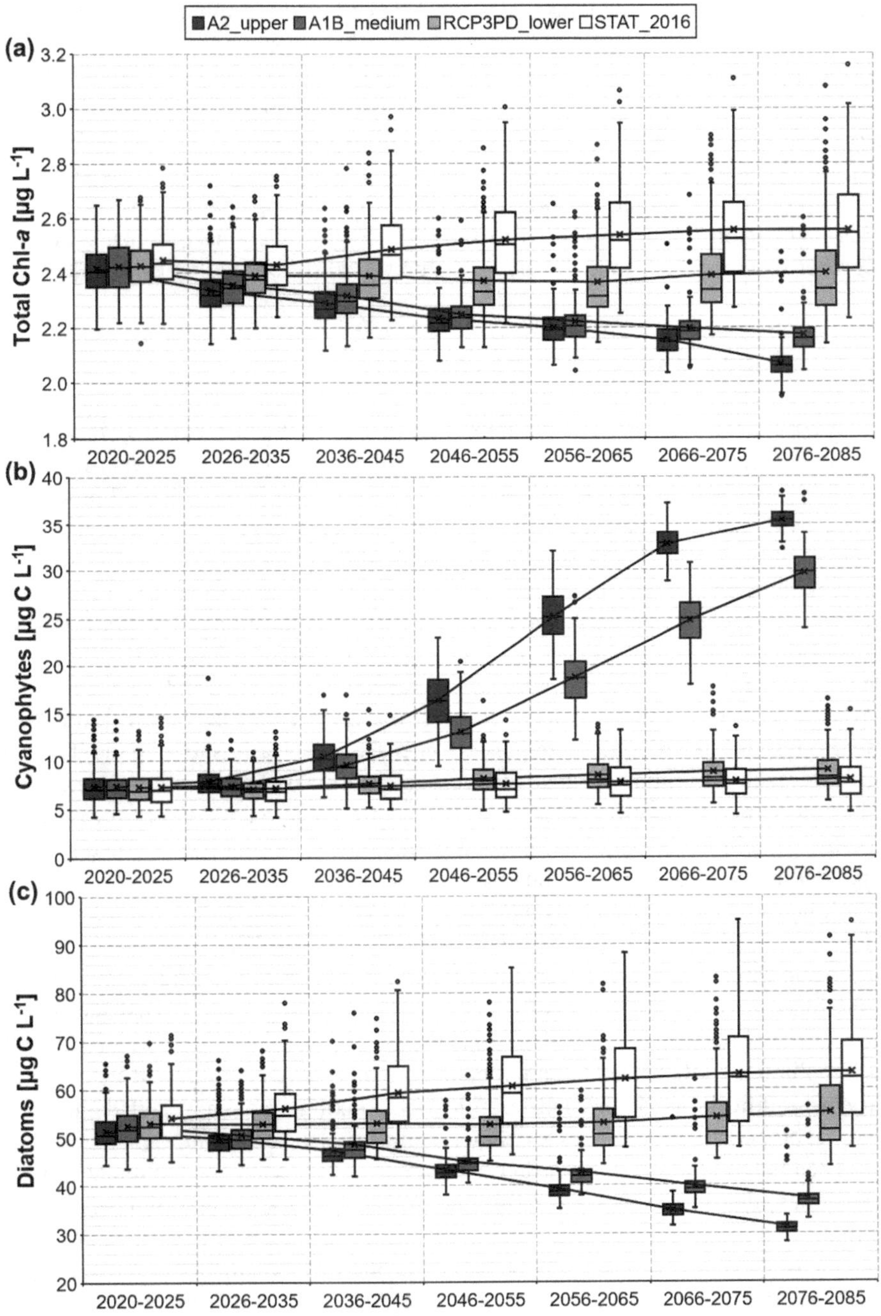

Fig. 7 Box plots of the distributions over the 200 model realizations of the volume-weighted mean annual concentrations for 2020–2085 averaged over decades of total Chl-a (**a**), cyanobacteria (**b**), and diatoms (**c**) in the 0–20 m layer for the A2_upper, A1B_medium, RCP3PD_lower, and STAT_2016 CH2011 climate scenarios and invariant current external-load pressures (modified from [103]; crosses and solid lines identify the mean values and their trends)

which would co-dominate with diatoms by the end of the century, the latter group being penalized (Fig. 7c) by warmer waters and less reactive Si availability in the photic layer. As an example, under A2_upper, simulated cyanobacteria would rise from 7% of the mean overall annual phytoplankton biomass in 2020–2025 up to 37% in 2076–2085, whereas diatoms would correspondingly drop from 47% down to 33%. Stabilization of present phytoplankton numerosity would be expected under RCP3PD_lower. It should be pointed out that as a modeling compromise cyanobacteria and diatoms were modelled in [103] as single functional groups, grouping together species which sometimes have some very different ecological traits.

4 Conclusions and Perspectives

Long-term data analysis and modelling results indicate that external loads and catchment characteristics are still important in determining nutrient levels and their variability in lakes [8, 18]. However, in deep lakes in particular, internal processes driven by climate change contribute to the trophic status and the oxygenation of lake waters [28, 112, 113]. Decreasing oxygenation of deep-water layers and accumulation of nutrients, with limited replenishment of the surface layers, have been described for other deep lakes that underwent oligotrophication in the 1990s thanks to restoration measures and are now facing the counteracting effects of climate change [42, 114–116]. Despite successful interventions to abate external loads that have led to re-oligotrophication or at least to a decrease of average P levels, the monitoring of Lake Maggiore and of the other DSL is clearly showing the increasing relevance of climate change in their long-term evolution: climate warming might stop the full recovery of the lakes from eutrophication or contrast the positive effects of previous or ongoing remediation actions, requiring more efficient management interventions to meet the restoration targets [114].

Climate change is profoundly altering nutrient cycling and DO dynamics in Lake Maggiore, as in many other deep lakes worldwide [9, 43]. These changes are leading to substantial deviations in key parameters traditionally used for the classification of lake water quality and ecological status, including average annual and spring overturn total phosphorus concentrations integrated over the water column, as well as hypolimnetic DO levels. Such indicators are increasingly controlled by climate-driven modifications of thermal stratification, mixing frequency and depth, and internal biogeochemical processes, rather than solely by external nutrient loads [43].

When the WFD assessment criteria were originally established, climate change was not explicitly considered as a relevant pressure, implicitly assuming stationary climatic and limnological conditions. In contrast, climate forcing has now emerged as a major driver of physical, chemical, and ecological dynamics in deep lake ecosystems, affecting both nutrient availability and oxygen consumption patterns. Similar conceptual limitations apply to other lake evaluation and classification schemes adopted at the national level, which are likewise based on reference conditions that do not explicitly account for long-term climate-induced shifts. As suggested by [117], there is therefore a clear need to incorporate a climate-informed assessment framework into WFD protocols, explicitly integrating climate-related pressures and their cascading effects on lake functioning. Such an approach would allow a more robust interpretation of long-term monitoring data, reduce the risk of misclassifying ecological status in climate-sensitive systems, and ultimately support adaptive management and conservation strategies under ongoing and future climate change.

Questions

- Which are the main characteristics of deep lakes, and which parameters are usually considered in their long-term monitoring?
- Which data quality control may be adopted to check the time consistency of lake monitoring data?
- What are the differences between statistical and process-based lake models?
- Why are long-term high-quality data fundamental for the implementation of solid lake numerical models?
- Why is dissolved oxygen decreasing and phosphorus increasing in the deep hypolimnion of Lake Maggiore?

Glossary

- Trophic status: the overall level of biological productivity within a lake, often used as an indicator of water quality. Based on their productivity levels, lakes can be classified as oligotrophic (low productivity), mesotrophic (moderate), eutrophic (high), and hypereutrophic (very high).
- Primary producers: organisms that can produce their own food through photosynthesis or chemosynthesis. In lakes, the main primary producers are phytoplanktonic organisms which represent the base of the aquatic food webs.
- Epilimnion (or trophogenic layer): the shallow, less dense, warm water layer in lakes, separated from the deeper, more-dense, colder-water layer (hypolimnion) by a zone of high rate of temperature change (metalimnion or thermocline).
- Thermal stability: in a lake, the work that would be required to transform a thermally stratified lake into a lake characterized by isothermal conditions. It is often assessed by the Schmidt stability value.
- Oligomictic: a lake characterized by a persistent and relatively stable thermal stratification, punctuated by only infrequent and often incomplete mixing events. This regime typically occurs in low-latitude regions with limited seasonal temperature variability, as well as in very deep lakes where complete overturn is inhibited.
- Interlaboratory comparisons (or proficiency tests): evaluations of the performances of a laboratory by means of comparisons with other laboratories; they consist in testing the same samples by different laboratories and in comparing the results.
- Relative standard deviation (RSD): a measure expressing the standard deviation of a dataset as a percentage of the mean of that dataset. In analytical chemistry, it represents a key metric for comparing method precision across different concentration levels or instruments.

Bibliography

1. Jones D, Smol P (2023) Wetzel's Limnology. Lake and river ecosystems.4th edn. Academic Press. https://doi.org/10.1016/C2019-0-04412-3
2. Bartoli M, Viaroli P (2015) Chapter 2: Lake ecology and chemistry. Surface water photochemistry. In: Calza P, Vione D (eds) Royal Society of Chemistry. Comprehensive series in photochemical & photobiological sciences. https://doi.org/10.1039/9781782622154-00017
3. Pierson DC (2012) Light and primary production in lakes. Encyclopedia of lakes and reservoirs. In: Bengtsson L, Herschy RW, Fairbridge RW (eds) Encyclopedia of earth sciences series. Springer, Dordrecht. https://doi.org/10.1007/978-1-4020-4410-6_261
4. Salmaso N, Buzzi F, Capelli C, Cerasino L, Leoni B, Lepori F, Rogora M (2020) Responses to local and global stressors in the large southern perialpine lakes: present status and challenges for research and management. J Great Lakes Res 46:752–766. https://doi.org/10.1016/j.jglr.2020.01.017
5. Marcé R, George G, Buscarinu P, Deidda M, Dunalska J, de Eyto E, Flaim G, Grossart HP, Istvanovics V, Lenhardt M, Moreno-Ostos E, Obrador B, Ostrovsky I, Pierson DC, Potuzak J, Poikane S, Rinke K, Rodirguez-Mozaz S, Staehr PA, Sumberova K, Waajen G, Weyhenmeyer GA, Weathers KC, Zion M, Ibelings BW, Jennings E (2016) Automatic high frequency monitoring for improved lake and reservoir management. Environ Sci Technol 50(20):10780–10794. https://doi.org/10.1021/acs.est.6b01604
6. Nava V, Chandra S, Aherne J, Alfonso MB, Antão-Geraldes AM, Attermeyer K, Bao R, Bartrons M, Berger SA, Biernaczyk M, Bissen R, Brookes JD, Brown D, Cañedo-Argüelles M, Canle M, Capelli C, Carballeira R, Cereijo JL, Chawchai S, Christensen ST, Christoffersen KS, de Eyto E, Delgado J, Dornan TN, Doubek JP, Dusaucy J, Erina O, Ersoy Z, Feuchtmayr H, Frezzotti ML, Galafassi S, Gateuille D, Gonçalves V, Grossart HP, Hamilton DP, Harris TD, Kangur K, Kankılıç GB, Kessler R, Kiel C, Krynak EM, Leiva-Presa À, Lepori F, Matias MG, Matsuzaki SS, McElarney Y, Messyasz B, Mitchell M, Mlambo MC, Motitsoe SN, Nandini S, Orlandi V, Owens C, Özkundakci D, Pinnow S, Pociecha A, Raposeiro PM, Rõõm EI, Rotta F, Salmaso N, Sarma SSS, Sartirana D, Scordo F, Sibomana C, Siewert D, Stepanowska K, Tavşanoğlu ÜN, Tereshina M, Thompson J, Tolotti M, Valois A, Verburg P, Welsh B, Wesolek B, Weyhenmeyer GA, Wu N, Zawisza E, Zink L, Leoni B (2023) Plastic debris in lakes and reservoirs. Nature 619:317–322. https://doi.org/10.1038/s41586-023-06168-4
7. Crini N, Lichtfouse E, Liu G, Balaram V, Lado Ribeiro A, Lu Z, Stock F, Carmona E, Teixeira M, Picos Corrales L, Moreno-Piraján J, Giraldo L, Li C, Pandey A, Hocquet D, Torri G, Crini G (2022) Worldwide cases of water pollution by emerging contaminants: a review. Environ Chem Lett 20:2311–2338. https://doi.org/10.1007/s10311-022-01447-4
8. Carpenter SR, Caraco NF, Correll DL, Howarth RW, Sharpley AN, Smith VH (1998) Nonpoint pollution of surface waters with phosphorus and nitrogen. Ecol Appl 8:559–568. https://doi.org/10.1890/1051-0761
9. O'Reilly CM, Sharma S, Gray DK, Hampton SE, Read JS, Rowley RJ, Schneider P, Lenters JD, McIntyre PB, Kraemer BM, Weyhenmeyer GA, Straile D, Dong B, Adrian R, Allan MG, Anneville O, Arvola L, Austin J, Bailey JL, Baron JS, Brookes JD, de Eyto E, Dokulil MT, Hamilton DP, Havens K, Hetherington AL, Higgins SN, Hook S, Izmest'eva LR, Jöhnk KD, Kangur K, Kasprzak P, Kumagai M, Kuusisto E, Leshkevich G, Livingstone DM, MacIntyre S, May L, Melack JM, Müller-Navarra DC, Naumenko M, Nõges P, Nõges T, North RP, Plisnier P, Rigosi A, Rimmer A, Rogora M, Rudstam LG, Rusak J, Salmaso N, Samal NR, Schindler DE, Schladow SG, Schmid M, Schmidt SR, Silow E, Soylu ME, Teubner K, Verburg P, Voutilainen A, Watkinson A, Williamson CE, Zhang G (2015) Rapid and highly variable warming of lake surface waters around the globe. Geophys Res Lett 42:10,773–10,781. https://doi.org/10.1002/2015GL066235
10. Ghanbari H, Gregory-Eaves I, Aulard C, Baud A, Zilkey D, Fradette M, Keller P, Giorgio P, Smol J, Huot Y, Antoniades D (2025) Widespread 20th-century increases in Canadian lake

primary production and the roles of climate warming, solar irradiance and human impacts. Commun Earth Environ 6:583. https://doi.org/10.1038/s43247-025-02569-7

11. Moss B, Kosten S, Meerhoff M, Battarbee RW, Jeppesen E, Mazzeo N, Havens K, Lacerot G, Liu Z, De Meester L, Paerl H, Scheffer M (2011) Allied attack: climate change and nutrient pollution. Inland Waters 1:101–105. https://doi.org/10.5268/IW-1.2.359
12. Meerhoff M, de los Ángeles González-Sagrario M (2022) Habitat complexity in shallow lakes and ponds: importance, threats, and potential for restoration. Hydrobiologia 849:3737–3760. https://doi.org/10.1007/s10750-021-04771-y
13. Jones JA, Driscoll CT (2022) Long-term ecological research on ecosystem responses to climate change. Bioscience 72:814–826. https://doi.org/10.1093/biosci/biac021
14. Morabito G, Mazzocchi M, Salmaso N, Zingone A, Bergami C, Flaim G, Accoroni S, Basset A, Bastianini M, Belmonte G, Bernardi Aubry F, Bertani I, Bresciani M, Buzzi F, Cabrini M, Camatti E, Caroppo C, Cataletto B, Castellano M, Pugnetti A (2018) Plankton dynamics across the freshwater, transitional and marine research sites of the LTER-Italy network. Patterns, fluctuations, drivers. Sci Total Environ 627:373–387. https://doi.org/10.1016/j.scitotenv.2018.01.153
15. Dirnböck T, Haase P, Mirtl M, Pauw J, Templer PH (2019) Contemporary international long-term ecological research (ILTER)—from biogeosciences to socio-ecology and biodiversity research. Reg Environ Chang 19:309–311. https://doi.org/10.1007/s10113-018-1445-0
16. Weathers KC, Hanson PC, Arzberger P, Brentrup J, Brookes JD, Carey CC, Gaiser E, Hamilton DP, Hong GS, Ibelings BW, Istvánovics V, Jennings E, Kim B, Kratz T, Lin F-P, Muraoka K, O'Reilly C, Rose KC, Ryder E, Zhu G (2013) The Global Lake Ecological Observatory Network (GLEON): the evolution of grassroots network science. Limnol Oceanogr Bull 22:71–73. https://doi.org/10.1002/lob.201322371
17. Hanson PC, Weathers KC, Kratz TK (2016) Networked Lake science: how the Global Lake Ecological Observatory Network (GLEON) works to understand, predict, and communicate lake ecosystem response to global change. Inland Waters 6(4):543–554. https://doi.org/10.1080/IW-6.4.904
18. Jeppesen E, Søndergaard M, Jensen JP, Havens KE, Anneville O, Carvalho L, Coveney MF, Deneke R, Dokulil MT, Foy B, Gerdeaux D, Hampton SE, Hilt S, Kangur K, Köhler J, Lammens EHHR, Lauridsen TL, Manca M, Miracle MR, Moss B, Nõges P, Persson G, Phillips G, Portielje R, Romo S, Schelske CL, Straile D, Tatrai I, Willén E, Winder M (2005) Lake responses to reduced nutrient loading—an analysis of contemporary long-term data from 35 case studies. Freshw Biol 50:1747–1771. https://doi.org/10.1111/j.1365-2427.2005.01415.x
19. Dove A, Chapra SC (2015) Long-term trends of nutrients and trophic response variables for the Great Lakes. Limnol Oceanogr 60:696–721. https://doi.org/10.1002/lno.10055
20. Qin B, Zhou J, Elser JJ, Gardner WS, Deng J, Brookes JD (2020) Water depth underpins the relative roles and fates of nitrogen and phosphorus in lakes. Environ Sci Technol 54(6):3191–3198. https://doi.org/10.1021/acs.est.9b05858
21. Guilizzoni P, Levine S, Manca M, Marchetto A, Lami A, Ambrosetti W, Brauer A, Gerli S, Carrara E, Guzzella L, Vignati D (2012) Ecological effects of multiple stressors on a deep lake (Lago Maggiore, Italy) integrating neo and palaeolimnological approaches. J Limnol 71:1–22. https://doi.org/10.4081/jlimnol.2012.e1
22. Hipsey MR, Bruce LC, Boon C, Busch B, Carey CC, Hamilton DP, Hanson PC, Read JS, de Sousa E, Weber M, Winslow LA (2019) A General Lake model (GLM 3.0) for linking with high-frequency sensor data from the global lake ecological observatory network (GLEON). Geosci Model Dev 12:473–523. https://doi.org/10.5194/gmd-12-473-2019
23. Trolle D, Elliott JA, Mooij WM, Janse JH, Bolding K, Hamilton DP, Jeppesen E (2014) Advancing projections of phytoplankton responses to climate change through ensemble modelling. Environ Model Softw 61:371–379. https://doi.org/10.1016/j.envsoft.2014.01.032
24. Bruce LC, Frassl MA, Arhonditsis GB, Gal G, Hamilton DP, Hanson PC, Hetherington AL, Melack JM, Read JS, Rinke K, Rigosi A, Trolle D, Winslow LA, Adrian R, Ayala AI, Bocaniov SA, Boehrer B, Boon C, Hipsey MR (2018) A multi-lake comparative analysis of

the General Lake Model (GLM): stress-testing across a global observatory network. Environ Model Softw 102:274–291. https://doi.org/10.1016/j.envsoft.2017.11.016
25. Dresti C, Fenocchi A, Copetti D (2021) Modelling physical and ecological processes in medium-to-large deep European perialpine lakes: a review. J Limnol 80(3). https://doi.org/10.4081/jlimnol.2021.2041
26. Salmaso N, Mosello R, Garibaldi L, Decet F, Brizzio MC, Cordella P (2002) Vertical mixing as a determinant of trophic status in Deep Lakes: a case study from two lakes south of the Alps (Lake Garda and Lake Iseo). J Limnol 62:33–41. https://doi.org/10.4081/jlimnol.2003.s1.33
27. Yankova Y, Neuenschwander S, Köster O, Posch T (2019) Abrupt stop of deep water turnover with lake warming: drastic consequences for algal primary producers. Sci Rep 7(1):13770. https://doi.org/10.1038/s41598-017-13159-9
28. Jansen J, Simpson GL, Weyhenmeyer GA, Härkönen LH, Paterson AM, del Giorgio P, Prairie YT (2024) Climate-driven deoxygenation of northern lakes. Nat Clim Chang 14:832–838. https://doi.org/10.1038/s41558-024-02058-3
29. Valerio G, Pilotti M (2025) Contribution of chemical stratification to the extent of water renewal in a deep lake. Water Resour Res 61:e2024WR039382. https://doi.org/10.1029/2024WR039382
30. Goldman CR, Jassby A Spring mixing depth as a determinant of annual primary production in lakes. In: Tilzer MM, Serruya C (eds) Large lakes. Ecological structure and function. Springer, Berlin, Heidelberg, pp 125–132. https://doi.org/10.1007/978-3-642-84077-7_6
31. Ambrosetti W, Barbanti L (1999) Deep water warming in lakes: an indicator of climatic change. J Limnol 58:1–9. https://doi.org/10.4081/jlimnol.1999.1
32. Salmaso N (2005) Effects of climatic fluctuations and vertical mixing on the interannual trophic variability of Lake Garda, Italy. Limnol Oceanogr 50:553–565. https://doi.org/10.4319/lo.2005.50.2.0553
33. Rogora M, Buzzi F, Dresti C, Leoni B, Patelli M, Lepori F, Mosello R, Salmaso N (2018) Climatic effects on vertical mixing and deep-water oxygen content in the subalpine lakes in Italy. Hydrobiologia 824:33–50. https://doi.org/10.1007/s10750-018-3623-y
34. Rogora M, Austoni M, Caroni R, Giacomotti P, Kamburska L, Marchetto A, Mosello R, Orrù A, Tartari G, Dresti C (2021) Temporal changes in nutrients in a deep oligomictic lake: the role of external loads versus climate change. J Limnol 80(3). https://doi.org/10.4081/jlimnol.2021.2051
35. European Commission Directive 2000/60/EC of the European Parliament and of the Council of 23 October 2000 establishing a framework for community action in the field of water policy. Off. J. 2000: L327/1, 2000. European Commission, Brussels. https://eur-lex.europa.eu/eli/dir/2000/60/oj/eng
36. Tanentzap AJ, Morabito G, Volta P, Rogora M, Yan ND, Manca M (2020) Climate warming restructures an aquatic food web over 28 years. Glob Chang Biol 26:6852–6866. https://doi.org/10.1111/gcb.15347
37. Soulignac F, Anneville O, Bouffard D, Chanudet V, Dambrine E, Guénand Y, Harmel T, Ibelings BW, Trevisan D, Uittenbogaard R, Danis P (2019) Contribution of 3D coupled hydrodynamic-ecological modeling to assess the representativeness of a sampling protocol for lake water quality assessment. Knowl Manag Aquat Ecosyst 42. https://doi.org/10.1051/kmae/2019034
38. Kiefer I, Odermatt D, Anneville O, Wüest A, Bouffard D (2015) Application of remote sensing for the optimization of in-situ sampling for monitoring of phytoplankton abundance in a large lake. Sci Total Environ 527–528:493–506. https://doi.org/10.1016/j.scitotenv.2015.05.011
39. Rupesh P, Hamilton D, Ausseil O, Kittridge M, Özkundakci D, McDowell RW (2024) Optimising sampling frequency for change detection of variables in lake monitoring programs. Ecol Indic 166:112321. https://doi.org/10.1016/j.ecolind.2024.112321
40. Carey CC (2023) Causes and consequences of changing oxygen availability in lakes: Kilham plenary lecture article. Inland Waters 13:316–326. https://doi.org/10.1080/20442041.2023.2239110

41. Stefan HG, Hondzo M, Fang X, Eaton JG, McCormick JH (1996) Simulated long-term temperature and dissolved oxygen characteristics of lakes in the north-Central United States and associated fish habitat limits. Limnol Oceanogr 41:1124–1135. https://doi.org/10.4319/lo.1996.41.5.1124
42. Posch T, Köster O, Salcher MM, Pernthaler J (2012) Harmful filamentous cyanobacteria favoured by reduced water turnover with lake warming. Nat Clim Chang 2:809–813. https://doi.org/10.1038/nclimate1581
43. Jane SF, Hansen GJA, Kraemer BM, Leavitt PR, Mincer JL, North RL, Pilla RM, Stetler JT, Williamson CE, Woolway RI, Arvola L, Chandra S, De Gasperi CL, Diemer L, Dunalska J, Erina O, Flaim G, Grossart HP, Hambright KD, Hein C, Hejzlar J, Janus LL, Jenny JP, Jones JR, Knoll LB, Leoni B, Mackay E, Matsuzaki SS, McBride C, Müller-Navarra DC, Paterson AM, Pierson D, Rogora M, Rusak JA, Sadro S, Saulnier-Talbot E, Schmid M, Sommaruga R, Thiery W, Verburg P, Weathers KC, Weyhenmeyer GA, Yokota K, Rose KC (2021) Widespread deoxygenation of temperate lakes. Nature 594:66–70. https://doi.org/10.1038/s41586-021-03550-y
44. Zhang Y, Shi K, Woolway RI, Wang X, Zhang Y (2025) Climate warming and heatwaves accelerate global lake deoxygenation. Sci Adv 11:eadt5369. https://doi.org/10.1126/sciadv.adt5369
45. Dillon PJ, Rigler FH (1974) The phosphorus-chlorophyll relationship in lakes. Limnol Oceanogr 19:767–773. https://doi.org/10.4319/lo.1974.19.5.0767
46. Schindler DW, Carpenter SR, Chapra SC, Hecky RE, Orihel DM (2016) Reducing phosphorus to curb Lake eutrophication is a success. Environ Sci Technol 50(17):8923–8929. https://doi.org/10.1021/acs.est.6b02204
47. Elser JJ, Bracken ME, Cleland EE, Gruner DS, Harpole WS, Hillebrand H, Ngai JT, Seabloom EW, Shurin JB, Smith JE (2007) Global analysis of nitrogen and phosphorus limitation of primary producers in freshwater, marine and terrestrial ecosystems. Ecol Lett 10:1135–1142. https://doi.org/10.1111/j.1461-0248.2007.01113.x
48. Søndergaard M, Lauridsen TL, Johansson LS, Jeppesen E (2017) Nitrogen or phosphorus limitation in lakes and its impact on phytoplankton biomass and submerged macrophyte cover. Hydrobiologia 795:35–48. https://doi.org/10.1007/s10750-017-3110-x
49. Gobler C, Burkholder J, Davis T, Harke M, Johengen T, Stow C, Van de Waal D (2016) The dual role of nitrogen supply in controlling the growth and toxicity of cyanobacterial blooms. Harmful Algae 54:87–97. https://doi.org/10.1016/j.hal.2016.01.010
50. Paerl HW, Scott JT, McCarthy MJ, Newell SE, Gardner WS, Havens KE, Hoffman DK, Wilhelm SW, Wurtsbaugh WH (2016) It takes two to tango: when and where dual nutrient (N & P) reductions are needed to protect lakes and downstream ecosystems. Environ Sci Technol 50:10805–10813. https://doi.org/10.1021/acs.est.6b02575
51. Paerl HW (2009) Controlling eutrophication along the freshwater–marine continuum: dual nutrient (N and P) reductions are essential. Estuar Coasts 32:593–601. https://doi.org/10.1007/s12237-009-9158-8
52. Analytical Methods Committee (1999) Principles of data quality control in chemical analysis. Analyst 114:1497–1503. https://doi.org/10.1039/AN9891401497
53. Rogora M, Minella M, Orrù A, Tartari GA (2006) A comparison between high-temperature catalytic oxidation and persulphate oxidation for the determination of total nitrogen in freshwater. Int J Environ Anal Chem 86(14):1065–1078. https://doi.org/10.1080/03067310600739632
54. Wilson HL, Ayala AI, Jones ID, Rolston A, Pierson D, de Eyto E, Grossart H-P, Perga M-E, Woolway RI, Jennings E (2020) Variability in epilimnion depth estimations in lakes. Hydrol Earth Syst Sci 24:5559–5577. https://doi.org/10.5194/hess-24-5559-2020
55. Jonas T, Stips A, Eugster W, Wuest A (2003) Observations of a quasi-shear-free lacustrine convective boundary layer: stratification and its implications on turbulence. J Geophys Res Oceans 108:3328. https://doi.org/10.1029/2002JC001440
56. Jennings E, Jones S, Arvola L, Staehr PA, Gaiser E, Jones ID, Weathers KC, Weyhenmeyer GA, Chiu C-Y, De Eyto E (2012) Effects of weather-related episodic events in lakes:

an analysis based on high-frequency data. Freshw Biol 57:589–601. https://doi.org/10.1111/j.1365-2427.2011.02729.x
57. Copetti D, Guyennon N, Buzzi F (2020) Generation and dispersion of chemical and biological gradients in a large-deep multi-basin lake (Lake Como, North Italy): the joint effect of external drivers and internal wave motions. Sci Total Environ 749:141587. https://doi.org/10.1016/j.scitotenv.2020.141587
58. Rogora M, Cancellario T, Caroni R, Kamburska L, Manca D, Musazzi S, Tiberti R, Lami A (2023) High frequency monitoring through in-situ fluorometric sensors: a supporting tool to long-term ecological research on lakes. Front Environ Sci 10. https://doi.org/10.3389/fenvs.2022.1058515
59. Fenocchi A, Rogora M, Sibilla S, Dresti C (2017) Relevance of inflows on the thermodynamic structure and on the modeling of a deep subalpine lake (Lake Maggiore, northern Italy/southern Switzerland). Limnologica. -Ecology and Management of Inland Waters 63:42–56. https://doi.org/10.1016/j.limno.2017.01.006
60. APHA AWWA WEF (2012) Standard methods for the examination of water and wastewater.22nd edn. American Public Health Association, Washington
61. ISO 8258 (1991) Shewhart control charts. International Organization for Standardization, Geneva
62. Tiberti R, Caroni R, Cannata M, Lami A, Manca D, Strigaro D, Rogora M (2021) Automated high frequency monitoring of Lake Maggiore through in situ sensors: system design, field test and data quality control. J Limnol 80(2). https://doi.org/10.4081/jlimnol.2021.2011
63. Dresti C, Rogora M, Fenocchi A (2023) Hypolimnetic oxygen depletion in a deep oligomictic lake under climate change. Aquat Sci 85:4. https://doi.org/10.1007/s00027-022-00902-2
64. Livingstone DM (1997) An example of the simultaneous occurrence of climate-driven "sawtooth" deep-water warming/cooling episodes in several Swiss lakes. SIL Proceedings 26(2):822–828. https://doi.org/10.1080/03680770.1995.11900832
65. Pilla RM, Williamson CE, Adamovich BV, Adrian R, Anneville O, Chandra S, Colom-Montero W, Devlin SP, Dix MA, Dokulil MT, Gaiser EE, Girdner SF, Hambright D, Hamilton DP, Havens K, Hessen DO, Higgins SN, Huttula TH, Huuskonen H, Isles PDF, Joehnk KD, Jones ID, Keller WB, Knoll LB, Korhonen J, Kraemer BM, Leavitt PR, Lepori F, Luger MS, Maberly SC, Melack JM, Melles SJ, Müller-Navarra DC, Pierson DC, Pislegina HV, Plisnier P, Richardson DC, Rimmer A, Rogora M, Rusak JA, Sadro S, Salmaso N, Saros JE, Saulnier-Talbot É, Schindler DE, Schmid M, Shimaraeva SV, Silow EA, Sitoki LM, Sommaruga R, Straile D, Strock KE, Thiery W, Timofeyev MA, Verburg P, Vinebrooke RD, Weyhenmeyer GA, Zadereev E (2020) Deeper waters are changing less consistently than surface waters in a global analysis of 102 lakes. Sci Rep 10:20514. https://doi.org/10.1038/s41598-020-76873-x
66. Finlay JC, Small GE, Sterner RW (2013) Human influences on nitrogen removal in lakes. Science 342:247–250. https://doi.org/10.1126/science.1242575
67. Müller B, Thoma R, Baumann KBL, Callbeck CM, Schubert CJ (2021) Nitrogen removal processes in lakes of different trophic states from on-site measurements and historic data. Aquat Sci 83:37. https://doi.org/10.1007/s00027-021-00795-7
68. Baron JS, Hall EK, Nolan BT, Finlay JC, Bernhardt ES, Harrison JA, Chan F, Boyer EW (2013) The interactive effects of excess reactive nitrogen and climate change on aquatic ecosystems and water resources of the United States. Biogeochemistry 114:71–92. https://doi.org/10.1007/s10533-012-9788-y
69. Fenocchi A, Pella N, Copetti D, Buzzi F, Magni D, Salmaso N, Dresti C (2025) Use of process-based coupled ecological-hydrodynamic models to support lake water ecosystem service protection planning at the regional scale. J Contam Hydrol 268:104469. https://doi.org/10.1016/j.jconhyd.2024.104469
70. Dresti C, Rogora M, Buzzi F, Beghi A, Magni D, Canziani A, Fenocchi A (2023) A modelling approach to evaluate the present and future effectiveness of hypolimnetic withdrawal for the restoration of eutrophic Lake Varese (Northern Italy). J Environ Manag 347:119042. https://doi.org/10.1016/j.jenvman.2023.119042

71. Hipsey MR, Boon C, Paraska D, Bruce LC, Huang P (2019) AquaticEcoDynamics/libaed2: v1.3.0-rc2, aquatic EcoDynamics (AED). Model Library & Science Manual. https://doi.org/10.5281/zenodo.2538495
72. Woolway RI, Merchant CJ (2019) Worldwide alteration of lake mixing regimes in response to climate change. Nat Geosci 12:271–276. https://doi.org/10.1038/s41561-019-0322-x
73. Fornarelli R, Galelli S, Castelletti A, Antenucci JP, Marti CL (2013) An empirical modeling approach to predict and understand phytoplankton dynamics in a reservoir affected by inter-basin water transfers. Water Resour Res 49:3626–3641. https://doi.org/10.1002/wrcr.20268
74. Huszar VLM, Caraco NF, Roland F, Cole J (2006) Nutrient–chlorophyll relationships in tropical–subtropical lakes: do temperate models fit? Biogeochemistry 79:239–250. https://doi.org/10.1007/s10533-006-9007-9
75. Zwart JA, Hararuk O, Prairie YT, Jones SE, Solomon CT (2019) Improving estimates and forecasts of lake carbon dynamics using data assimilation. Limnol Oceanogr Methods 17:97–111. https://doi.org/10.1002/lom3.10302
76. Vollenweider RA (1968) Scientific fundamentals of the eutrophication of lakes and flowing waters, with particular reference to nitrogen and phosphorus as factors in eutrophication. OECD Technical Report DA 5/SCI/68.27, Organisation for Economic Cooperation and Development, Paris (France)
77. Vollenweider RA, Kerekes J (1982) Eutrophication of waters: monitoring, assessment and control. Report of the OECD cooperative program on eutrophication, organisation for economic cooperation and development, Paris (France)
78. Bryhn AC, Håkanson L (2007) A somparison of predictive phosphorus load-concentration models for lakes. Ecosystems 10:1084–1099. https://doi.org/10.1007/s10021-007-9078-z
79. Steward JS, Lowe EF (2009) General empirical models for estimating nutrient load limits for Florida's estuaries and inland waters. Limnol Oceanogr 55(1):433–445. https://doi.org/10.4319/lo.2010.55.1.0433
80. Jensen JP, Pedersen AR, Jeppesen E, Søndergaard M (2006) An empirical model describing the seasonal dynamics of phosphorus in 16 shallow eutrophic lakes after external loading reduction. Limnol Oceanogr 51(1):791–800. https://doi.org/10.4319/lo.2006.51.1_part_2.0791
81. Livingstone DM (2003) Impact of secular climate change on the thermal structure of a large temperate Central European lake. Clim Chang 57:205–225. https://doi.org/10.1023/A:1022119503144
82. Fenocchi A, Rogora M, Sibilla S, Ciampittiello M, Dresti C (2018) Forecasting the evolution in the mixing regime of a deep subalpine lake under climate change scenarios through numerical modelling (Lake Maggiore, Northern Italy/southern Switzerland). Clim Dyn 51:3521–3536. https://doi.org/10.1007/s00382-018-4094-6
83. Håkanson L (1999) On the principles and factors determining the predictive success of ecosystem models, with a focus on lake eutrophication models. Ecol Model 121(2–3):139–160. https://doi.org/10.1016/S0304-3800(99)00083-6
84. Matzinger A, Müller B, Niederhauser P, Schmid M, Wüest A (2010) Hypolimnetic oxygen consumption by sediment-based reduced substances in former eutrophic lakes. Limnol Oceanogr 55. https://doi.org/10.4319/lo.2010.55.5.2073
85. Lepori F, Lucchini B, Capelli C, Rotta F (2022) Mesotrophy is not enough: re-assessing phosphorus objectives for the restoration of a deep Alpine lake (Lake Lugano, Switzerland and Italy). Adv Oceanogr Limnol 13(2):104–115. https://doi.org/10.4081/aiol.2022.11061
86. Park J, Patel K, Lee WH (2024) Recent advances in algal bloom detection and prediction technology using machine learning. Sci Total Environ 938:173546. https://doi.org/10.1016/j.scitotenv.2024.173546
87. Paturi UMR, Ramesh C, Muppala M, Mekala RR, Kasu SR, Reddy NS (2025) Artificial neural networks for modeling harmful algal blooms: a review. Mar Ecol 46(4):e70037. https://doi.org/10.1111/maec.70037
88. Ndong M, Bird D, Nguyen-Quang T, de Boutray M-L, Zamyadi A, Vinçon-Leite B, Lemaire BJ, Prévost M, Dorner S (2014) Estimating the risk of cyanobacterial occurrence using an

index integrating meteorological factors: application to drinking water production. Water Res 56:98–108. https://doi.org/10.1016/j.watres.2014.02.023

89. Vinçon-Leite B, Casenave C (2019) Modelling eutrophication in lake ecosystems: a review. Sci Total Environ 651(p2):2985–3001. https://doi.org/10.1016/j.scitotenv.2018.09.320
90. Recknagel F, Michener WK (eds) (2017) Ecological informatics: data management and knowledge discovery. Springer
91. Bruce LC, Hamilton D, Imberger J, Gal G, Gophen M, Zohary T, Hambright KD (2006) A numerical simulation of the role of zooplankton in C, N and P cycling in Lake Kinneret, Israel. Ecological Modelling 193(3–4):412–436. https://doi.org/10.1016/j.ecolmodel.2005.09.008
92. Dietzel A, Reichert P (2012) Calibration of computationally demanding and structurally uncertain models with an application to a lake water quality model. Environ Model Softw 38:129–146. https://doi.org/10.1016/j.envsoft.2012.05.007
93. Dietzel A, Mieleitner J, Kardaetz S, Reichert P (2013) Effects of changes in the driving forces on water quality and plankton dynamics in three Swiss lakes—long-term simulations with BELAMO. Freshw Biol 58:10–35. https://doi.org/10.1111/fwb.12031
94. Perroud M, Goyette S, Martynov A, Beniston M, Anneville O (2009) Simulation of multiannual thermal profiles in deep Lake Geneva: a comparison of one-dimensional lake models. Limnol Oceanogr 54:1574–1594. https://doi.org/10.4319/lo.2009.54.5.1574
95. Rinke K, Yeates P, Rothhaupt K-O (2010) A simulation study of the feedback of phytoplankton on thermal structure via light extinction. Freshw Biol 55:1674–1693. https://doi.org/10.1111/j.1365-2427.2010.02401.x
96. Fragoso CR, Motta Marques DML, Collischonn W, Tucci CEM, van Nes EH (2008) Modelling spatial heterogeneity of phytoplankton in Lake Mangueira, a large shallow subtropical lake in South Brazil. Ecol Model 219:125–137. https://doi.org/10.1016/j.ecolmodel.2008.08.004
97. Fenocchi A, Petaccia G, Sibilla S (2016) Modelling flows in shallow (fluvial) lakes with prevailing circulations in the horizontal plane: limits of 2D compared to 3D models. J Hydroinf 18:928–945. https://doi.org/10.2166/hydro.2016.033
98. Salvadè G, Stocker K, Trösch J, Zamboni F (1992) Hydrodynamics of Lake Lugano. Aquat Sci 54:187–204. https://doi.org/10.1007/BF00878136
99. Amadori M, Piccolroaz S, Giovannini L, Zardi D, Toffolon M (2018) Wind variability and Earth's rotation as drivers of transport in a deep, elongated subalpine lake: the case of Lake Garda. J Limnol 77(3):505–521. https://doi.org/10.4081/jlimnol.2018.1814
100. Valerio G, Cantelli A, Monti P, Leuzzi G (2017) A modeling approach to identify the effective forcing exerted by wind on a prealpine lake surrounded by a complex topography. Water Resour Res 53:4036–4052. https://doi.org/10.1002/2016WR020335
101. Amadori M, Giovannini L, Toffolon M, Piccolroaz S, Zardi D, Bresciani M, Giardino C, Luciani G, Kliphuis M, van Haren H, Dijkstra HA (2021) Multi-scale evaluation of a 3D lake model forced by an atmospheric model against standard monitoring data. Environ Model Softw 139:105017. https://doi.org/10.1016/j.envsoft.2021.105017
102. Piccolroaz S, Amadori M, Toffolon M, Dijkstra HA (2019) Importance of planetary rotation for ventilation processes in deep elongated lakes: evidence from Lake Garda (Italy). Sci Rep 9:8290. https://doi.org/10.1038/s41598-019-44730-1
103. Fenocchi A, Rogora M, Marchetto A, Sibilla S, Dresti C (2020) Model simulations of the ecological dynamics induced by climate and nutrient load changes for deep subalpine Lake Maggiore (Italy/Switzerland). J Limnol 79(3):221–237. https://doi.org/10.4081/jlimnol.2020.1963
104. CH2011 Swiss climate change scenarios CH2011. C2SM, MeteoSwiss, ETH, NCCR Climate, and OcCC, 2011, Zurich, Switzerland: 88 pp.
105. Hamilton DP, Schladow SG (1997) Prediction of water quality in lakes and reservoirs. Part I—model description. Ecol Model (96, 1–3):91–110. https://doi.org/10.1016/S0304-3800(96)00062-2
106. Fenocchi A, Rogora M, Morabito G, Marchetto A, Sibilla S, Dresti C (2019) Applicability of a one-dimensional coupled hydrodynamic-ecological numerical model to future projections

in a very deep large lake (Lake Maggiore, Northern Italy/Southern Switzerland). Ecol Model 392:38–51. https://doi.org/10.1016/j.ecolmodel.2018.11.005
107. Cui Y, Zhu G, Li H, Luo L, Cheng X, Jin Y, Trolle D (2016) Modeling the response of phytoplankton to reduced external nutrient load in a subtropical Chinese reservoir using DYRESM-CAEDYM. Lake Reserv Manag 32(2):146–157. https://doi.org/10.1080/10402381.2015.1136365
108. Gal G, Makler-Pick V, Shachar N (2014) Dealing with uncertainty in ecosystem model scenarios: application of the single-model ensemble approach. Environ Model Softw 61:360–370. https://doi.org/10.1016/j.envsoft.2014.05.015
109. Valerio G, Pilotti M, Barontini S, Leoni B (2015) Sensitivity of the multiannual thermal dynamics of deep pre-alpine lake to climatic change. Hydrol Process 29:767–779. https://doi.org/10.1002/hyp.10183
110. Schlabing D, Frassl MA, Eder MM, Rinke K, Bàrdossy A (2014) Use of a weather generator for simulating climate change effects on ecosystems: a case study on Lake Constance. Environ Model Softw 61:326–338. https://doi.org/10.1016/j.envsoft.2014.06.02
111. Jöhnk KD, Huisman J, Sharples J, Sommeijer B, Visser PM, Stroom JM (2008) Summer heatwaves promote blooms of harmful cyanobacteria. Glob Chang Biol 14(3):495–512. https://doi.org/10.1111/j.1365-2486.2007.01510.x
112. Schwefel R, Müller B, Boisgontier H, Wüest A (2019) Global warming affects nutrient upwelling in deep lakes. Aquat Sci 81:50. https://doi.org/10.1007/s00027-019-0637-0
113. Woolway RI, Sharma S, Smol JP (2022) Lakes in hot water: the impacts of a changing climate on aquatic ecosystems. Bioscience 72:1050–1061. https://doi.org/10.1093/biosci/biac052
114. Lepori F, Roberts JJ, Schmidt TS (2018) A paradox of warming in a deep peri-Alpine lake (Lake Lugano, Switzerland and Italy). Hydrobiologia 824:215–228. https://doi.org/10.1007/s10750-018-3649-1
115. Salmaso N, Boscaini A, Capelli C, Cerasino L (2018) Ongoing ecological shifts in a large lake are driven by climate change and eutrophication: evidences from a three-decade study in Lake Garda. Hydrobiologia 824:177–195. https://doi.org/10.1007/s10750-017-3402-1
116. Soares L, Thouillot M, Frossard V, Desgue O, Barouillet C, Baulaz Y, Clement J, Domaizon I, Dorioz J-M, Goulon C, Guillard J, Jacquet S, Réalis-Doyelle E, Tran Khac V, Jenny JP (2025) Expanding the European water framework directive indicators to address long-term climate change impacts on lakes using mechanistic lake models. Ecol Indic 172:113220. https://doi.org/10.1016/j.ecolind.2025.113220
117. Free G, Poikane S, Solheim AL, Bussettini M, Bradley C, Smith J, Caroni R, Bresciani M, Pinardi M, Giardino C, van de Bund W (2024) Climate change and ecological assessment in Europe under the WFD—hitting moving targets with shifting baselines? J Environ Manag 370:122884. https://doi.org/10.1016/j.jenvman.2024.122884

Chemometrics in Environmental Analysis: Case Studies

Mery Malandrino, Stefano Bertinetti, and Riccardo Cecire

Abstract

This chapter sketches, in a deliberately practical way, how chemometric data exploration can assist environmental analysis when datasets become too multivariate, too noisy, or simply too difficult to read with univariate logic. Environmental questions often arrive as multivariable data, where useful structure is mixed with redundancy and measurement noise; the chapter treats this as the normal condition rather than an exception. Against this backdrop, the guiding idea is that chemometrics, used carefully, can help environmental chemists move from complicated measurements to decision support, while accepting that interpretation remains a human responsibility.

After outlining common data structures and the need for preprocessing, the chapter introduces the core families of chemometric methods and focuses on widely used unsupervised approaches. Principal Component Analysis (PCA), Factor Analysis (FA), and Cluster Analysis (CA) are described as complementary lenses: dimensionality reduction and correlation patterns on one side, grouping and similarity structures on the other, with some unavoidable information loss in the former and a fuller retention of variance in the latter. A simple pipeline is proposed as a repeatable sequence of operations, though it is implied that real datasets rarely behave as neatly as flowcharts suggest.

Several case studies across contrasting matrices are presented as examples: contaminated soils, moss biomonitoring, urban PM_{10} (including a focused look near an incinerator startup), remote Arctic PM_{10}, Antarctic lacustrine waters, and the elemental fingerprinting of Piedmont Freisa wines. In each example, multivariate patterns are used to suggest plausible sources, clusters, or seasonal dynamics, sometimes with clean separations, sometimes with overlap that leaves

M. Malandrino (✉) · S. Bertinetti · R. Cecire
Department of Chemistry, University of Turin, Torino, Italy
e-mail: mery.malandrino@unito.it

M. Minella (ed.), *Complexity in Environmental Chemistry*,
https://doi.org/10.1007/978-3-032-31774-2_14

room for doubt. Still, a recurring outcome is that chemometrics can reduce complexity into interpretable structures that point toward targeted hypotheses and, potentially, practicable actions.

Keywords

Chemometrics · Environmental analysis · Data exploration · Source apportionment · Decision support · Sustainability

Chapter Learning Goals
In this chapter some case studies concerning the application of the main chemometric methods at environmental analysis will be discussed to highlight how they can address complex environmental problems for reaching shared and practicable solutions.

The first question an environmental chemist asks himself when using chemometric methods is the following: "What do we expect from Chemometrics in Environmental Analysis?". The use of chemometric methods allows us to obtain support of decisions. So, with the aid of decisions, we come to conclusions concerning reality. This is what we really need! Chemometric methods allow multivariate data treatment. Since environmental problems are usually described by multiple variables, and taking into consideration multiple samples, this way of operating is essential to reach the solution to the problem.

1 Chemometric Processing

Before digging into the theory, it is worth noting that a chemometric analysis is not a single operation but a sequence of interconnected steps. From the organization of the data matrix to preprocessing, model selection, and interpretation, each phase contributes to the reliability of the final outcome. The steps described in the following paragraphs should therefore be considered as parts of a coherent workflow rather than independent procedures. Although often represented schematically, this workflow requires critical evaluation at every stage, since choices made during data preparation and modelling can substantially influence the conclusions drawn from environmental datasets.

1.1 Data Structure

We should remember that an ordered set of data can be written in the form of a matrix, which usually contains the values of measured or known properties (variables, features) listed in columns. The rows of the matrix are then associated with the objects (e.g., samples at different locations or times) under study. Hence, an environmental analysis usually allows to obtain a dataset composed of n objects and *p* variables (Fig. 1).

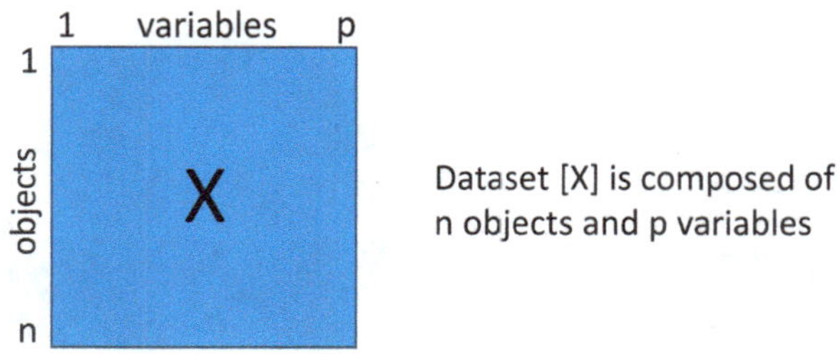

Fig. 1 Example of typical environmental dataset obtained considering p variables determined in *n* samples (objects)

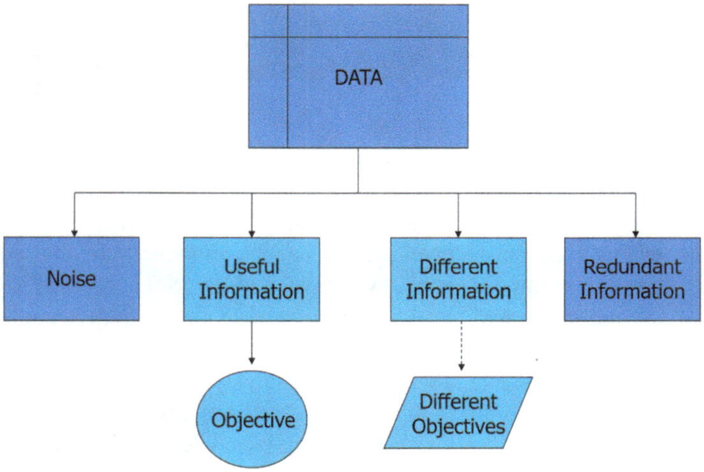

Fig. 2 Conceptual scheme of data information content in chemometric analysis

This dataset includes (Fig. 2) noise, useful information to achieve our goal; different information, useful to achieve other goals; and redundant information which can make it confusing [1]. It is, therefore, difficult to locate the useful information we are looking for. The chemometric methods allow to distinguish these different components present in the dataset in order to select only the useful information to solve our problem. Chemometric methods can be used for data exploration, which is to find relationships and correlations between the variables measured and know their influence on samples, to evaluate the synergic or antagonistic effect among the variables and their effect on the samples [2, 3].

1.2 Chemometric Methods

The chemometric methods are divided into two important families (Fig. 3): Unsupervised Pattern Recognition Methods and Supervised Pattern Recognition Methods [1, 2]. For the first methods, we do not have, or do not want, to use additional or a priori information about the data measured. We use these methods to explore relationships among samples and variables and to identify possible patterns

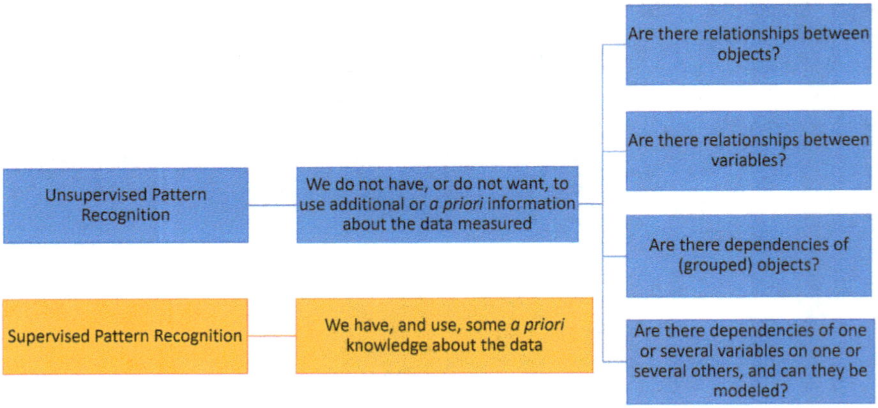

Fig. 3 Schematic representation of unsupervised and supervised pattern recognition

or groupings within the data without relying on prior assumptions. Moreover, it is possible to evaluate if these dependencies can be modelled. In the Supervised Pattern Recognition methods, instead, we have, and use, some a priori knowledge about the data to develop qualitative and/or quantitative models that can be used to predict the behavior or the value of an unknown sample or variable [2, 4]. The most used chemometric methods in environmental analysis belong to the first category, i.e., they are Unsupervised Pattern Recognition Methods, also known as Exploration Data methods [3]. The main exploration methods used in environmental analysis are Principal Component Analysis and Cluster Analysis, while Factor Analysis is often applied to support and refine the interpretation of multivariate data.

1.3 Principal Component Analysis

Briefly, Principal Component Analysis (PCA) allows to evaluate the correlations existing between the variables and their relevance, to visualize the objects in a low-dimensional geometric space for the identification of outliers and clusters, to provide a simplified representation of complex datasets, and to define a model for representing the data on orthogonal axes reducing the dimensionality of the data [2, 5]. Finally, the PCA allows to consider variables and objects simultaneously, but it does not allow to retain all the information contained in the original dataset [5]. In general, in the PCA, the experimental variables are transformed into new variables, called principal components, which are linear combinations of the original variables and are orthogonal to each other [1, 5]. The samples can be represented on these new axes, which are the principal components.

1.4 Factor Analysis

Factor Analysis (FA) is a multivariate statistical method closely related to Principal Component Analysis, which is used to identify latent variables, called factors, which explain the correlations observed among the original variables. In environmental analysis, FA allows to reduce the complexity of large datasets by describing the system in terms of a limited number of factors that represent common sources of variability [1, 2]. Unlike PCA, which aims at maximizing the explained variance, FA is based on a statistical model that assumes that the observed variables are linear combinations of a smaller number of underlying factors plus a residual error term [5]. The factors obtained can often be interpreted in terms of physical, chemical, or environmental processes, such as pollution sources or geochemical controls, making FA particularly useful for source identification studies [3]. In many applications, a rotation of the factor axes, such as varimax rotation, is applied in order to obtain a simpler and more interpretable structure of the factor loadings. Varimax rotation, in particular, maximizes the variance of the squared loadings within each factor, thereby enhancing high loadings and reducing low ones, which facilitates the identification of groups of variables associated with specific sources or processes [1]. As for PCA, FA does not allow to retain all the information contained in the original dataset, but it provides a simplified representation that facilitates the interpretation of complex environmental data. For this reason, Factor Analysis is often used in combination with PCA to corroborate and refine the interpretation of multivariate environmental datasets [2].

1.5 Cluster Analysis

Cluster Analysis (CA) allows the identification of non-random structures in the data in an objective way, providing evidence for the presence of clusters based on similarity measures, most commonly Euclidean distance [2, 6]. Depending on the aim of the study, CA can be applied in Q-mode, where objects (samples) are grouped according to their similarity, or in R-mode, where variables are clustered in order to highlight associations among them. In this way, CA can focus either on relationships among samples or among variables, though not simultaneously. Unlike PCA, Cluster Analysis retains the full information contained in the original dataset [1]. The graphical output of CA is typically a dendrogram, in which samples or variables are progressively linked to form clusters according to their degree of similarity [6].

1.6 Chemometric Pipeline

The chemometric treatment of environmental experimental data usually requests the application of the following operations: the first step is to calculate any missing data using different methods taking into account the characteristic of the missing

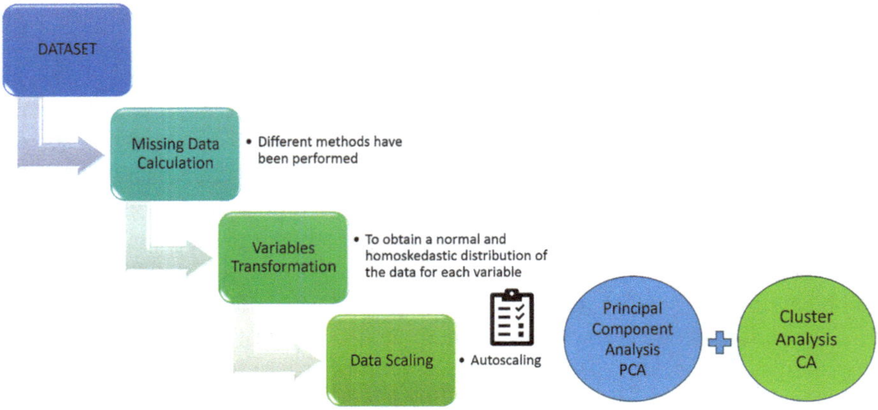

Fig. 4 Schematic diagram of the steps to be taken for a correct chemometric elaboration of environmental data

data, for example, a value lower than detection limit or a value too high for a possible contamination [3, 4]; after, it is necessary to evaluate the distribution of each variable and to apply a transformation to obtain normal and homoscedastic distribution for all the variables considered [1, 2]; the last step of the data pre-treatment consists in the scaling of data, usually using the autoscaling method [1, 2]. After performing these steps, the experimental data is ready to be treated with the most common data exploration methods, PCA and CA. As we have seen, these methods are complementary and are often both performed in order to corroborate the conclusions reached [2, 3]. Figure 4 shows the flowchart of the steps to be taken for a correct chemometric treatment of environmental data.

2 Case Studies

The case studies concern different environmental problems relative to several matrices. In all the selected case studies, a chemometric approach allowed to identify the causes of the environmental problem in order to propose targeted solutions. Precisely, exploration data methods, mainly PCA and CA, have been used to study the causes of the soil contamination [7] and the possibility to use mosses as environmental metal pollution indicators [8]; to identify the possible sources of urban atmospheric particulate matter, precisely the fraction with a dimension less to 10 μm, in the metropolitan area of Turin city (NW, Italy) [9, 10]; to evaluate the sources of atmospheric particulate matter collected in polar areas, precisely in the Svalbard Islands (Arctic region) [11]; to understand the element dynamics in Antarctic natural waters [12, 13]; and, finally, to evaluate the possibility to distinguish a type of Piedmont red wine from other Italian and non-Italian red wines in order to ensure its traceability [14].

2.1 Contaminated Soils

In the first case study, a soil contaminated by heavy metals was considered. This site (Fig. 5) is located near the city of Borgomanero in the Province of Novara [7]. In this site 28 soil samples have been collected to evaluate the extension of the contamination. Precisely, 11 samples are collected along the horizontal profile, 11 samples are collected along the vertical profile, and 6 samples are collected at increasing distances from the contaminated site. In all the samples, the content of 15 elements, including geogenic and potentially toxic elements, has been quantified by spectroscopic techniques (inductively coupled plasma-optical emission spectrometer, ICP-OES, and electrothermal atomic absorption spectrometer equipped with Zeeman-effect background correction, ET-AAS). The final dataset was composed of 28 samples arranged along the rows and 15 elements, namely, Al, Cd, Cu, Cr, Fe, La, Mn, Ni, Pb, Sc, Ti, V, Y, Zn, and Zr, arranged along the columns.

The dataset was processed by PCA and CA in order to obtain a visual representation of the soil samples and gain insight into the distribution of the pollutants. In Fig. 6, the biplot, obtained combining PC1 and PC2, is shown. The total information contained in the biplot is equal to 55%, so it is evident that all the information is not conserved, but this partial information is the most significant because it contains the greatest variance. The biplot shows that the soil samples can be divided into three groups: in the first group, enclosed by the blue circle, there are the soil samples belonging to the horizontal profile collected in the site core or just outside it, together with the first two layers of the vertical profile, B12 and B13 samples; these samples

Fig. 5 Geographical map of Piedmont region (NW, Italy), with the indication of the sampling places

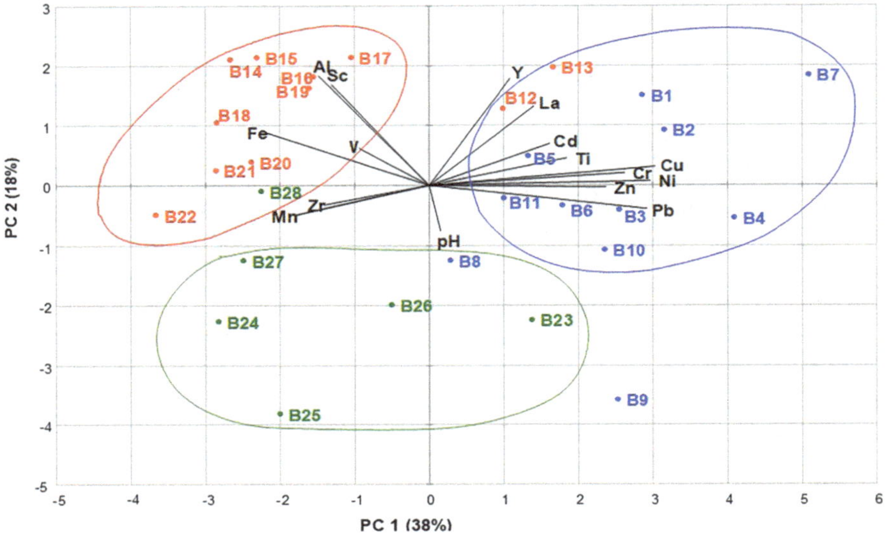

Fig. 6 Combined plot of scores and loadings obtained by PCA (PC1 vs PC2). (Reprinted with permission, Copyright 2002, Environmental Pollution, Elsevier)

are characterized by high contaminant content; the second cluster is composed of the soil samples relative to deeper layers of the vertical profile; finally the samples collected in the surroundings of the site, excluding B23, and in the city center are grouped in a larger cluster. Samples B8 and B9, collected under the vegetation grown just outside the site core, even if included in horizontal profile, have a stronger similarity to this group. The PCA allowed to establish the extension of the contamination. Concerning the variables, their loadings show a correlation among the elements identified as pollutants (Cd, Cr, Cu, Ni, Pb, and Zn). These are anticorrelated to Mn and Fe, typically geogenic elements. Finally, pH shows a negative loading on PC2, whereas all the samples characterized by higher contamination levels are located at positive PC2 values. This is confirmation that contaminated soils are usually characterized by lower pH.

Dendrogram, obtained using agglomerative hierarchical clustering procedure, in which Euclidean distances are calculated between objects to evaluate the similarity among soil samples, shows a similar clustering to PCA (Fig. 7). Indeed, two clusters formed by the samples collected along the horizontal and vertical profile, respectively, are evident. The first two samples of vertical profile are included in the cluster colored in blue. Finally, most of the soil samples collected in the surroundings of the site are grouped in a third cluster. It is evident that CA confirms the clustering evidenced in biplot.

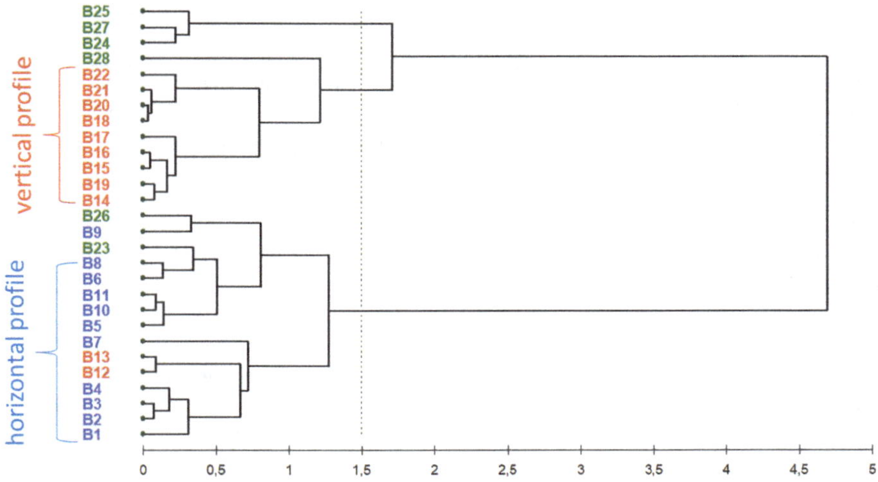

Fig. 7 Dendrogram for horizontal profile samples at Borgomanero site

2.2 Mosses as Environmental Metal Pollution Indicators

The second case study concerns the evaluation of the possible use of mosses as environmental metal pollution indicators [8]. Mosses were chosen for their ability in accumulating potentially toxic elements and for their features particularly suitable for environmental monitoring. *Bryum argenteum* is a moss species able to survive in scarcely fertile and polluted zones, growing mostly on low walls and on road borders. For this reason, it fits particularly for environmental monitoring. Nine sites were selected, located in Piedmont region and characterized by different types of area and environmental conditions (Table 1). Mosses were sampled monthly in all the sites for 1 year, collecting five different samples from which analysis a mean value was obtained. 20 elements, ranging from major (K, P, Al, Ca, Fe, and Mg) to minor (Mn, Na, Ti, and Zn) and trace (As, Ba, Cd, Co, Cr, Cu, Li, Ni, Pb, and Sr), have been quantified by atomic spectroscopy (ET-AAS and ICP-OES). The dataset is composed of 9 moss samples arranged along the rows and 20 elements arranged along the columns.

In order to check possible correlations between element content and pollution grade, PCA was applied to the data obtained. In the biplot, shown in Fig. 8, the results obtained combining PC1 and PC2 are reported. The total information contained in the biplot is equal to 57%. It is apparent that macronutrient elements, like K, Na, and P, are correlated and more abundant in sites where pollution can be considered lower, that is, Borgomanero Street, Ferrera, and Turin Botanical Garden. These sites tend to form a distinct group. Borgomanero site is strongly characterized by its high content of metals of anthropogenic origin, precisely Cr, Cu, and Ni. Finally, the remaining sites tend to cluster together, identifying intermediate pollution conditions with sources different from those of Borgomanero.

Table 1 Description of sampling sites

Site	Type of area	Environmental conditions
Borgomanero	Polluted rural	Moss grows on a highly polluted soil
Borgomanero Street	Rural	Moss grows on a low wall near an unsurfaced road
Ferrera	Mountain	Moss grows on a rock beside a mountain lake, at 1400 m a.s.l.
Giaveno	Pedemountain	Moss grows beside a low traffic road
Murisenghi	Rural	Moss grows beside a low traffic road
Turin Botanical Garden	Urban	Moss grows inside a greenhouse
Turin Corso Lepanto	Urban	Moss grows on a low wall, 3 m from the road
Turin Corso Orbassano	Urban	Moss grows on a low wall, 3 m from the road
Turin via Nizza	Urban	Moss grows on a seeded patch, 5 m from the road

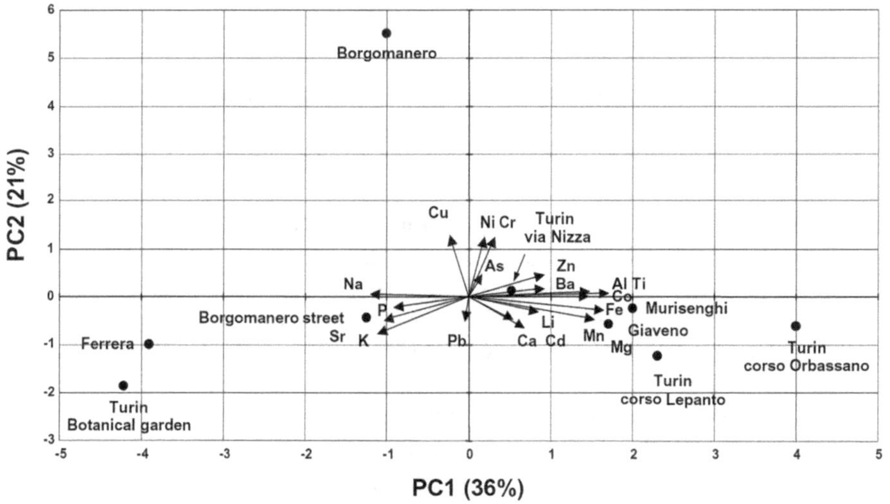

Fig. 8 PCA biplot (PC1 vs PC2). Each point represents a sampling site described by the annual mean metal concentrations (average of 12 monthly samplings). (Reprinted with permission, Copyright 2003, Chemosphere, Elsevier)

2.3 Urban Atmospheric Particulate Matter (PM_{10})

In this case study, the composition of atmospheric particulate matter, precisely the fraction less than 10 μm, collected in five urban sampling sites was evaluated [9]. Three sites are selected in the Turin area, namely, TO1 is in the historical center; TO2 is in the northern out-skirts of the town, in an area of high vehicular traffic; DR is located in the small town of Druento, within a regional park. Two sampling sites are selected in the Cuneo province: BSD is close to Borgo San Dalmazzo, and SA is in Saliceto (SA), a village on the border with Liguria region. The distribution of the sampling sites is shown in Fig. 9. The aim of this study was to identify the

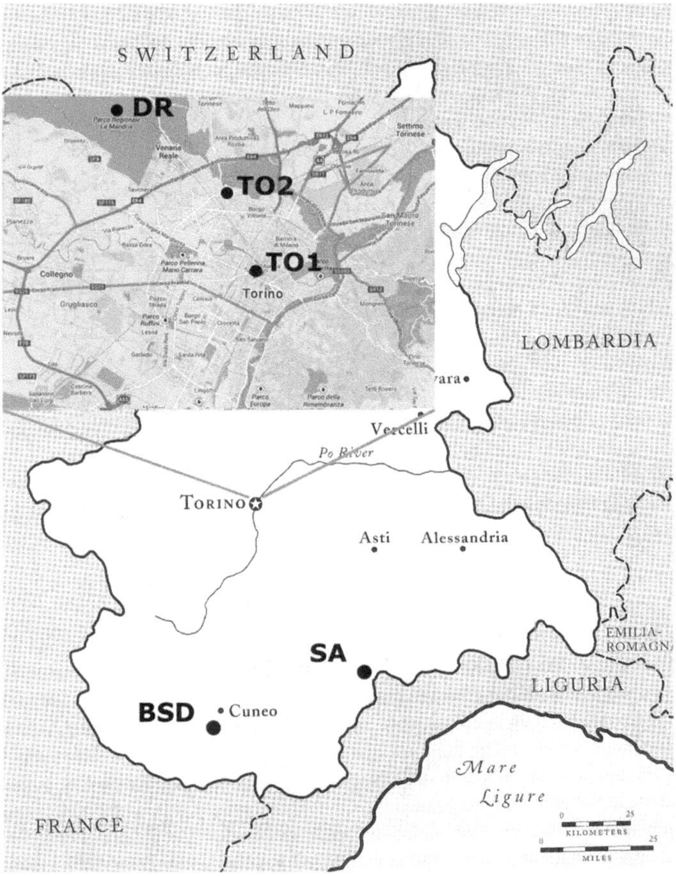

Fig. 9 Map of Piedmont and detail of sampling areas. (Reprinted with permission, Copyright 2016, Chemosphere, Elsevier)

emission sources influencing the air quality in these urban sites. To this end, 28 samples were collected at each sampling site, corresponding to 1 week of sampling in February, May, August, and November, thus covering one representative month for each season. 23 elements, including main elements, potentially toxic elements, and rare earth elements, have been quantified by spectroscopic and spectrometric techniques (ICP-OES and inductively coupled plasma-high resolution mass spectrometer, HR-ICP-MS). Overall, the dataset obtained is composed of 135 PM_{10} samples arranged along the rows and 23 elements arranged along the columns.

A statistical-chemometric elaboration was applied in order to identify the principal sources of pollution and to highlight similarities and differences between the five sampling sites. In the score plot, the results obtained combining PC1 and PC2 are reported (Fig. 10). The total information contained in the biplot is equal to 61.46%. An evident differentiation exists on PC2 among the sites, forming clusters with centroids separated from each other. PC2 is influenced by the variables characterizing SA samples; therefore these samples have high scores on PC2 and are all

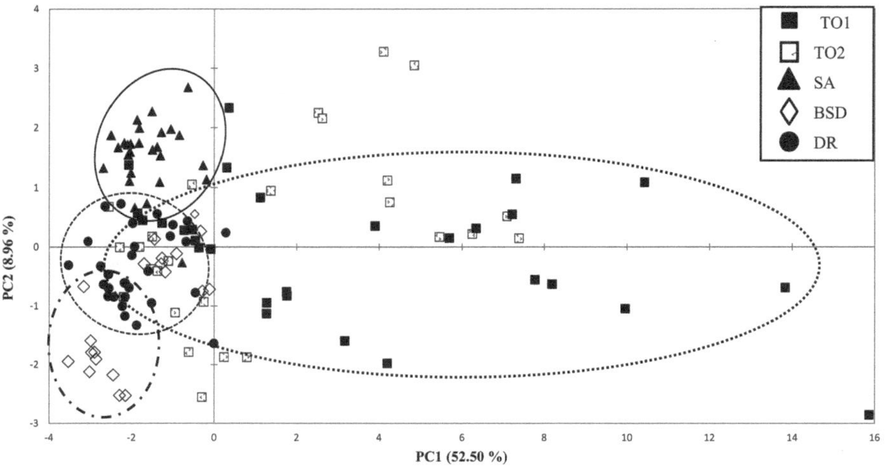

Fig. 10 Score and loading plot obtained by PCA (PC1 vs PC2). (Reprinted with permission, copyright 2016, Chemosphere, Elsevier)

grouped in a compact cluster. At the opposite side of the plot, we find BSD samples. PM_{10} from this site does not display a particular pattern, with the exception of the high content of Ca. At the center there are samples from TO1, TO2, and DR. The samples collected in Turin are arranged along PC1, which represent variables related to anthropic pollution; these samples, in particular from TO2 site, are more scattered, owing to a clearer seasonal pattern in their elemental content: the PM_{10} samples collected in the cold seasons show higher score values on PC1 because they are in greater extent influenced by pollutant and geogenic variables. Instead, the PM_{10} from DR site is the least influenced by the pollution variables; indeed, all the samples are contained in a very compact cluster, characterized by negative values on PC1.

As the complex dataset available for this study, it was possible to evaluate the emission sources using the Varimax rotation of loadings calculated by PCA. In Table 2 are the reported factor loadings obtained for the PM_{10} samples from all the five sites.

Six factors were obtained summing about 78% of the total variance in the entire dataset. The first factor, accounting for 43% of the total variance, presents highest loadings for many elements (e.g., Ba, Cu, Fe, Mo, and Zn) that have often been reported as deriving from brake abrasion and tire wear. Therefore, this factor describes non-exhaust vehicular emissions and road dust resuspension. The second factor may be associated with different mineralogy in SA because of the high loadings of Ce and La. The third factor is characterized by high loadings of Mg, Ca, and Ti: it can be associated with crustal contribution and, for BSD, with typical contribution from a cement plant. As and Sb have high loadings in the fourth factor: this factor can be associated with a possible industrial source even if it is also likely that the correlation between these two elements arises from a common trend in atmosphere, due to their similar chemical properties. Finally, the last two factors present high loadings for K and Ni and for V, respectively; therefore, they can be associated

Table 2 Factor loadings obtained by Varimax rotation for PM10 samples at five monitoring sites

Element	Factor 1	Factor 2	Factor 3	Factor 4	Factor 5	Factor 6
Mg	0.680	0.094	**0.701**	0.030	−0.060	0.088
Al	**0.624**	0.127	0.491	0.047	−0.200	0.251
K	0.397	0.283	0.045	0.052	**0.764**	−0.012
Ca	0.228	0.016	**0.776**	−0.014	0.168	−0.131
Fe	**0.966**	0.135	0.183	0.116	0.020	0.033
Ti	0.368	−0.018	**0.488**	0.082	−0.091	0.227
Mn	**0.842**	0.167	0.265	0.058	0.122	0.063
Cu	**0.899**	0.259	0.185	0.207	−0.042	−0.036
Zn	**0.870**	0.062	0.166	0.101	0.352	0.027
Sb	0.129	−0.012	0.007	**0.835**	−0.017	−0.060
Ba	**0.912**	−0.010	0.102	0.129	0.190	0.013
Pb	**0.748**	0.070	−0.023	0.112	0.558	0.097
Cr	**0.827**	0.069	0.279	0.103	0.021	0.083
V	0.286	0.308	0.018	−0.026	0.004	**0.527**
Co	**0.745**	−0.059	0.098	0.047	0.357	0.334
Ni	0.092	−0.048	0.033	−0.008	**0.116**	0.073
As	0.450	0.088	0.022	**0.813**	0.115	0.076
Sr	**0.598**	−0.018	0.244	0.049	0.093	0.282
Zr	**0.920**	0.156	0.102	0.216	0.059	0.047
Mo	**0.883**	0.003	0.060	0.097	0.261	0.275
Cd	**0.759**	0.259	0.087	0.085	0.165	−0.083
Ce	0.137	**0.959**	0.175	0.038	0.002	−0.113
La	0.173	**0.799**	−0.209	0.017	0.291	0.359

The main factor loading for each variable is reported in bold

with fossil fuel combustion processes as well as biomass burning in residential heating systems and with ship and industrial emissions. It is evident that the chemometric treatment of the dataset allows to identify the emission sources of PM_{10} collected in these areas.

With regard to urban atmospheric particulate matter, it is interesting to take into account the study concerning the evaluation of the impact on air quality of the installation of an incinerator built in the city of Turin between 2010 and 2013 [10]. PM_{10} sampling was carried out in spring and summer 2013 in one site near the incinerator, in the area with the highest probability that PM emitted from the plant is falling to the ground. Twenty-two PM_{10} samples were chosen for this study in the period between May and July 2013. The content of 24 elements, namely, Ca, Al, Na, K, Mg, Mn, Fe, Ti, V, Cr, Cu, Zn, As, Zr, Mo, Cd, Sn, Ba, Pb, Co, Ni, La, Ce, and Tl, was determined using spectroscopic and spectrometric techniques (ICP-OES and HR-ICP-MS). The final dataset was composed of 22 PM_{10} samples arranged along the rows and 24 elements arranged along the columns.

With the aim to gain insight into the possible sources of PM_{10}, a chemometric treatment of the experimental data was carried out through the Q-mode CA in which

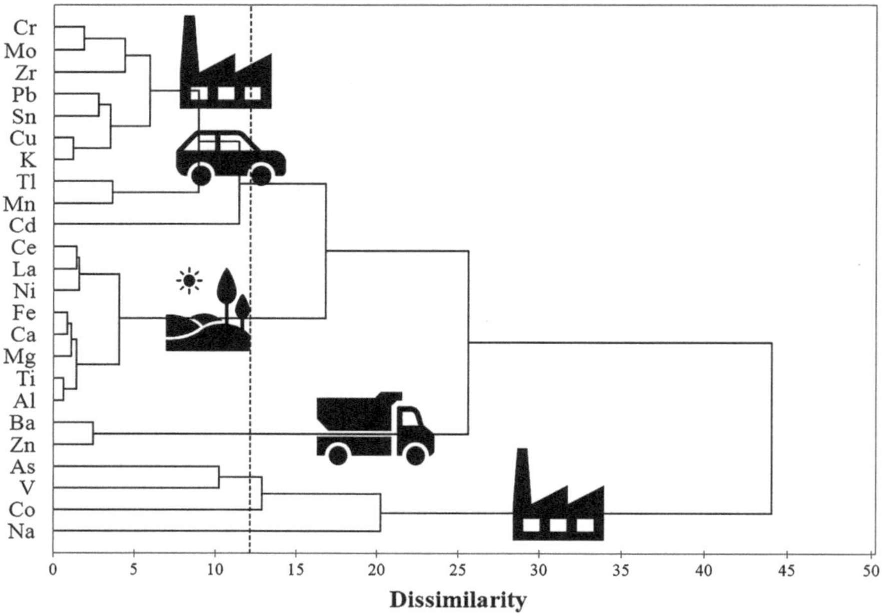

Fig. 11 Dendrogram obtained by Q-mode HCA. (Reprinted with permission, Copyright 2021, Chemosphere, Elsevier)

we assume that distinct levels of dissimilarity among elements are indicative of different emission sources. The dendrogram obtained is reported in Fig. 11.

Starting from the bottom of the figure, the first cluster is composed by As, Co, Na, and V: they are linked at high levels of dissimilarity, suggesting that they do not derive from a single emission source but they are likely all emitted in atmosphere from several anthropogenic sources, probably attributable also to activities that have occurred in the start-up phase of the incinerator, such as the temporary commissioning of the second combustion line. A second cluster is characterized by only two elements, namely, Ba and Zn; several researchers identified these elements as typical markers of road dust source because the former is added to lubricating oils to prevent smoke and diesel engine abrasion and the latter is also generated by dust caused by vehicular movement or exhaust emission. Generally, these elements are also associated with crustal elements in identifying the road dust source, while here they are linked at a high level of dissimilarity with these elements grouped together in a third cluster. This could be due to the fact that while crustal elements are enriched when the environmental conditions are favorable to the resuspension of soil dust, the resuspension of road dust occurred due to the heavier traffic of trucks carrying the wastes to the incinerator. The third cluster is characterized by several elements, namely, Al, Ca, Ce, Fe, La, Mg, Ni, and Ti, which are linked at low levels of dissimilarity, suggesting a common source associated with soil dust. Finally, the last cluster groups are formed by many elements that are linked at different levels of dissimilarity, suggesting that they are emitted in atmosphere from several anthropogenic sources. In particular, the clustering of K, Cu, Pb, and Sn could represent direct incinerator emissions since Pb, Sn, and Cu are commonly associated with this source and K is a common marker of biomass burning events, while the other elements probably derive from city vehicular traffic and other industrial activities. Overall, it is likely that the increase of concentrations for Ba, Cd, Cr, Cu, K, Mo, Pb, Sn, and Zn during start-up phase of municipal solid waste incinerator (MSWI) plant was caused by a combination of factors: the vehicular traffic presumably increased in this area due to the commissioning of the incinerator and a possible direct contribution deriving from the incinerator emissions; however, a contribution from the city vehicular traffic and industrial activities cannot be excluded.

2.4 Atmospheric Particulate Matter (PM_{10}) in Remote Areas

Chemometric data treatment can also be applied to understand emission sources influencing PM_{10} in uncontaminated areas, as the Arctic. Concerning this polar area, in this study the composition of atmospheric particulate matter collected at Ny-Ålesund, on Svalbard Islands, was determined [11]. Ny-Ålesund is a small settlement acting as a center for international Arctic scientific research and environmental monitoring. The aerosol was sampled at Gruvebadet research station, not far away from Ny-Ålesund. PM_{10} samples were collected with a 4-day resolution from 2010 to 2013. In particular, 2012 Arctic Campaign started in April and ended in September. 36 Arctic PM_{10} samples were collected in 2012 Campaign, and the content of 29 elements was analyzed by spectroscopic and spectrometric techniques. Overall, the dataset was composed of 36 Arctic PM_{10} samples, arranged along the rows, and 29 elements, arranged along columns (total number of experimental data 1044).

The complexity of the matrices and of the phenomena under study makes interpretation of experimental results difficult. Exploration data methods can be a valuable tool for this aim, thanks to their multivariate approach, which takes into account the behavior of multiple variables simultaneously. Score and loading plots obtained by PCA for 2012 samples are shown in Fig. 12, considering the combination of PC1 and PC2. The first PC is mainly associated with analytes having a geogenic origin, except Ce. By looking at the score plot, it is possible to distinguish samples collected in April and May, having positive scores on PC1, from samples collected in June, August, and September, having negative scores; samples collected in July, though, are spread among the two groups. Considering that, in spring, the soil of Svalbard Islands is still covered by snow and ice, it is likely that long-range transport of PM from midlatitudes took place in this period, thus influencing the composition of PM_{10} samples collected near to Ny-Ålesund; this result is coherent with previous findings, evidencing the so-called "Arctic haze" in spring. The position of Ce with respect to the other lanthanides suggests that this element might have a non-geogenic, i.e., anthropogenic, origin.

The dendrogram obtained by Q-mode HCA (Fig. 13) highlights three main clusters of objects, that is, groups of samples characterized by similar elemental

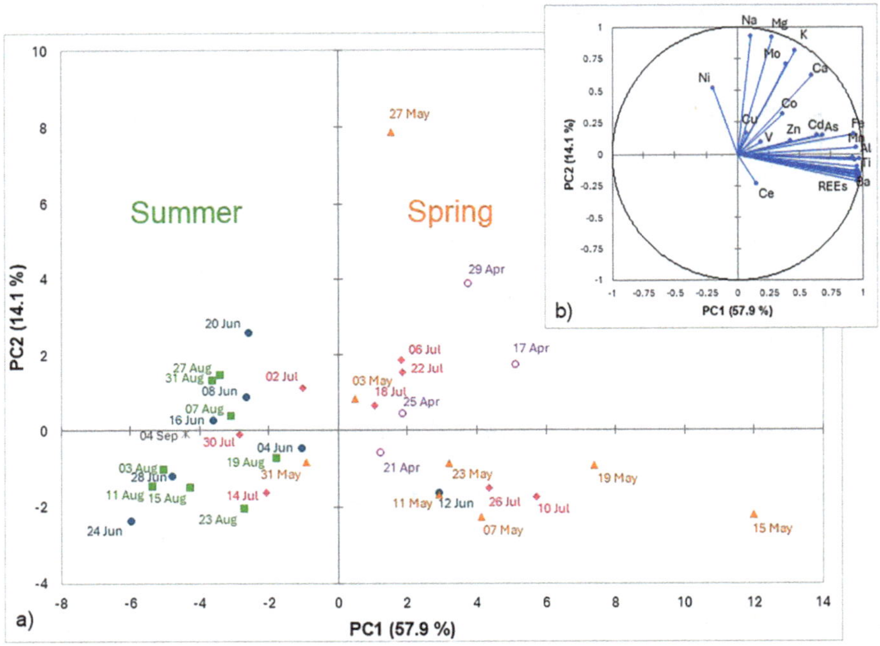

Fig. 12 Score (a) and loading (b) plot obtained by PCA (PC1 vs PC2). (Reprinted with permission, Copyright 2019, Atmospheric Environment, Elsevier)

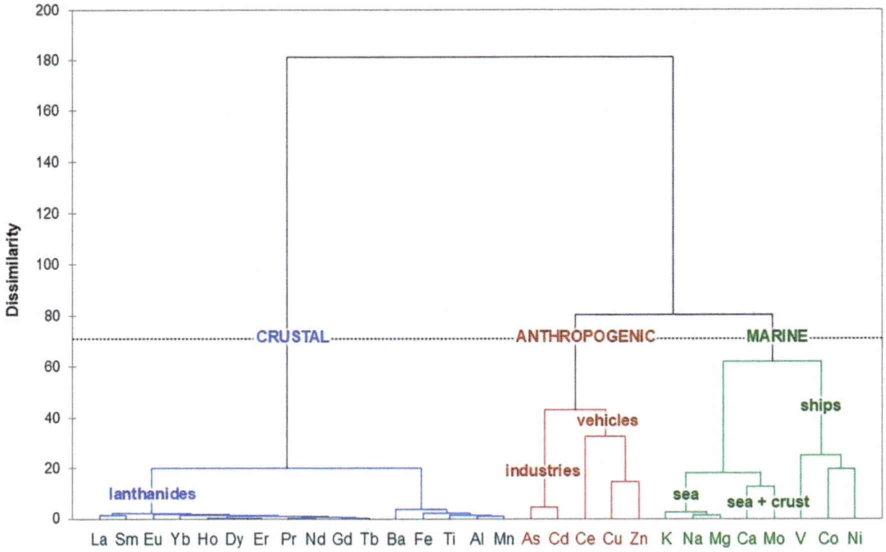

Fig. 13 Dendrogram obtained by Q-mode CA. (Reprinted with permission, Copyright 2019, Atmospheric Environment, Elsevier)

composition. The first cluster is composed by analytes typically geogenic; as expected, at a lower level of dissimilarity, it is possible to see that lanthanides form a distinct cluster from the other five analytes. The second cluster is composed by analytes probably having an anthropogenic origin; at a lower level of dissimilarity, it is possible to distinguish a small cluster composed of As and Cd, a small cluster composed of Cu and Zn, and Ce as an outlier. As and Cd are commonly emitted during combustion and industrial production processes, which are likely occurring at mid-latitudes; Cu and Zn are likely associated with non-exhaust vehicle emissions, as these elements are common indicators of brake and tire wear; considering its position in the dendrogram, Ce might also derive from vehicular emissions, as this element is often used in catalytic converters. The third cluster is composed by analytes deriving from marine aerosol. This cluster also includes Co, Ni, and V; therefore, airborne pollution derives from ship emissions.

2.5 Natural Waters

In the present case study, a distinct environmental matrix was addressed, namely, lacustrine waters from Terra Nova Bay in Antarctica, with the intention of examining, at least in part, the behavior of inorganic constituents across the summer period as discussed by Conca et al. (2017) [12]. Antarctic lake systems of this type have been investigated from 2002 to 2013, though the resulting datasets remain uneven in both temporal and spatial density. Within Terra Nova Bay, attention was largely directed toward six shallow lakes, Carezza (CA), Edmonson Point sites No. 14 and 15a, Inexpressible Island (II), Gondwana (GW), and Tarn Flat (TF), each of which is subject to somewhat different degrees of marine and cryospheric influence. A total of 55 water samples was obtained during the XXVII Italian Antarctic Expedition in the austral summer of 2011 to 2012. Sampling was repeated at intervals ranging from about 6 to 18 days, a frequency considered sufficient to capture at least the broader seasonal oscillations in solute concentrations. The inorganic composition of each sample was characterized through the determination of 21 elemental concentrations, giving rise to a data matrix of 55 rows and 21 columns.

The experimental data were subsequently explored by Principal Component Analysis. The first two components, which together account for roughly 60% of the total variance, are shown in Fig. 14. Although some overlap remains, the PCA output tends to arrange the samples into lake-specific groupings, suggesting that local controls on chemistry are not negligible. A dominant factor separating the lakes appears to be the relative contribution of marine aerosol deposition. In particular, II10b and EP14 samples occupy markedly positive positions along PC1, which implies a strong imprint of marine-derived inputs on their chemical composition. By contrast, CA and EP15a samples show a narrower spread and form comparatively compact clusters, consistent with more limited seasonal shifts. Taken as a whole, the data suggest that each lake exhibits its own chemical fingerprint along PC1, while superimposed on this pattern is a progressive modification of analyte levels through the summer, plausibly linked to increasing meltwater influx.

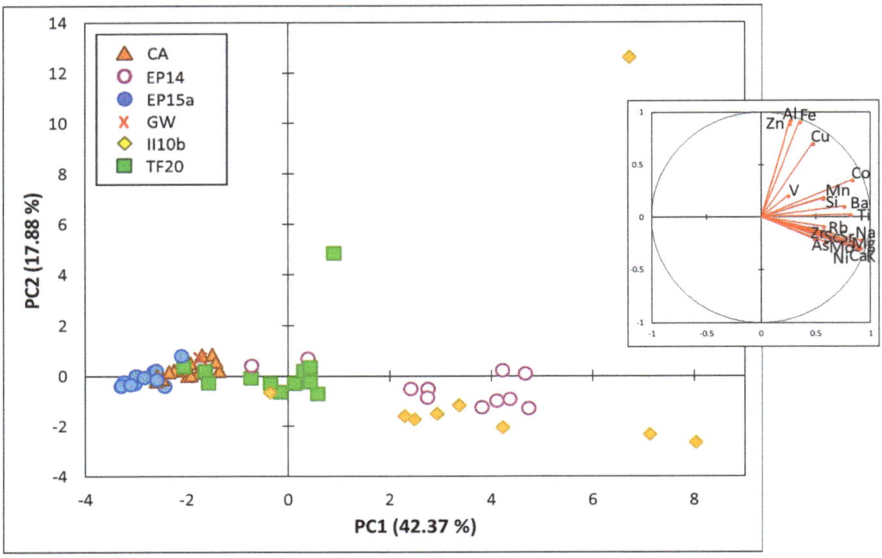

Fig. 14 Score and loading plot obtained by PCA (PC1 vs PC2). (Reprinted with permission, Copyright 2017, Chemosphere, Elsevier)

A second case study considered the same general lacustrine system but focusing on a different issue: evaluating whether the observed variability in inorganic and trace metal concentrations reflects primarily natural geochemical processes or whether signals of anthropogenic contamination might be discerned [13]. Over the period between 2002 and 2011, several Antarctic summer campaigns produced multivariate geochemical datasets for Terra Nova Bay lakes, interpreted mainly through Principal Component Analysis and hierarchical clustering. The focus here was on small ice influenced basins, notably Carezza (CA), Edmonson Point sites EP14 and EP15a, and Tarn Flat (TF). In total, 11 water samples were available, with a denser sampling effort during December 2010 and January 2011 for EP14 and EP15a, where short-term variability was expected to be more pronounced. Elemental concentrations for 21 inorganic and trace metal species were measured, again yielding a matrix of 11 samples by 21 variables.

These data were evaluated by a combination of PCA and Hierarchical Cluster Analysis (Fig. 15). The first principal component, explaining about 66% of the total variance, together with the second component, was used to visualize the structure of the dataset. Samples tend to cluster according to the lake of origin, though the boundaries are not always sharp. The primary separation among lakes reflects differing geochemical regimes driven by varying contributions from marine spray, bedrock weathering, and seasonal meltwater. EP14 samples plot toward the positive end of PC1, which points to waters enriched in marine and geogenic components. In contrast, CA and TF samples occupy the opposite side of PC1 and partially overlap, implying broadly similar yet not identical chemical controls. EP15a samples are more scattered along PC1, a pattern that may be taken as evidence of enhanced

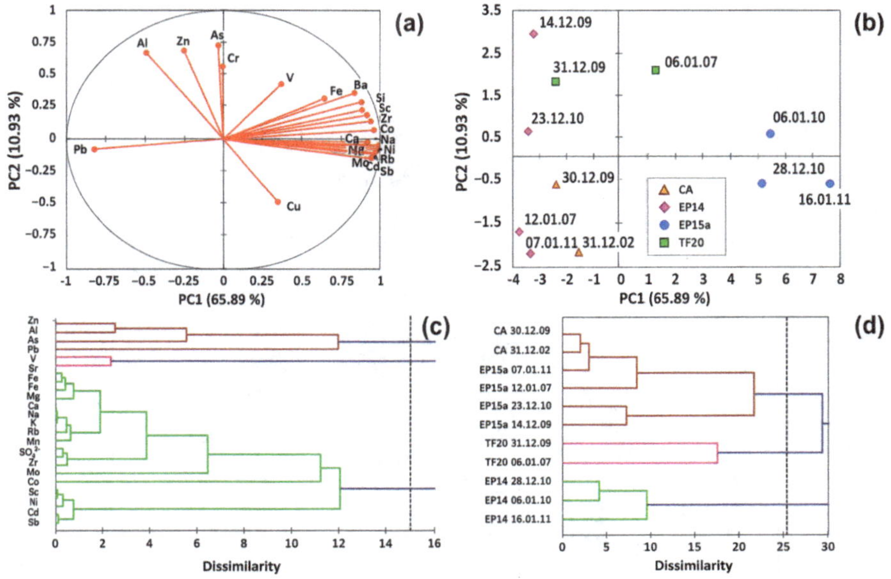

Fig. 15 Element concentrations in lake water samples: plots of (**a**) loadings and (**b**) scores obtained by PCA for PC1 vs PC2; (**c**) sample and (**d**) variable classification obtained by HCA. (Reprinted with permission, Copyright 2017, Chemosphere, Elsevier)

short-term variability. Overall, the covariation of potentially toxic metals with lithogenic elements suggests that the temporal and spatial changes in lake chemistry are, to a considerable extent, governed by natural processes rather than by a systematic accumulation of anthropogenic contaminants, even if minor external inputs cannot be entirely excluded.

2.6 Wine Samples

Among Italian red wines, Freisa is almost exclusively produced in the Piedmont region. Five denominations of controlled origin of Piedmont Freisa wines are recognized (Fig. 16): Freisa di Chieri (FC), Freisa d'Asti (FA), Monferrato Freisa (MF), Langhe Freisa (LF), and Colli Tortonesi Freisa (CTF), which are related to the area of production [14]. In this study the concentration of 15 elements was determined in 55 wines purchased from different wineries and shops, belonging to the 5 different DOC typologies. Overall, the dataset obtained is composed of 55 Freisa wine samples, arranged along the rows, and 15 elements, arranged along the columns.

A multivariate study on the analytical data was carried out to obtain a visual representation of the metal distribution in the Freisa wine samples, to find out similarities and correlations among variables, and to identify possible clusters among objects. Total metal concentrations in wines are useful for characterizing them and classifying them according to geographical origin and assessment of authenticity.

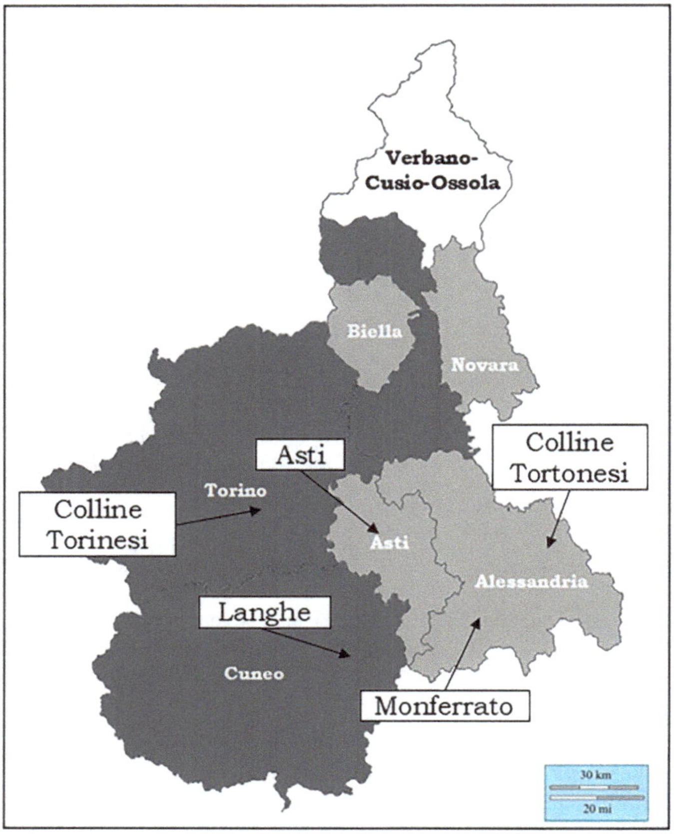

Fig. 16 Areas of production of the five different Freisa wines in Piedmont. (Original figure published under Creative Common License CC-BY 4.0)

Indeed, it is known that the information on metal contents in finished wines is useful for their "fingerprinting," mostly because metals are stable and represent factors affecting wine composition. In the dendrogram obtained by CA, shown in Fig. 17, the identification of three clusters is possible: the first composed by all the CTF samples and most of the LF samples; the second, larger in size and composed by most of the FA and FC samples; the third in which three Freisa wine samples are grouped, belonging to the MF, LF, and CTF sites and characterized by a higher content of elements. Overall, CA only allows to distinguish the CTF and LF samples from FA and FC samples. The high similarity found for all the other samples allows to hypothesize that Freisa wine has similar content of elements regardless of the Piedmont collection area.

In order to assess whether the elemental profile of Freisa wines provides a distinctive chemical fingerprint, a comparative analysis was performed including both Freisa samples and other Italian red wines. The dataset, composed of the mean elemental concentrations measured in these wines, was subsequently treated by

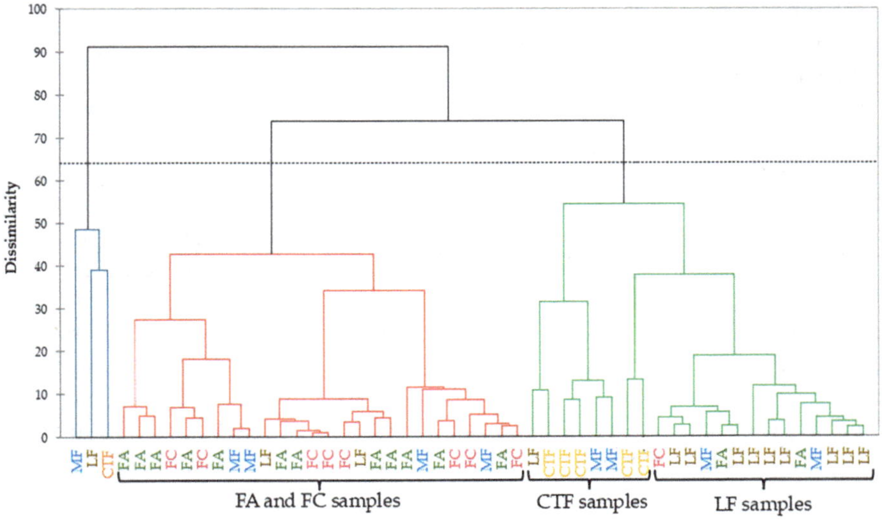

Fig. 17 Dendrogram obtained by CA. (Original figure published under Creative Common License CC-BY 4.0)

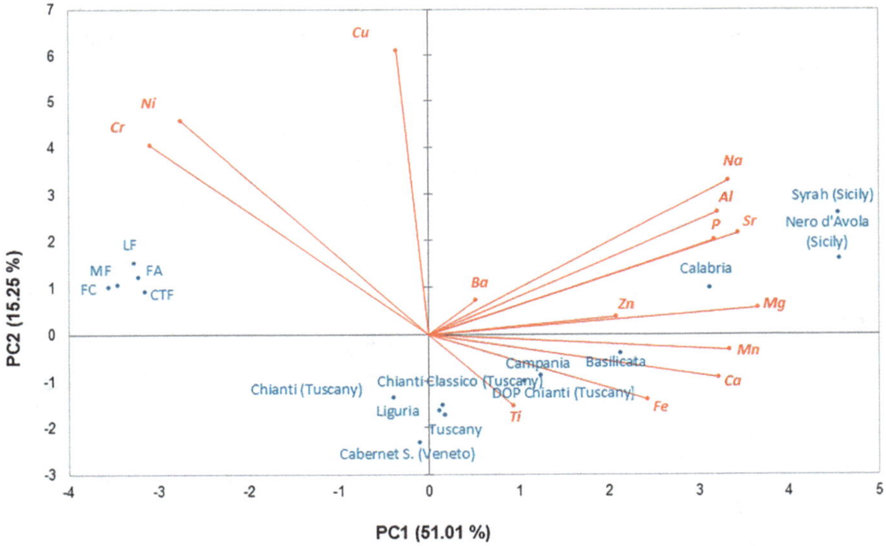

Fig. 18 Combined plot of scores and loadings obtained by PCA (PC1 vs PC2). (Original figure published under Creative Common License CC-BY 4.0)

PCA (Fig. 18) to evaluate their possible classification according to geographical origin.

From the PCA plot, it is apparent how most wine samples taken from the same geographical area tend to be grouped in well-defined clusters. In particular, Freisa wine samples have negative scores on PC1 and are all grouped in a very compact cluster, meaning that these wine samples have characteristic elemental composition. On the opposite side of the plot, we find Southern and Sicilian wine samples characterized by positive scores on PC1. Finally, the Northern wine samples are more scattered and arranged along PC1 and are characterized by negative scores on PC2. It is interesting to observe that the Freisa wine cluster is strongly influenced by Cr and Ni variables showing high and negative loadings on PC1. It is possible, therefore, to assess Cr and Ni as possible markers for Freisa wine for their high discriminant capacity to distinguish this wine from other red Italian wines. These two elements are also strongly correlated to each other, showing the highest Pearson correlation coefficient (0.980) in the correlation matrix calculated. This correlation could be explained on the basis of their common origin deriving from the high content of these elements in the Piedmont soils of the investigated sites. Finally, the Sicilian and Calabrian wines are mainly influenced by Na, Al, Sr, and Mg variables, while most of the Northern wine samples are negatively influenced by Cu.

3 Conclusion

This chapter provides a broad overview of the possible applications of chemometric data exploration methods to environmental analysis, highlighting the added value of the application of these chemometric tools. In particular, unsupervised chemometric methods, such as PCA and HCA, allow to deal with highly complex multivariate data and to provide decision criteria that can be useful to achieve the proposed objective, which generally consists in solving a complex problem. It is however necessary to remember that chemometrics helps to deal with multivariate data and make decisions, but the understanding of the results obtained and the responsibility for the decision is of the chemist!

Questions

- How does the structure of an environmental dataset as an $n \times p$ matrix influence the choice and interpretation of chemometric methods?
- In what way do PCA, Factor Analysis, and Cluster Analysis complement each other when exploring complex environmental data?
- Why is data pre-treatment (handling missing values, transformation, scaling) a critical step in the chemometric pipeline, and what could happen if it is neglected?
- How did chemometric tools help distinguish between geogenic and anthropogenic sources in at least two different case studies presented in the chapter?
- In the PM10 investigations, how were specific emission sources inferred from factor loadings and clustering patterns, and what assumptions underlie these interpretations?

Glossary

- **CA**: Cluster Analysis
- **ET-AAS**: Electrothermal Atomic Absorption Spectrometry
- **FA**: Factor Analysis
- **HCA**: Hierarchical Cluster Analysis
- **HR-ICP-MS**: High-Resolution Inductively Coupled Plasma Mass Spectrometry
- **ICP-OES**: Inductively Coupled Plasma Optical Emission Spectrometry
- **MSWI**: Municipal Solid Waste Incinerator
- **PCA**: Principal Component Analysis
- **PC**: Principal Component
- **PM10**: Particulate Matter with aerodynamic diameter ≤10 μm

Bibliography

1. Massart DL, Vandeginste BGM, Buydens LMC, De Jong S, Lewi PJ, Smeyers-Verbeke J (1997) Handbook of chemometrics and qualimetrics, Part A. Elsevier, Amsterdam
2. Brereton RG (2003) Chemometrics: data analysis for the laboratory and chemical plant. Wiley, Chichester. https://doi.org/10.1002/0470863242
3. Einax JW, Zwanziger HW, Geiss S (1997) Chemometrics in environmental analysis. VCH, Weinheim
4. Vandeginste BGM, Massart DL, Buydens LMC, De Jong S, Lewi PJ, Smeyers-Verbeke J (1998) Handbook of chemometrics and qualimetrics, Part B. Elsevier, Amsterdam
5. Jolliffe IT (2011) Principal component analysis. In: Lovric M (ed) International encyclopedia of statistical science. Springer, Berlin, Heidelberg, pp 1094–1096. https://doi.org/10.1007/978-3-642-04898-2_455
6. Everitt BS, Landau S, Leese M, Stahl D (2001) Cluster analysis. 4th edn. Arnold, London
7. Abollino O, Aceto M, Malandrino M, Mentasti E, Sarzanini C, Barberis R (2002) Distribution and mobility of metals in contaminated sites: chemometric investigation of pollutant profiles. Environ Pollut 119:177–193. https://doi.org/10.1016/S0269-7491(01)00333-5
8. Aceto M, Abollino O, Conca R, Malandrino M, Mentasti E, Sarzanini C (2003) The use of mosses as environmental metal pollution indicators. Chemosphere 50:333–342. https://doi.org/10.1016/S0045-6535(02)00533-7
9. Padoan E, Malandrino M, Giacomino A, Grosa MM, Lollobrigida F, Martini S, Abollino O (2016) Spatial distribution and potential sources of trace elements in PM10 monitored in urban and rural sites of Piedmont region. Chemosphere 145:495–507. https://doi.org/10.1016/j.chemosphere.2015.11.094
10. Ziegler D, Malandrino M, Barolo C, Adami G, Sacco M, Pitasi F, Abollino O, Giacomino A (2021) Influence of start-up phase of an incinerator on inorganic composition and lead isotope ratios of the atmospheric PM10. Chemosphere 266:129091. https://doi.org/10.1016/j.chemosphere.2020.129091
11. Conca E, Abollino O, Giacomino A, Buoso S, Traversi R, Becagli S, Grotti M, Malandrino M (2019) Source identification and temporal evolution of trace elements in PM10 collected near Ny-Ålesund (Norwegian Arctic). Atmos Environ 203:153–165. https://doi.org/10.1016/j.atmosenv.2019.02.001
12. Conca E, Malandrino M, Giacomino A, Buoso S, Berto S, Verplanck PL, Magi E, Abollino O (2017) Dynamics of inorganic components in lake waters from Terra Nova Bay, Antarctica. Chemosphere 183:454–470. https://doi.org/10.1016/j.chemosphere.2017.05.104

13. Zelano I, Malandrino M, Giacomino A, Buoso S, Conca E, Sivry Y, Benedetti M, Abollino O (2017) Element variability in lacustrine systems of Terra Nova Bay (Antarctica) and concentration evolution in surface waters. Chemosphere 180:343–355. https://doi.org/10.1016/j.chemosphere.2017.04.005
14. Rabezzana R, Malandrino M, Abollino O, Bonometti E, Giordana A, Turco F, Volpi G, Operti L (2023) Characterization of Freisa wines from Piedmont (Italy) by aroma and element profile. Appl Sci 13:7425. https://doi.org/10.3390/app13137425

FSC
www.fsc.org
MIX
Papier | Fördert
gute Waldnutzung
FSC® C083411

Kolibri 360 GmbH
Ferdinand-Jühlke-Straße 7
99095 Erfurt, Deutschland
produktsicherheit@kolibri360.de